CAMBRIDGE STUDIES IN
ADVANCED MATHEMATICS 1

Algebraic automata theory

Algebraic automata theory

W.M.L.HOLCOMBE
Department of Pure Mathematics, The Queen's University of Belfast

CAMBRIDGE UNIVERSITY PRESS
CAMBRIDGE
LONDON NEW YORK NEW ROCHELLE
MELBOURNE SYDNEY

Shaftesbury Road, Cambridge CB2 8EA, United Kingdom

One Liberty Plaza, 20th Floor, New York, NY 10006, USA

477 Williamstown Road, Port Melbourne, VIC 3207, Australia

314–321, 3rd Floor, Plot 3, Splendor Forum, Jasola District Centre, New Delhi – 110025, India

103 Penang Road, #05–06/07, Visioncrest Commercial, Singapore 238467

Cambridge University Press is part of Cambridge University Press & Assessment, a department of the University of Cambridge.

We share the University's mission to contribute to society through the pursuit of education, learning and research at the highest international levels of excellence.

www.cambridge.org
Information on this title: www.cambridge.org/9780521604925

First published 1982
First paperback edition 2004, 2024

A catalogue record for this publication is available from the British Library

Library of Congress catalogue card number: 81-18169

ISBN 978-0-521-60492-5 Paperback

To Jill, Lucy, and my mother, and in fond memory of my father and grandfather

Contents

Introduction

In recent years there has been a growing awareness that many complex processes can be regarded as behaving rather like machines. The theory of machines that has developed in the last twenty or so years has had a considerable influence, not only on the development of computer systems and their associated languages and software, but also in biology, psychology, biochemistry, etc. The so-called 'cybernetic view' has been of tremendous value in fundamental research in many different areas. Underlying all this work is the mathematical theory of various types of machine. It is this subject that we will be studying here, along with examples of its applications in theoretical biology, etc.

The area of mathematics that is of most use to us is that which is known as modern (or abstract) algebra. For a hundred years or more, algebra has developed enormously in many different directions. These all had origins in difficult problems in the theory of equations, number theory, geometry, etc. but in many areas the subject has taken on its own momentum, the problems arising from within the subject, and as a result there has been a general feeling that much of abstract algebra is of little practical value. The advent of the theory of machines, however, has provided us with new motivation for the development of algebra since it raises very real practical problems that can be examined using many of the abstract tools that have been developed in algebra. This, to me, is the most exciting aspect of the subject, the ability of using algebra in a useful and meaningful way to tackle some of the fundamental questions facing us today: What can machines do? How do we think and speak? How do cells repair themselves? How do biochemical systems function? How do organisms grow and develop? etc. We will not be able to answer these questions here, that is neither possible nor the aim of this book. What we will be doing is to lay the foundations for the algebraic study of machines, by looking at various types of machines, their properties, and ways in which complex machines can be simulated by simpler machines joined together in some way. This will then provide a theoretical framework for the more detailed analysis of the applications of machine theory in these subjects, with the ultimate aim of explaining many natural and artificial phenomena. Perhaps a later work devoted to the applications of machine theory would make use of the developments outlined here.

We begin with some elementary material concerning the theory of semigroups. This is presented as concisely as possible; it may be omitted by those readers familiar with the material. Others could easily start by reading the first few pages of chapter 2, which introduces the state machine, before returning (hopefully better motivated) to the details of chapter 1.

The second chapter examines many elementary properties of the state machine, the ways in which it can be connected together with others, and finishes with some applications. I have tried to include as wide a variety as possible and I have not treated them in great depth because the required background knowledge in biology, biochemistry, etc. may not be available. For those interested, the references provide sources of further reading.

Chapter 3 develops the idea of a covering, by which state machines can be simulated by other, perhaps simpler, state machines in various configurations. This area represents a major change in philosophy in algebra since we do not attempt to describe the machine exactly but rather what it can do. There are some general results that enable us to start with an arbitrary state machine and simulate it with simpler machines constructed from finite simple groups and elementary 'two-state' machines connected up suitably. The best known method for doing this, the holonomy decomposition, is examined in chapter 4. However, this process leads to simulating machines that can be very large and thus relatively inefficient. In specific situations it is possible to develop much better simulators and a variety of techniques for doing this are examined in chapter 3.

The theory of recognizers is intimately connected with the theory of state machines and is of considerable importance in the theory of computers. This area is studied in chapter 5.

Finally we end with a more practical and realistic type of machine and apply the previous results to this situation in chapter 6.

Some of this material would be suitable for an advanced undergraduate course on applied algebra or automata theory and I have indeed given such a course for some years at Queen's University, Belfast. The more advanced material (chapters 4, 6) would be suitable for a graduate course.

I hope that this book can help forge links between pure mathematicians, computer scientists and theoretical biologists. There are great benefits in a dialogue between practitioners of these subjects and although I realize that the approach here is rather mathematical, I hope that it will not prevent others from making use of the material. With this in mind I have included as an appendix a computer program for evaluating the semigroup of a state machine. This has been developed for me by Dr A. W. Wickstead (Pure Mathematics, Q.U.B.) and I would like to take this opportunity of thanking him for his help. The program is suitable for use on a microcomputer, something that is becoming readily available these days.

My other thanks go to many of my colleagues at Queen's who have helped me with various questions. As usual, though, I have to take responsibility for any errors that may occur in this work.

Sheila O'Brien (Q.U.B.) made an excellent job of typing my manuscript and I would like to record my gratitude here.

I must also thank Dr E. Dilger (Tübingen) for reading the manuscript and Dr B. McMaster for helping me with the proofs.

Michael Holcombe (*July 1981*)

1

Semigroups and their relatives

We may as well begin at the beginning and this will involve us in a brief excursion through some of the fundamental concepts essential for any algebraic subject. It will also enable us to become acquainted with the notation used, although the experienced reader could easily skip through this chapter. We will assume that the reader has a knowledge of elementary set theory.

1.1 Relations

One of the fundamental concepts in mathematics is that of a relation. It can be introduced in a variety of ways but the most useful one for us is the following abstract approach.

Let A be a non-empty set. A *relation*, $\mathcal{R}$, on A is a subset $\mathcal{R} \subseteq A \times A$. If $(a, a') \in A \times A$ and $(a, a') \in \mathcal{R}$ we say that a is *$\mathcal{R}$-related* to a'. Sometimes a natural notation is used in mathematics to express this relationship between two elements of a set, for example if $A = \mathbb{Z}$, the set of all integers, then there is a relation $\leq$ that can be defined on $\mathbb{Z}$. We write $a \leq a'$ if the number $a' - a$ is not negative and the set $\mathcal{R} \subseteq \mathbb{Z} \times \mathbb{Z}$ defining this relation consists of all ordered pairs $(a, a') \in \mathbb{Z} \times \mathbb{Z}$ such that $a \leq a'$.

A relation $\mathcal{R}$ on the set A is an *equivalence relation* if:

(i) $(a, a) \in \mathcal{R}$ for all $a \in A$
(ii) $(a, a') \in \mathcal{R} \Rightarrow (a', a) \in \mathcal{R}$ for $a, a' \in A$
(iii) $(a, a') \in \mathcal{R}$ and $(a', a'') \in \mathcal{R} \Rightarrow (a, a'') \in \mathcal{R}$ for $a, a', a'' \in A$.

The existence of an equivalence relation is a very useful fact because it means that we can partition the set A into a disjoint union of subsets in a natural way.

Let $\mathcal{R}$ be an equivalence relation on the set A, for each $a \in A$ define the set $[a]_{\mathcal{R}} = \{a' \in A | (a, a') \in \mathcal{R}\}$, so $[a]_{\mathcal{R}}$ is the subset of all elements of A that are related to a under $\mathcal{R}$. It is called the *equivalence class* defined by a. If the relation $\mathcal{R}$ is understood we just write $[a]$.

Consider now the collection of all the *distinct* subsets of the form $[a]$ where $a \in A$. If we denote this collection by the notation $A/\mathcal{R}$ we can establish the following:

Theorem 1.1.1

Let $\mathcal{R}$ be an equivalence relation on the set A. The set $A/\mathcal{R}$ consists of a collection of pairwise disjoint subsets of A that cover A. By this we mean that if the collection $A/\mathcal{R}$ is indexed by some set I so that

$$A/\mathcal{R} = \{H_i | i \in I\} \quad \text{where each } H_i \text{ is of the form } [a]$$

for some $a \in A$ then

(i) $\bigcup_{i \in I} H_i = A$

(ii) $H_i \cap H_j = \varnothing$ if $i \neq j$ $(i, j \in I)$.

Proof Let $a \in A$, then $(a, a) \in \mathcal{R}$ and so $a \in [a]$ and thus there is $i \in I$ such that $[a] = H_i$. Hence $a \in H_i$ for some $i \in I$. This proves (i). Now suppose that $a \in H_i$ for some $i \in I$, then there exists $a' \in A$ such that $a \in [a']$ where $H_i = [a']$. Let $b \in [a]$, then $(a, b) \in \mathcal{R}$, but $(a', a) \in \mathcal{R}$ and so $(a', b) \in \mathcal{R}$, which means $b \in [a']$. Hence $[a] \subseteq [a']$. Let $c \in [a']$, then $(a', c) \in \mathcal{R}$, however $(a', a) \in \mathcal{R}$ implies $(a, a') \in \mathcal{R}$ and then $(a, c) \in \mathcal{R}$ and $c \in [a]$. Hence $[a] = [a']$. Finally choose $d \in H_i \cap H_j$ where $H_i = [a']$ and $H_j = [a'']$. Since $[a'] = [d]$ and $[a''] = [d]$ from the above it is clear that $H_i = H_j$ and so $i = j$. This proves (ii). □

The set $A/\mathcal{R}$ is called the *quotient set of A with respect to* $\mathcal{R}$.

If A is a set and $\pi = \{H_i | i \in I\}$ is a collection of subsets of A satisfying (i) $\bigcup_{i \in I} H_i = A$ and (ii) $H_i \cap H_j = \varnothing$ if $i \neq j$, $(i, j \in I)$; then π is called a *partition* of A. We call the subsets H_i $(i \in I)$, the *π-blocks*. Clearly an equivalence relation defines a partition based upon the distinct equivalence classes. Conversely, given a partition $\pi = \{H_i | i \in I\}$ we can define an equivalence relation $\mathcal{R}$ in the following way:

$$a\mathcal{R}a' \Leftrightarrow \text{there exists } i \in I \text{ such that } a \in H_i \text{ and } a' \in H_i.$$

So two elements are equivalent precisely when they belong to the same

π-block. It will sometimes be convenient to identify an equivalence relation with its partition.

Associated with a general relation $\mathscr{R}$ on a set A are two subsets of A defined as follows:

$$\mathfrak{D}(\mathscr{R}) = \{a \in A \mid (a, b) \in \mathscr{R} \text{ for some } b \in A\}$$
$$\mathfrak{R}(\mathscr{R}) = \{a \in A \mid (b, a) \in \mathscr{R} \text{ for some } b \in A\}.$$

We call $\mathfrak{D}(\mathscr{R})$ the *domain* of $\mathscr{R}$ and $\mathfrak{R}(\mathscr{R})$ the *range* of $\mathscr{R}$. If $\mathscr{R}$ is an equivalence relation then $\mathfrak{D}(\mathscr{R}) = \mathfrak{R}(\mathscr{R}) = A$.

One of the benefits of taking this approach to relations is the ease of generalizing ideas to relations *between* sets.

Let X and Y be sets, a *relation $\mathscr{R}$ from X to Y* is a subset $\mathscr{R} \subseteq X \times Y$. As before we will say that elements (x, y) belonging to $\mathscr{R}$ are *$\mathscr{R}$-related*, $(x \in X, y \in Y)$. Let us denote this relation by the symbols $\mathscr{R} : X \rightsquigarrow Y$.

There are certain 'extreme' situations that we will briefly examine now. We no longer prevent the sets X and Y from being empty. If either X or Y or both are empty then $X \times Y$ is *defined* to be the empty set. So that we can certainly define a relation from the empty set $\varnothing$ to a set Y, or from a set X to the empty set $\varnothing$, in both cases $\mathscr{R} = \varnothing$ is the only possible relation. In fact the *empty relation* $\mathscr{R} = \varnothing$ can be defined from any set X to any set Y, we write it as $\boldsymbol{\theta}: X \to Y$.

As before we can define the concepts of domain and range, thus if $\mathscr{R} : X \rightsquigarrow Y$ is a relation then

$$\mathfrak{D}(\mathscr{R}) = \{x \in X \mid (x, y) \in \mathscr{R} \text{ for some } y \in Y\}$$
$$\mathfrak{R}(\mathscr{R}) = \{y \in Y \mid (x, y) \in \mathscr{R} \text{ for some } x \in X\}$$

Clearly $\mathfrak{D}(\mathscr{R}) \subseteq X$, $\mathfrak{R}(\mathscr{R}) \subseteq Y$ and one or both may be $\varnothing$.

Suppose, now, that $\mathscr{R} : X \rightsquigarrow Y$ is a relation, so that $\mathscr{R} \subseteq X \times Y$. Define a relation $\mathscr{R}^{-1}: Y \rightsquigarrow X$ as follows:

$$\mathscr{R}^{-1} = \{(y, x) \in Y \times X \mid (x, y) \in \mathscr{R}\} \subseteq Y \times X.$$

We call $\mathscr{R}^{-1}$ the *inverse relation* of $\mathscr{R}$. Then $\mathfrak{D}(\mathscr{R}^{-1}) = \mathfrak{R}(\mathscr{R})$ and $\mathfrak{R}(\mathscr{R}^{-1}) = \mathfrak{D}(\mathscr{R})$.

A function (or mapping) is a special type of relation. Let $\mathscr{R} : X \rightsquigarrow Y$ be a relation such that

$$\text{if } (x, y) \in \mathscr{R} \text{ and } (x, y') \in \mathscr{R} \text{ then } y = y',$$

where $x \in X$; $y, y' \in Y$. We call $\mathscr{R}$ a *partial function* and write it as $\mathscr{R} : X \to Y$. A relation is thus a partial function if each element of the domain $\mathfrak{D}(\mathscr{R})$ is related to exactly one element of the range $\mathfrak{R}(\mathscr{R})$.

A *function* is a partial function $\mathscr{R} : X \to Y$ such that $\mathfrak{D}(\mathscr{R}) = X$.

Example 1.1

Let $X=\{a, b, c\}$, $Y=\{w, x, y, z\}$. If $\mathcal{R}_1: X \rightsquigarrow Y$ is defined by $\mathcal{R}_1=\{(a, w), (a, x), (c, w)\}$ then $\mathcal{R}_1$ is a relation, it is not a partial function since $(a, w)\in\mathcal{R}_1$ and $(a, x)\in\mathcal{R}_1$. Also note that

$$\mathfrak{D}(\mathcal{R}_1)=\{a, c\}, \quad \mathfrak{R}(\mathcal{R}_1)=\{w, x\}$$

$\mathcal{R}_1^{-1}: Y\rightsquigarrow X$ is given by $\mathcal{R}_1^{-1}=\{(w, a), (x, a), (w, c)\}$.

Now let $R_2: X\rightsquigarrow Y$ be defined by $\mathcal{R}_2=\{(a, w), (b, w)\}$ then $\mathcal{R}_2$ is a partial function, $\mathfrak{D}(\mathcal{R}_2)=\{a, b\}\neq X$, $\mathfrak{R}(\mathcal{R}_2)=\{w\}\neq Y$ and $\mathcal{R}_2^{-1}: Y\rightsquigarrow X$ is given by $\mathcal{R}_2^{-1}=\{(w, a), (w, b)\}$. Note that $\mathcal{R}_2^{-1}$ is not a partial function.

Finally define $\mathcal{R}_3: X\to Y$ by $\mathcal{R}_3=\{(a, w), (b, w), (c, x)\}$, then $\mathcal{R}_3$ is a function. $\mathfrak{D}(\mathcal{R}_3)=X$, $\mathfrak{R}(\mathcal{R}_3)=\{x, w\}\neq Y$. The relation $\mathcal{R}_3^{-1}: Y\rightsquigarrow X$ is not a partial function.

Suppose that $\mathcal{R}: X\rightsquigarrow Y$ is a relation. We say that $\mathcal{R}$ is *surjective* if $\mathfrak{R}(\mathcal{R})=Y$, and $\mathcal{R}$ is called *injective* if, given

$$(x, y)\in\mathcal{R} \text{ and } (x', y)\in\mathcal{R} \text{ then } x=x'.$$

Theorem 1.1.2

Let $\mathcal{R}: X\rightsquigarrow Y$ be a relation.

(i) If $\mathcal{R}$ is injective then $\mathcal{R}^{-1}$ is a partial function.

(ii) If $\mathcal{R}$ is injective and surjective then $\mathcal{R}^{-1}$ is a function.

Proof (i) We must show that $\mathcal{R}^{-1}: Y\rightsquigarrow X$ is a partial function, so that if

$$(y, x)\in\mathcal{R}^{-1} \text{ and } (y, x')\in\mathcal{R}^{-1} \text{ then } x=x'$$

but $(y, x)\in\mathcal{R}^{-1}$ is equivalent to $(x, y)\in\mathcal{R}$ and similarly $(y, x')\in\mathcal{R}^{-1}$ yields $(x', y)\in\mathcal{R}$. The injectivity of $\mathcal{R}$ gives us $x=x'$ and so $\mathcal{R}^{-1}$ is a partial function.

(ii) Since $\mathfrak{D}(\mathcal{R}^{-1})=\mathfrak{R}(\mathcal{R})=Y$ we see that $\mathcal{R}^{-1}: Y\to X$ is now a function. □

Examples 1.2

Let $X=\{a, b, c\}$, $Y=\{w, x, y, z\}$. Define $\mathcal{R}_4: X\rightsquigarrow Y$ by $\mathcal{R}_4=\{(a, x), (a, y), (b, z)\}$ then $\mathcal{R}_4$ is injective and $\mathcal{R}_4^{-1}=\{(x, a), (y, a), (z, b)\}$ is a partial function $\mathcal{R}_4^{-1}: Y\to X$.

If $\mathcal{R}_5: X\rightsquigarrow Y$ is given by $\mathcal{R}_5=\{(a, x), (a, y), (b, z), (a, w)\}$ then $\mathcal{R}_5^{-1}: Y\to X$ becomes $\mathcal{R}_5^{-1}=\{(w, a), (x, a), (y, a), (z, b)\}$ which is a function.

In many cases we will indicate the definition of a relation by using a diagram, for example the relation $\mathscr{R}_5$ is represented by the arrows in the following:

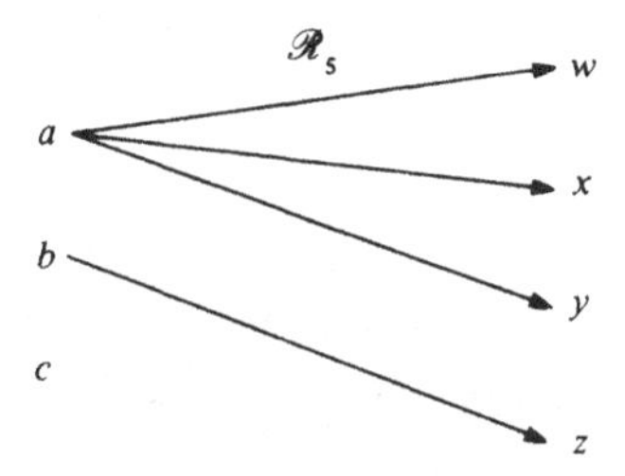

and $\mathscr{R}_5^{-1}$ is given by:

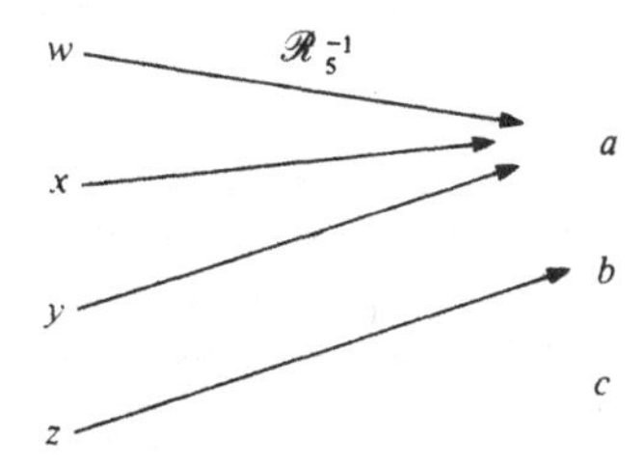

Theorem 1.1.3

Let $\mathscr{R} : X \to Y$ be a function, then $\mathscr{R}^{-1} : Y \to X$ is a function if and only if $\mathscr{R}$ is surjective and injective.

Proof If $\mathscr{R}$ is an injective, surjective function then $\mathscr{R}^{-1}$ is a function by 1.1.2. If $\mathscr{R}^{-1}$ is a function and $\mathscr{R}$ is a function then $\mathfrak{D}(\mathscr{R}^{-1}) = Y = \mathfrak{R}(\mathscr{R})$ and so $\mathscr{R}$ is surjective. Suppose that $(x, y) \in \mathscr{R}$ and $(x', y) \in \mathscr{R}$ for some $x, x' \in X$, $y \in Y$, then $(y, x) \in \mathscr{R}^{-1}$ and $(y, x') \in \mathscr{R}^{-1}$. But $\mathscr{R}^{-1}$ is a function and so $x = x'$, which means that $\mathscr{R}$ is injective. □

There is a natural concept of inclusion that can be defined between relations. If $\mathscr{R} : X \rightsquigarrow Y$ and $\mathscr{S} : X \rightsquigarrow Y$ are relations then $\mathscr{R} \subseteq X \times Y$ and $\mathscr{S} \subseteq X \times Y$. Suppose that $\mathscr{R} \subseteq \mathscr{S}$ then we see that

given $(x, y) \in \mathscr{R}$ then $(x, y) \in \mathscr{S}$.

This inclusion of relations may also be applied to partial functions in the natural way.

Example 1.3
The relation $\mathcal{R}_6 : X \rightsquigarrow Y$ given by

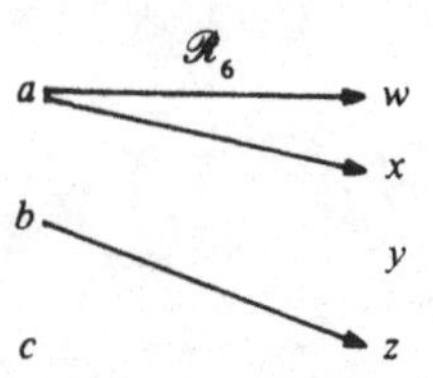

is such that $\mathcal{R}_6 \subseteq \mathcal{R}_5$ where $\mathcal{R}_5$ is defined in example 1.2.
The partial function $\mathcal{R}_2 : X \to Y$ defined in example 1.1 is

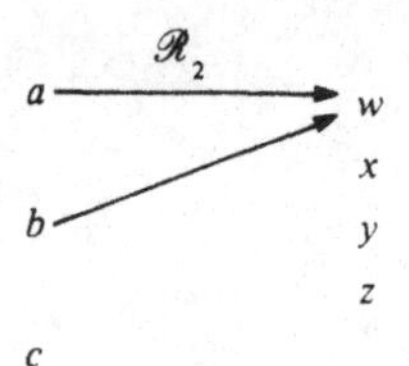

and $\mathcal{R}_3 : X \to Y$ is the function

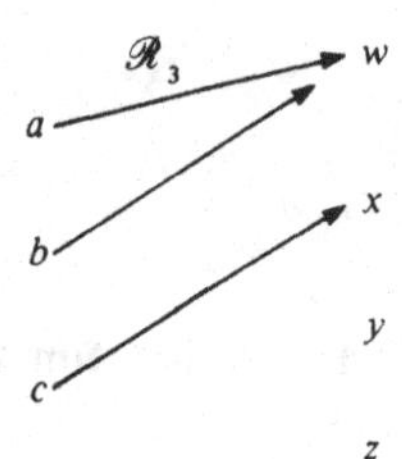

and so $\mathcal{R}_2 \subseteq \mathcal{R}_3$.

Finally we examine the problem of defining functions between empty sets. We have already noted that $\varnothing \times Y$ and $X \times \varnothing$ both equal the empty set $\varnothing$, consequently there are relations $\varnothing : \varnothing \rightsquigarrow Y$ and $\varnothing : X \rightsquigarrow \varnothing$, where X and Y are sets. Both of these relations are in fact partial functions since the condition for a partial function is satisfied vacuously. However $\varnothing : \varnothing \to Y$ is a function whereas $\varnothing : X \to \varnothing$ is not if $X \neq \varnothing$, since $\mathfrak{D}(\varnothing) = \varnothing \neq X$ in this case.

(Notice that a relation $\mathcal{R} : X \rightsquigarrow Y$ will *not* be a partial function if we can find $x \in X$, $y, y' \in Y$ such that $(x, y) \in \mathcal{R}$ and $(x, y') \in \mathcal{R}$ and $y \neq y'$. In neither of these last two cases can this be done and so both $\varnothing : \varnothing \rightsquigarrow Y$ and $\varnothing : X \to \varnothing$ are partial functions. The relations $\varnothing : \varnothing \to \varnothing$, $\boldsymbol{\theta} : X \to Y$ are in fact partial functions.)

The relation notation used here is sometimes a little cumbersome in practice and we propose to adjust it slightly especially when we are dealing with functions and partial functions.

Let $\mathcal{R}:X \rightarrow Y$ be a partial function and suppose that $x \in X$, then either there exists a $y \in Y$ such that $(x, y) \in \mathcal{R}$ or no such y exists. In the first case we will write $y = \mathcal{R}(x)$ and in the second case $\emptyset = \mathcal{R}(x)$. By a natural extension of this we will use the notation $\mathcal{R}(X)$ for $\mathfrak{R}(\mathcal{R})$, the range of $\mathcal{R}$. Generally if $X' \subseteq X$ then we write

$$\mathcal{R}(X') = \{y \in Y \mid (x', y) \in \mathcal{R} \text{ for some } x' \in X'\}.$$

Now let X_i $(i \in I)$ be a family of subsets of X. Then

$$\begin{aligned}\mathcal{R}\Big(\bigcup_{i \in I} X_i\Big) &= \Big\{y \in Y \mid (x', y) \in \mathcal{R} \text{ for some } x' \in \bigcup_{i \in I} X_i\Big\} \\ &= \bigcup_{i \in I} \{y \in Y \mid (x_i, y) \in \mathcal{R} \text{ for some } x_i \in X_i\} \\ &= \bigcup_{i \in I} \mathcal{R}(X_i).\end{aligned}$$

We describe this situation by saying that the relation $\mathcal{R}$ is *completely additive*.

If $\mathcal{R}:X \rightsquigarrow Y$ is a relation and $x \in X$ then we define

$$\mathcal{R}(x) = \{y \in Y \mid (x, y) \in \mathcal{R}\}.$$

In some circumstances $\mathcal{R}(x)$ is a singleton (for example when $\mathcal{R}$ is a function) and it is convenient to identify this singleton subset with the element it contains and in this way we may establish a coherent notation. So that if $\mathcal{R}:X \rightsquigarrow Y$ and $\mathcal{S}:X \rightsquigarrow Y$ are relations and $x, x' \in X$ then the phrase $\mathcal{R}(x) \subseteq \mathcal{S}(x')$ will be meaningful whether $\mathcal{R}$ and $\mathcal{S}$ are functions or not.

From example 1.3 we have $\mathcal{R}_2(a) \subseteq \mathcal{R}_3(a)$ meaning $\mathcal{R}_2(a) = \mathcal{R}_3(a)$ and $\mathcal{R}_2(c) \subseteq \mathcal{R}_3(c)$ which means $\emptyset \subseteq \{x\}$.

If $\mathcal{R}:X \rightsquigarrow Y$ and $\mathcal{S}: Y \rightsquigarrow Z$ are relations then we define the *composition* or *product* relation $\mathcal{S} \circ \mathcal{R}:X \rightsquigarrow Z$ by $(x, z) \in \mathcal{S} \circ \mathcal{R}$ if and only if there exists $y \in Y$ such that $(x, y) \in \mathcal{R}$ and $(y, z) \in \mathcal{S}$. Clearly if $\mathfrak{R}(\mathcal{R}) \cap \mathfrak{D}(\mathcal{S}) = \emptyset$ then $\mathcal{S} \circ \mathcal{R} = \emptyset$. When $\mathcal{R}:X \rightarrow Y$ and $\mathcal{S}: Y \rightarrow Z$ are functions then $\mathcal{S} \circ \mathcal{R}:X \rightarrow Z$ is also a function (except in the case when $\mathcal{S} = \emptyset$ and $X \neq \emptyset$).

In all cases $(\mathcal{S} \circ \mathcal{R})(x) = \mathcal{S}(\mathcal{R}(x))$ where $\mathcal{S}(\mathcal{R}(x))$ is defined to be $\bigcup\{\mathcal{S}(y) \mid y \in \mathcal{R}(x)\}$.

Since relations are defined as subsets it is possible to consider the intersection of two relations. So that if $\mathcal{R}:X \rightsquigarrow Y$ and $\mathcal{S}: X \rightsquigarrow Y$ are relations then $\mathcal{R} \subseteq X \times Y$ and $\mathcal{S} \subseteq X \times Y$. The *intersection relation*

$\mathcal{R} \cap \mathcal{S}: X \rightsquigarrow Y$ is then defined to be the subset $\mathcal{R} \cap \mathcal{S} \subseteq X \times Y$. One particular example of interest is the case where $X = Y$ and $\mathcal{R}$ and $\mathcal{S}$ are equivalence relations. If $a \in X$ then $[a]_{\mathcal{R} \cap \mathcal{S}} = [a]_{\mathcal{R}} \cap [a]_{\mathcal{S}}$, where $[a]_{\mathcal{R} \cap \mathcal{S}}$ means the equivalence class of a with respect to the equivalence relation $\mathcal{R} \cap \mathcal{S}$.

1.2 Semigroups and homomorphisms

On many occasions we will be dealing with a set S which has some additional structure. This will often take the form of a rule for 'combining' certain elements of S to 'produce' a *unique* element of S. We shall refer to such processes as multiplications on S.

If s and s_1 are elements of S and they can be combined to form a new element of S we would write this as $s \cdot s_1 = s_2$ where $s_2 \in S$. This process is somewhat more precisely stated if we use the concept of a relation. Then we are considering a relation $\mathcal{R}: S \times S \rightsquigarrow S$ defined by $((s, s_1), s_2) \in \mathcal{R}$ if and only if $s \cdot s_1 = s_2$. In fact $\mathcal{R}$ is a partial function since we want s_2 to be a *unique* element. The domain $\mathfrak{D}(\mathcal{R})$ may be a proper subset of $S \times S$ in which case we cannot multiply an arbitrary pair of elements of S. $\mathcal{R}$ is said to define a *closed partial binary operation* on S. If $\mathfrak{D}(\mathcal{R}) = S \times S$ then $\mathcal{R}$ is a *closed binary operation* on S, in this case we can multiply any two elements of S to obtain an element of S. We will usually drop the multiplication symbol when dealing with products of elements in S, writing ss_1 in place of $s \cdot s_1$.

If S has a closed binary operation satisfying the *associativity* condition $a(bc) = (ab)c$ for every $a, b, c \in S$ we call S a *semigroup*. If we need to specify the operation involved we will write '$(S, \cdot)$ is a semigroup'. Semigroups are very common in many branches of mathematics and they certainly play a central role in the theory of automata so we had better look at some of their more important properties. First of all we will examine some examples.

Examples 1.4

(i) The set of natural numbers $\{1, 2, \ldots\}$ is a semigroup with respect to both the binary operations of addition and multiplication.

(ii) The set of all integers is a semigroup with respect to the binary operations of addition and multiplication.

(iii) Let A be a set and S the set of all relations that can be defined on A. Suppose that $\mathcal{R}: A \rightsquigarrow A$ and $\mathcal{R}': A \rightsquigarrow A$ are both members of S, as before we define a new relation $\mathcal{R} \circ \mathcal{R}': A \rightsquigarrow A$ by

$$(a, b) \in \mathcal{R} \circ \mathcal{R}' \Leftrightarrow \exists c \in A \text{ such that } (a, c) \in \mathcal{R}' \textit{ and } (c, b) \in \mathcal{R}.$$

This process defines a closed binary operation on the set S which is associative and so S is a semigroup.

(iv) If A is a set and S is the set of all functions from A to A then we can define a closed binary operation on S in the same way as in (iii). So if $\mathcal{R}: A \to A$ and $\mathcal{R}': A \to A$ are functions we define a new function $\mathcal{R} \circ \mathcal{R}': A \to A$ by:

$$\mathcal{R} \circ \mathcal{R}'(a) = b \Leftrightarrow \exists c \in A \text{ such that } \mathcal{R}'(a) = c \textit{ and } \mathcal{R}(c) = b.$$

$$\text{i.e. } \mathcal{R} \circ \mathcal{R}'(a) = b \Leftrightarrow \mathcal{R}(\mathcal{R}'(a)) = b.$$

Therefore the binary operation is just function composition.

The set of all partial functions from A to A is a semigroup under this operation, the partial function $\boldsymbol{\theta}: A \to A$ belongs to the set; we denote it by $\boldsymbol{PF}(A)$.

(v) Let Σ be any non-empty set. A *string* or *word* from Σ is any finite sequence of elements from Σ. Define a closed binary operation on the set Σ^+ of all words from Σ as follows. Let $\sigma_1 \ldots \sigma_n, \sigma'_1 \ldots \sigma'_m \in \Sigma^+$, then $\sigma_1 \ldots \sigma_n \cdot \sigma'_1 \ldots \sigma'_m$ is called their *concatenation* and is clearly a word from Σ. It is easy to see that Σ^+ is a semigroup with respect to this operation. We shall call Σ^+ the *free semigroup generated by the set* Σ. We regard Σ as being embedded in Σ^+.

(vi) The empty set is a semigroup, the binary operation is the empty function $\varnothing: \varnothing \times \varnothing \to \varnothing$ which is associative – a set S with a closed binary operation is a semigroup *unless* we can find elements $a, b, c \in S$ such that

$$a(bc) \neq (ab)c.$$

If $(S, \cdot)$ is a semigroup and $A \subseteq S$, $B \subseteq S$ are subsets we define

$$AB = \{s \in S \mid s = ab \text{ for some } a \in A, b \in B\}.$$

Closely associated with the idea of a semigroup is the concept of a monoid. For this we need another definition.

Let S be a semigroup, an element $e \in S$ is called a *unit element* of S if $ae = ea = a$ for every $a \in S$. A simple exercise shows that if S possesses a unit element then that element is unique.

A semigroup possessing a unit element is called a *monoid*.

Examples 1.5

(i) Both semigroups in example 1.4(i) are monoids, in fact all the semigroups in that example with the exception of 1.4(v) and 1.4(vi) are monoids.

(ii) Let Σ be a non-empty set. We have already examined the set Σ^+ consisting of all words from Σ. Let us adjoin an extra element, called

the *null word* and denoted by Λ, this is just the empty sequence and has the following formal properties:
if

$$a \in \Sigma^+$$

then

$$\Lambda a = a\Lambda = a$$

and so Λ acts like a unit. If we define $\Sigma^* = \Sigma^+ \cup \{\Lambda\}$ then Σ^* is a monoid. It is called the *free monoid generated by the set* Σ.

This procedure gives us a general method for transforming a semigroup into a monoid.

Let S be a semigroup. Suppose that S is not a monoid, choose any element $e \notin S$ and form the set $S^{\cdot} = S \cup \{e\}$, we define a multiplication denoted by $*$ on $S^{\cdot}$ by extending the multiplication already on S so that

$$a * b = \begin{cases} ab & \text{if } a, b \in S \\ b & \text{if } a = e \\ a & \text{if } b = e \end{cases}$$

where $a, b \in S^{\cdot}$.

Clearly $S^{\cdot}$ is a monoid with unit element e. If S is already a monoid we will define $S^{\cdot} = S$.

Let $(S, \cdot)$ and $(T, *)$ be semigroups and $f: S \to T$ be a mapping. We call f a *semigroup homomorphism* if

$$f(a) * f(b) = f(ab) \quad \text{for all } a, b \in S.$$

If $(S, \cdot)$ and $(T, *)$ are monoids with identities e and e' respectively and $f: S \to T$ is a semigroup homomorphism such that

$$f(e) = e'$$

then f is called a *monoid homomorphism*.

Semigroup homomorphisms forge a strong link between the semigroup structures concerned and are of great importance in the development of the theory. Sometimes, however, it is necessary to consider slightly more general relationships between semigroups. If $\mathcal{R}: S \rightsquigarrow T$ is a relation then it is called a *semigroup relation* if

$$\mathcal{R}(a) * \mathcal{R}(b) \subseteq \mathcal{R}(ab) \quad \text{for } a, b \in S.$$

What does this mean? Well suppose that $c \in T$ with $(a, c) \in \mathcal{R}$ and $d \in T$ with $(b, d) \in \mathcal{R}$, then

$$c * d \in \mathcal{R}(a) * \mathcal{R}(b) \subseteq \mathcal{R}(ab)$$

which means that $(ab, c * d) \in \mathcal{R}$.

Example 1.6

Let $S = \{0, 1\}$ and $T = \{a, b, c, d\}$ with multiplications defined by the following tables:

S	0	1
0	0	1
1	1	1

T	a	b	c	d
a	a	b	c	d
b	b	c	a	d
c	c	a	b	d
d	d	d	d	d

so that, for example $a * c = c$, where c is the common entry in the row labelled a and the column labelled c, etc.

We will define various relations and functions between these semigroups:

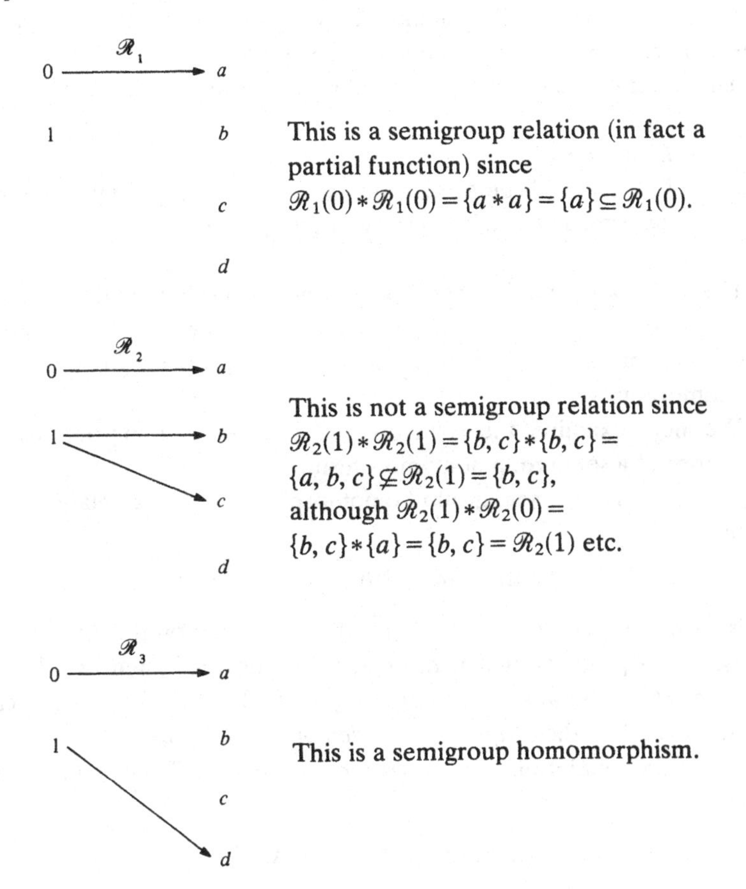

This is a semigroup relation (in fact a partial function) since $\mathcal{R}_1(0) * \mathcal{R}_1(0) = \{a * a\} = \{a\} \subseteq \mathcal{R}_1(0)$.

This is not a semigroup relation since $\mathcal{R}_2(1) * \mathcal{R}_2(1) = \{b, c\} * \{b, c\} = \{a, b, c\} \not\subseteq \mathcal{R}_2(1) = \{b, c\}$, although $\mathcal{R}_2(1) * \mathcal{R}_2(0) = \{b, c\} * \{a\} = \{b, c\} = \mathcal{R}_2(1)$ etc.

This is a semigroup homomorphism.

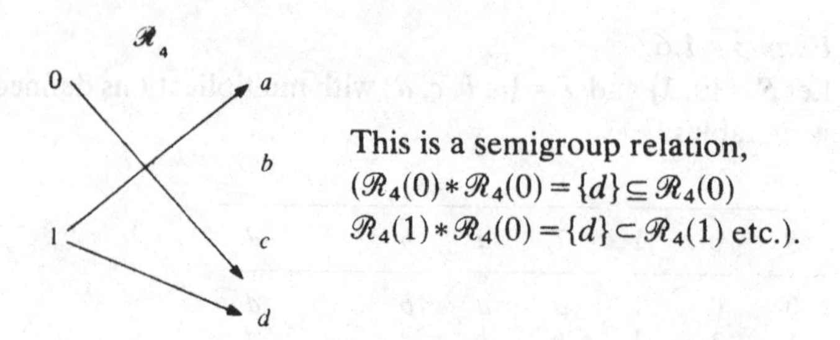

$\mathcal{R}_4$
0
1
a
b
c
d
This is a semigroup relation,
$(\mathcal{R}_4(0) * \mathcal{R}_4(0) = \{d\} \subseteq \mathcal{R}_4(0)$
$\mathcal{R}_4(1) * \mathcal{R}_4(0) = \{d\} \subset \mathcal{R}_4(1)$ etc.).

Let S be a semigroup and S' a subset of S such that if $a, b \in S'$ then $ab \in S'$, then we say that S' is a *subsemigroup* of S. Usually we denote this by $S' \subseteq S$ and rely on the context to indicate that S' is a subsemigroup of S rather than just a subset. If X is a subset of S define $\langle X \rangle$ to be the intersection of all subsemigroups of S that contain X. Then $\langle X \rangle$ is the *subsemigroup generated by X.*

In the case where S is a monoid with unit element $e \in S$ then a subsemigroup $S' \subseteq S$ is a *submonoid* of S if e also belongs to S'.

The proof of the following elementary fact is left to the reader.

Lemma 1.2.1

Let S and T be semigroups and $f: S \rightarrow T$ a partial semigroup homomorphism. Then $f(S)$ is a subsemigroup of T.

If S and T are monoids then $f(S)$ may not be a submonoid of T unless $f(e)$ is the unit element of T, where e is the unit of S. Semigroup homomorphisms possessing this unit-preserving property are the monoid homomorphisms.

We next examine the structure induced on a semigroup by the existence of a semigroup homomorphism.

Let $f: S \rightarrow T$ be a semigroup homomorphism, define a relation $\sim$ on S by

$$a \sim b \text{ if and only if } f(a) \doteq f(b), \qquad \text{for } a, b \in S.$$

It is easily verified that $\sim$ is an equivalence relation on the set S. It also satisfies the property that if $a, b \in S$, $a \sim b$ and $s \in S$ then $as \sim bs$ and $sa \sim sb$. This is because $f(as) = f(a) * f(s) = f(b) * f(s) = f(bs)$ etc. Such a relation is called the *congruence relation on S defined by f.*

Given any semigroup S a *congruence relation* on S is an equivalence relation $\sim$ satisfying:

$$a \sim b \Rightarrow as \sim bs \text{ and } sa \sim sb \qquad \text{for all } s \in S.$$

We shall shortly see that every congruence relation is defined by some suitable semigroup homomorphism.

First let S be a semigroup and $\sim$ a congruence relation on S. Consider the set of all equivalence classes defined on S by $\sim$, denote this set by $S/\sim$. We define a multiplication on $S/\sim$ as follows:

let $[a], [b] \in S/\sim$

put $[a]*[b]=[ab] \quad (a, b \in S)$.

This operation is well defined for if

$[a]=[c]$ and $[b]=[d]$ then

$a \sim c$ and $b \sim d$, consequently

$ab \sim cb$ and $cb \sim cd$ and so $ab \sim cd$

that is $[ab]=[cd]$ and thus $[a]*[b]=[c]*[d]$.

Furthermore $S/\sim$ under the operation $*$ is a semigroup. We call $(S/\sim, *)$ the *quotient semigroup of S with respect to* $\sim$. There is a semigroup homomorphism $f: S \to S/\sim$ defined by $f(a)=[a]$, $a \in S$. This is called the *natural homomorphism onto* $S/\sim$.

Notice further that the congruence relation on S defined by f is just the relation $\sim$ that we started with. There is thus a precise correspondence between semigroup homomorphisms and congruences.

We will finish this section with two important but elementary results that are of fundamental importance.

Theorem 1.2.2

Let $f: S \to T$ be a surjective semigroup homomorphism and $\sim$ the congruence induced on S by f. There exists a bijective homomorphism $f^*: S/\sim \to T$ such that $f^*([a])=f(a)$ for each $a \in S$.

Proof The definition of f^* specified in the hypothesis is well-defined and using it we will just establish that f^* is bijective and a semigroup homomorphism. First let $f^*([a])=f^*([b])$ where $[a], [b] \in S/\sim$ then $f(a)=f(b)$ and so $a \sim b$ which means $[a]=[b]$. Thus f^* is injective. Next let $t \in T$, then there exists $a \in S$ such that $t=f(a)$ and consequently $f^*([a])=f(a)=t$ giving the surjectivity of f^*.

Finally let $[a], [b] \in S/\sim$, then $[a]*[b]=[ab]$ and

$$f^*([a]*[b])=f^*([ab])=f(ab)=f(a)f(b)=f^*([a])f^*([b])$$

proving that f^* is a semigroup homomorphism. □

We usually call a bijective semigroup homomorphism $f: S \to T$ an *isomorphism* and write $S \approx T$ to indicate that an isomorphism exists between S and T.

Theorem 1.2.3

Let Σ be a non-empty set, T a semigroup and $f: \Sigma \to T$ a mapping. There is a mapping $g: \Sigma^+ \to T$ which is a semigroup homomorphism and such that $g(\sigma) = f(\sigma)$ for all $\sigma \in \Sigma$.

Proof Let $a \in \Sigma^+$ then $a = \sigma_1\sigma_2 \ldots \sigma_n$ for some $\sigma_i \in \Sigma$; $i = 1, 2, \ldots, n$. Define $g(a) = f(\sigma_1)f(\sigma_2) \ldots f(\sigma_n)$ (using the multiplication in T) and so $g: \Sigma^+ \to T$ is defined. For $a, b \in \Sigma^+$ it is immediate that

$$g(ab) = g(a)g(b)$$

and so g is a semigroup homomorphism. □

In fact g is the unique homomorphism satisfying $g(\sigma) = f(\sigma)$ for $\sigma \in \Sigma$. This property of the semigroup Σ^+ is what gives it the name 'free semigroup'.

Now let S be a semigroup, a semigroup homomorphism $f: S \to S$ will be called an *endomorphism*. For any semigroup S the set of all endomorphisms will be denoted by **End**(S). The set **End**(S) has a natural semigroup structure with respect to 'composition of mappings'. Since the identity mapping $1_S: S \to S$ defined by $1_S(a) = a$ for all $a \in S$ is clearly an endomorphism we see that **End**(S) is actually a monoid.

If S is a semigroup and X is a non-empty subset of S we define the subsemigroup of S generated by X to be the intersection of all the subsemigroups of S that contain the subset X and denote this by $\langle X \rangle$.

Let n be a positive integer and write

$$\mathbf{n} = \{0, 1, \ldots, n-1\}.$$

Consider the set S of all functions of $\mathbf{n}$ into itself. This is a semigroup under the operation of function composition. The semigroup S has order n^n and is in fact a monoid with identity the identity map $1_\mathbf{n}$.

Let $s \in S$ be defined by:

$$s(x) = x + 1 \quad \text{for } 0 \le x < n - 1$$
$$s(n-1) = 0.$$

The subsemigroup generated by the set $\{s\}$ is the set

$$\langle\{s\}\rangle = \{1_\mathbf{n}, s, s^2, \ldots, s^{n-1}\}$$

which is a group satisfying $s^n = 1_\mathbf{n}$.

Another subsemigroup of interest is the subset

$$R = \{t \in S \mid |t(\mathbf{n})| = 1\}$$
$$= \{\bar{k} \mid k \in \mathbf{n}\}$$

where $\bar{k}$ is the function defined by

$$\bar{k}(x) = k$$

for all $x \in \mathbf{n}$.

1.3 Products

Semigroups can be joined together in various ways to produce more semigroups. We will examine here some important methods of doing this.

Let S and T be semigroups. Consider the set $S \times T$, the cartesian product of S and T, and define a multiplication on $S \times T$ as follows:

$$(a, x) \cdot (b, y) = (ab, xy)$$

where $a, b \in S$; $x, y \in T$.

The result is a semigroup $(S \times T, \cdot)$ which is called the *direct product* of S and T, written $S \times T$. Associated with the direct product $S \times T$ are two important functions:

$$p_1 : S \times T \to S \text{ defined by } p_1(a, x) = a \quad (a \in S, x \in T)$$

and

$$p_2 : S \times T \to T \text{ defined by } p_2(a, x) = x \quad (a \in S, x \in T).$$

These are called the *projections* onto the first and second factors respectively and are easily seen to be surjective semigroup homomorphisms. They satisfy an important property:

Theorem 1.3.1

Let S, T and W be semigroups and $f_1 : W \to S$, $f_2 : W \to T$ semigroup homomorphisms. There exists a unique semigroup homomorphism $g : W \to S \times T$ such that $p_i \circ g = f_i$ for $i = 1, 2$.

Proof Define $g(w) = (f_1(w),\ f_2(w)) \in S \times T$ for each $w \in W$. Then for $w, w' \in W$

$$\begin{aligned} g(ww') &= (f_1(ww'), f_2(ww')) \\ &= (f_1(w)f_1(w'), f_2(w)f_2(w')) \\ &= (f_1(w), f_2(w)) \cdot (f_1(w'), f_2(w')) \\ &= g(w) \cdot g(w') \end{aligned}$$

and so g is a semigroup homomorphism.

Clearly

$$p_1(g(w)) = p_1(f_1(w), f_2(w)) = f_1(w)$$

and

$$p_2(g(w)) = p_2(f_1(w), f_2(w)) = f_2(w).$$

Finally let $h: W \to S \times T$ also be a semigroup homomorphism with the property that $p_i \circ h = f_i$, $i = 1, 2$. Now let $h(w) = (a, x)$ where $w \in W$, $a \in S$, $x \in T$. Then $p_1(h(w)) = a = f_1(w)$, $p_2(h(w)) = x = f_2(w)$ and so $h(w) = (f_1(w), f_2(w)) = g(w)$. □

Given three semigroups S, T, W we can form $S \times T$ and then $(S \times T) \times W$; similarly $S \times (T \times W)$ can be constructed and it is natural to ask about the relationship between these two semigroups.

Lemma 1.3.2
$(S \times T) \times W \approx S \times (T \times W)$.
Proof The isomorphism is $f: (S \times T) \times W \to S \times (T \times W)$ defined by

$$f((a, b), c) = (a, (b, c))$$

where $a \in S$, $b \in T$, $c \in W$. □

Now let S and T be semigroups and suppose that $\theta: T \to \mathbf{End}\, S$ is a semigroup homomorphism. We will define another product on the set $S \times T$, let $t, t' \in T$; s, $s' \in S$ put $(s, t) \circledast (s', t') = (s\theta(t)(s'), tt')$, where $\theta(t): S \to S$ and so $\theta(t)(s') \in S$. Clearly $S \times T$ is a semigroup with respect to this product, since

$$\begin{aligned}
&((s, t) \circledast (s', t')) \circledast (s'', t'') \\
&\quad = (s\theta(t)(s'), tt') \circledast (s'', t'') \\
&\quad = (s\theta(t)(s')\theta(tt')(s''), tt't'') \\
&\quad = (s\theta(t)(s')\theta(t)(\theta(t')(s'')), tt't''), \quad \text{as } \theta \text{ is a homomorphism,} \\
&\quad = (s\theta(t)(s'\theta(t')(s'')), tt't''), \quad \text{as } \theta(t) \in \mathbf{End}\, S, \\
&\quad = (s, t) \circledast (s'\theta(t')(s''), t't'') \\
&\quad = (s, t) \circledast ((s', t') \circledast (s'', t''))
\end{aligned}$$

This semigroup is called the *semidirect product* of S and T with respect to θ and denoted by $S \times_\theta T$.

Our third product is constructed in the following way, let $S^{T^\cdot}$ denote the set of all functions from the monoid $T^\cdot$ to the semigroup S. We define a multiplication $\circ$ on the set $S^{T^{\cdot}} \times T$ by putting

$$(f, t) \circ (g, t') = (f \circ g, tt')$$

where $f \circ g \in S^{T^{\cdot}}$ is defined by $(f \circ g)(x) = f(x)g(xt)$ for $x \in T^{\cdot}$, $f, g \in S^{T^{\cdot}}$, $t, t' \in T$. If $f, g, h \in S^{T^{\cdot}}$ and $t, t', t'' \in T$ then

$$\begin{aligned}((f, t) \circ (g, t')) \circ (h, t'') &= (f \circ g, tt') \circ (h, t'') \\ &= ((f \circ g) \circ h, tt't'')\end{aligned}$$

and

$$(f, t) \circ ((g, t') \circ (h, t'')) = (f \circ (g \circ h), tt't'').$$

Now if $x \in T^{\cdot}$ then

$$\begin{aligned}((f \circ g) \circ h)(x) &= (f \circ g)(x)h(xtt') \\ &= f(x)g(xt)h(xtt')\end{aligned}$$

and

$$\begin{aligned}(f \circ (g \circ h))(x) &= f(x)(g \circ h)(xt) \\ &= f(x)g(xt)h(xtt')\end{aligned}$$

and thus we have established the associativity of the multiplication on the set $S^{T^{\cdot}} \times T$. We call the semigroup $S^{T^{\cdot}} \times T$ the *wreath product of S and T* and write it as $S \circ T$.

Now let S and T be semigroups and consider the set $S^{T^{\cdot}}$ of all mappings from $T^{\cdot}$ into S. Suppose that $f: T^{\cdot} \to S$ and $t \in T$ then we may define a mapping $f_t : T^{\cdot} \to S$ by $f_t(x) = f(xt)$ where $x \in T^{\cdot}$. The set $S^{T^{\cdot}}$ is a semigroup under the multiplication induced by the semigroup S, for example if $f, g \in S^{T^{\cdot}}$ then

$$(fg)(x) = f(x)g(x) \in S \quad \text{where } x \in T^{\cdot} \tag{$*$}$$

Furthermore the mapping $\theta_t : S^{T^{\cdot}} \to S^{T^{\cdot}}$ defined by

$$\theta_t(f) = f_t, \quad f \in S^{T^{\cdot}}$$

is an endomorphism, for if $f, g \in S^{T^{\cdot}}$ then $\theta_t(fg) = (fg)_t$ where $(fg)_t(x) = fg(xt) = f(xt)g(xt)$ by $(*)$ and $\theta_t(f) \cdot \theta_t(g) = f_t g_t$ where $(f_t g_t)(x) = f_t(x)g_t(x) = f(xt)g(xt)$ and thus

$$\theta_t(fg) = \theta_t(f)\theta_t(g).$$

Consequently there exists a mapping $\theta: T \to \mathbf{End}(S^{T^{\cdot}})$ defined by $\theta(t) = \theta_t$ for $t \in T$. It is now possible to define the semidirect product $S^{T^{\cdot}} \times_\theta T$ and we note that if $(f, t), (g, t') \in S^{T^{\cdot}} \times T$ then their semidirect multiplication is given by

$$(f, t) \circledast (g, t') = (f\theta_t(g), tt')$$

where

$$(f\theta_t(g))(x) = f(x)g(xt) \quad \text{for all } x \in T^{\cdot}$$
$$= (f \circ g)(x) \quad \text{for all } x \in T^{\cdot}$$

and so

$$(f, t) \circledast (g, t') = (f, t) \circ (g, t').$$

We thus have:

Theorem 1.3.3
If S and T are semigroups then there exists $\theta : T \to \mathbf{End}(S^{T^{\cdot}})$ such that

$$S \circ T = S^{T^{\cdot}} \times_\theta T.$$

Before we examine the last method for combining two semigroups we must introduce the idea of a zero.

If S is a semigroup an element $s \in S$ is called a *zero* of S if

$$sa = as = s \quad \text{for all } a \in S.$$

The empty partial function $\boldsymbol{\theta} : A \to A$ is a zero for $\mathbf{PF}(A)$.

A zero element, if it exists, is necessarily unique. Not all semigroups possess a zero but one can easily be adjoined, so that if S is an arbitrary semigroup without a zero element we define $S^0 = S \cup \{0\}$ where $0 \notin S$ and define

$$a * b = \begin{cases} ab & \text{if } a, b \in S \\ 0 & \text{otherwise,} \end{cases}$$

then S^0 is a semigroup under $*$ and 0 is the zero element of S^0.

Now let S and T be semigroups and suppose that $S \neq \emptyset$ and $T \neq \emptyset$. Consider their disjoint union $S \cup T$: this is not in general a semigroup but if we adjoint a zero element in a suitable way we can construct a semigroup multiplication.

Let $0 \notin S \cup T$ and put

$$S \vee T = S \cup T \cup \{0\}$$

and define

$$a * b = \begin{cases} ab & \text{if } a, b \in S \\ ab & \text{if } a, b \in T \\ 0 & \text{otherwise.} \end{cases}$$

Then $S \vee T$ is a semigroup under $*$ with a zero. We call it the *join of S and T.*

Occasionally we need to consider the case when either S or T is the empty semigroup $\varnothing$ in which case we define

$$S \vee \varnothing = \varnothing \vee S = S.$$

1.4 Groups

Let G be a monoid with identity e. Suppose that for each $g \in G$ there exists an element $\hat{g} \in G$ such that $g\hat{g} = e = \hat{g}g$. We say that G is a *group* and usually write the element $\hat{g}$ as g^{-1} and call it the *inverse* of g.

Groups are an important mathematical concept, they arise in many situations and we now briefly state some of their important properties that we will need later.

A group G is *abelian* if $gg_1 = g_1g$ for any $g, g_1 \in G$.

The *cyclic group of order n* is the set

$$\mathbf{n} = \{0, 1, \ldots, n-1\}$$

with multiplication defined by

$$xy = r$$

where $x + y = qn + r$ and $0 \le r \le n-1$. This group is abelian and is usually written as $\mathbf{Z}_n$.

A *subgroup* of a group G is a submonoid $H \subseteq G$ such that $hh_1^{-1} \in H$ for each $h, h_1 \in H$. A subgroup must contain $\{e\}$. The identity singleton $\{e\}$ and the group G are both subgroups of G. If they are the only subgroups then $G \approx \mathbf{Z}_p$ for some prime number p. (Groups are said to be isomorphic if they are isomorphic as monoids.)

Let H be a subgroup of G, we say that H is *normal in* G, written $H \lhd G$, if $g^{-1}hg \in H$ for all $h \in H$, $g \in G$. This is equivalent to saying that a semigroup homomorphism $f: G \to G'$ exists where G and G' are groups, $H = f^{-1}(\{e'\})$ and e' is the identity of G'.

Given any finite group G and a subgroup H we form the *right cosets* of H in G. These are all subsets of the form $Hg = \{hg \mid h \in H\}$. It is easily verified that the set G/H of *distinct* right cosets forms a partition of the set G. The equivalence relation defined by this partition is given by

$$g \sim g_1 \Leftrightarrow g_1 = hg$$

for some $h \in H$ where $g, g_1 \in G$.

If H is normal in G we can define a multiplication on the set, G/H, of all distinct right cosets of H in G by

$$Hg \cdot Hg_1 = Hgg_1, \quad Hg, Hg_1 \in G/H.$$

This turns G/H into a group with identity $He = H$.

The function $f_H: G \to G/H$ defined by

$$f_H(g) = Hg \quad \text{for } g \in G$$

is a homomorphism onto G/H and $H = f_H^{-1}(\{H\})$. We call f_H the *natural* (or canonical) homomorphism onto G/H.

Theorem 1.4.1

Let $f: G \to G_1$ be a homomorphism of the group G onto G_1. If $H = f^{-1}(\{e_1\})$ then $G_1 \approx G/H$.

Proof Construct a function $\phi: G/H \to G_1$ by

$$\phi(Hg) = f(g) \quad \text{for } Hg \in G/H.$$

This is well-defined for if $Hg = Hg'$ then $g' = hg$ for some $h \in H$ and $\phi(Hg') = f(g') = f(hg) = f(h)f(g) = e_1 f(g) = f(g) = \phi(Hg)$.

Furthermore it is easy to establish that ϕ is an isomorphism. □

Theorem 1.4.2

Let $H \lhd G$. There is a one-one correspondence between the subgroups of G/H and subgroups of G that contain H.

Proof Let K be a subgroup of G/H, so that K is a collection of cosets of the form Hg, $(g \in G)$. Recall that $f_H: G \to G/H$ is an onto homomorphism. Let $L = f_H^{-1}(K)$, then $L \subseteq G$. Since He, the identity of G/H, belongs to K we see that $H = f_H^{-1}(He) \subseteq L$. If $l, l_1 \in L$ then

$$f_H(ll_1^{-1}) = Hll_1^{-1} = Hl(Hl_1)^{-1} \in K$$

so L is a subgroup. Similarly given a subgroup $L \subseteq G$ with $H \subseteq L$ then $f_H(L)$ is a subgroup of G/H. □

A group G, with $|G| > 1$, is *simple* if its only normal subgroups are $\{e\}$ and G. A normal subgroup $H \lhd G$ is a *maximal proper normal subgroup* if $H \neq G$ and whenever $H \subseteq K \subseteq G$ with $K \neq G$ and $K \lhd G$ then $K = H$.

Let G be a group and $H \lhd G$. H is a maximal proper normal subgroup if and only if G/H is simple.

Our next result is of particular importance.

Theorem 1.4.3

Let G be a finite group. A sequence of subgroups

$$G = G_n \supset G_{n-1} \supset \ldots \supset G_1 \supset G_0 = \{e\}$$

exists such that

(i) $G_i \lhd G_{i+1}$ for $i = 0, \ldots, n-1$

(ii) G_{i+1}/G_i is simple for $i = 0, \ldots, n-1$.

Proof Choose first a maximal proper normal subgroup of G. If this is $\{e\}$ then G is simple and the result holds. If not, let this subgroup be H, then $|H| < |G|$ and G/H is simple. Now put $G_{n-1} = H$. Consider G_{n-1} and choose a maximal proper normal subgroup of G_{n-1}, call it G_{n-2}, then G_{n-1}/G_{n-2} is simple. We may continue this process, the finiteness of the set G will force an end after a finite number of steps. □

We call such a sequence a *composition series for G of length n.* The following theorem, known as the Jordan–Hölder theorem, is proved in most text-books on group theory.

Theorem 1.4.4
Let

$$G = G_n \supset G_{n-1} \supset \ldots \supset G_1 \supset G_0 = \{e\}$$

and

$$G = K_m \supset K_{m-1} \supset \ldots \supset K_1 \supset K_0 = \{e\}$$

be composition series for the finite group G. Then $m = n$ and for each $j \in \{0, \ldots, m-1\}$ there exists a distinct $i \in \{0, \ldots, n-1\}$ such that

$$K_{j+1}/K_j \approx G_{i+1}/G_i, \quad \text{and conversely.}$$

1.5 Permutation groups

Let Q be a finite non-empty set with $|Q| > 1$. The set A of all bijective functions of Q onto Q can be given the structure of a group, by using the composition of functions as a multiplication. We will write the operation of the function on the right hand side, so that if $q \in Q$ and $\alpha \in A$ then $q\alpha$ will denote the image of q under α.

Now let $\alpha, \alpha' \in A$ and define $\alpha\alpha'$ by $q(\alpha\alpha') = (q\alpha)\alpha'$ for all $q \in Q$. Then $\alpha\alpha' \in A$ and under this multiplication A becomes a group with identity 1_Q.

We call A the *group of all permutations of* Q. If $Q = \mathbf{n} = \{0, 1, \ldots, n-1\}$ we denote A by $\mathbf{S}_n$. The subgroups G of A are called *permutation groups on* Q. Notice that if G is a permutation group on Q then the following conditions are satisfied:

(i) There exists a function $F: Q \times G \to Q$ defined by $F(q, g) = qg$, $q \in Q$; $g \in G$, called the *action of G on Q*.

(ii) $(qg)g_1 = q(gg_1)$ for $q \in Q$; $g, g_1 \in G$.

(iii) If $qg = qg_1$ for all $q \in Q$ then $g = g_1$ $(g, g_1 \in G)$.

If G is a permutation group on Q we call G *transitive on Q* if given $q, q' \in Q$ there exists $g \in G$ such that $q' = qg$. If G equals A, the set of all bijective mappings of Q onto Q, then it is transitive. If G is a subgroup of A it may not be transitive. Given $q \in Q$ the subset $qG = \{qg \mid g \in G\}$ is called the *orbit of q*. The set of distinct orbits of Q (with respect to G) partitions the set Q, for if $qG \cap q'G \neq \varnothing$ then $x = qg = q'g'$ for some $g, g' \in G$. Then $q' = qg(g')^{-1}$ so $q'G \subseteq qG$ and similarly $qG \subseteq q'G$. Finally for $q \in Q$ we have $q = q1_Q \in qG$. This partition is called the *orbit decomposition* of Q (with respect to G). There is an equivalence relation on Q associated with this partition, it is defined by

$$q \sim q' \Leftrightarrow q' = qg \quad \text{for some } g \in G.$$

A transitive permutation group yields an orbit decomposition involving one orbit, namely $Q = qG$ for any $q \in Q$.

Let G be a transitive permutation group on Q, a subset $P \subseteq Q$ such that $|P| > 1$, $P \neq Q$ and $P \cap Pg = P$ or $\varnothing$ for each $g \in G$ is called a *primitive block* of Q with respect to G. Thus each permutation either fixes P, i.e. $Pg = P$, or moves it away from P $(P \cap Pg = \varnothing)$. A *primitive* permutation group is a transitive permutation group with no primitive blocks. An *imprimitive* permutation group is a transitive permutation group with primitive blocks.

1.6 Exercises

1.1 Let $\mathbb{Z}$ be the set of all integers and let n be any positive integer. Define a relation $\mathcal{R}_n$ by

$$(a, a') \in \mathcal{R}_n \Leftrightarrow n \text{ divides } a - a'.$$

Prove that $\mathcal{R}_n$ is an equivalence relation. Describe the set $\mathbb{Z}/\mathcal{R}_n$.

1.2 Prove that a relation $\mathcal{R}: X \rightsquigarrow Y$ may be identified with a function $f: \mathcal{P}(X) \to \mathcal{P}(Y)$ satisfying the condition $f(\bigcup_{i \in I} A_i) = \bigcup_{i \in I} f(A_i)$ where $\{A_i \mid i \in I\}$ is any collection of subsets of X. (Note that $\mathcal{P}(X)$ is the set of subsets of X.)

1.3 Let A be a non-empty set and $\mathcal{R}$ an equivalence relation on A. Define the relation $\bar{\mathcal{R}}: A \rightsquigarrow A/\mathcal{R}$ by

$$\bar{\mathcal{R}} = \{(a, [a]) \mid a \in A\}.$$

Prove that $\bar{\mathcal{R}}$ is a surjective function.

1.4 If A is any non-empty set show that no surjective function $f: A \to \mathcal{P}(A)$ can exist. Can a surjective relation $\mathcal{R} : A \rightsquigarrow \mathcal{P}(A)$ exist?

1.5 Let $\mathcal{R} : X \rightsquigarrow Y$ be any relation, show that

$$1_{\mathfrak{R}(\mathcal{R})} \subseteq \mathcal{R} \circ \mathcal{R}^{-1}, \quad 1_{\mathfrak{D}(\mathcal{R})} \subseteq \mathcal{R}^{-1} \circ \mathcal{R}.$$

If $\mathcal{R}$ is injective establish

$$\mathcal{R}^{-1} \circ \mathcal{R} = 1_{\mathfrak{D}(\mathcal{R})}.$$

If $\mathcal{R}$ is a partial function prove that

$$\mathcal{R} \circ \mathcal{R}^{-1} = 1_{\mathfrak{R}(\mathcal{R})}.$$

If $\mathcal{R}$ is an injective function then

$$\mathcal{R}^{-1} \circ \mathcal{R} = 1_X, \quad \mathcal{R} \circ \mathcal{R}^{-1} = 1_{\mathfrak{R}(\mathcal{R})}.$$

If $\mathcal{R}$ is a surjective partial function show that

$$\mathcal{R} \circ \mathcal{R}^{-1} = 1_Y.$$

If $\mathcal{R}$ is a surjective function establish that

$$\mathcal{R} \circ \mathcal{R}^{-1} = 1_Y, \quad 1_X \subseteq \mathcal{R}^{-1} \circ \mathcal{R}.$$

If $\mathcal{R}$ is a surjective and injective function prove that

$$\mathcal{R} \circ \mathcal{R}^{-1} = 1_Y, \quad \mathcal{R}^{-1} \circ \mathcal{R} = 1_X.$$

1.6 If $\mathcal{R} : S \rightsquigarrow T$ is a semigroup relation show that

$$\mathcal{R}^{-1} : T \rightsquigarrow S$$

is also a semigroup relation.

1.7 Investigate the monoid analogues of theorems 1.2.2 and 1.2.3.

1.8 Let S be a finite semigroup. Let $s \in S$, consider the set $\{s, s^2, \ldots, s^n, \ldots\}$. Since S is finite we must have integers p and r such that $s^{p+r} = s^p$. Show that, if p and r are chosen suitably the set $s^p, s^{p+1}, \ldots, s^{p+r-1}$ is a subgroup of S.

1.9 Show that a finite semigroup S contains an element $s \in S$ satisfying $s^2 = s$.

1.10 A semigroup S is called *free* on Σ if a set Σ exists such that $S = \Sigma^+$. Prove that S is free on Σ if and only if $\Sigma \subseteq S$ and every element of S can be expressed uniquely as a finite product of elements of Σ.

1.11 If $S = \Sigma^+$ then $\Sigma = S \backslash S^2$ where $S^2 = \{s \cdot s_1 \mid s, s_1 \in S\}$.

1.12 S is a free semigroup if and only if

(i) $sa = sb \Rightarrow a = b$,

(ii) $as = bs \Rightarrow a = b$,

(iii) s has no identity element,

(iv) if $as = bt$ then either $a = b$, $a = bc$ or $b = ad$ for c, $d \in S$,

(v) each element has a finite number of left divisors.

1.13 Let T be a subsemigroup of Σ^+ then T is a free semigroup if and only if

$$Ts \cap T \neq \varnothing \quad \text{and} \quad sT \cap T \neq \varnothing$$

implies $s \in T$.

1.14 Find an example of a subsemigroup $T \subseteq \Sigma^+$ such that T is not a free semigroup.

2

Machines and semigroups

One of the achievements of modern science has been the realization that very few things in the world are completely static. The behaviour of many systems, both organic and synthetic, is influenced greatly by environmental changes. This interaction between a system and its environment can be vastly complicated and yet it is an area that we must try to understand if we are going to be in a position to predict the behaviour of the system and its effect on its environment.

The particular type of analysis that we present here is based on techniques that are generally referred to as *algebraic*. In some cases we will draw on established algebraic results but in general it is a new type of algebra that has arisen from a desire to understand the behaviour of a system in an environment. This is perhaps the most refreshing aspect of the theory. Here, for a change, is a subject whose motivation can be linked to very real problems in the modern world, a subject that has a short but dramatic history and one which has played a large role in the development of the fundamentals of computer science. However its achievements have not been restricted to this case alone and we hope to illustrate this when we examine the examples at the end of this chapter.

In many of the systems environmental changes alter the behaviour of the system and these changes in behaviour then affect the environment in some way. In other examples the only thing altered in the system is some internal quality. These latter systems are easier to analyse mathematically and so we shall start our considerations with them, although, as we see later, the other types of system can also be brought into this discussion in a meaningful and elementary way.

2.1 State machines

Suppose that we have a system which is reacting to certain changes in its immediate environment and suppose, further, that this reaction is entirely one of changes in the internal qualities of the system. First of all we identify within the system the set of these internal qualities which we will call *internal states.* If we denote this set of internal states by the set Q we can then agree that at any given time, t, the system is in a particular internal state, $q(t)$, which is an element of the set Q. What these internal states are is not of great importance in general, we deliberately keep the definition fairly vague in order that we can then apply our model to a great many distinct situations.

Some examples of systems and possible sets of internal states may help to give a more intuitive idea of what we mean. Consider an electronic system, which involves various electrical components, such as transistors, connected together in a complex electrical circuit. As currents flow through the system some of these transistors 'fire', while others do not. If at a given time, t, we have a complete knowledge of what the various components of the circuit are doing, either firing or not firing, then we say that we know the state, $q(t)$, of the system at the time t. The set Q will then be the set of all the states $q(t)$ that are possible at some time or other. Obviously the larger the system, the more difficult will the definition of an internal state be, and the larger the set Q of all internal states. However, the total number of internal states will always be finite in examples of this kind.

For a biological example consider a single cell from a biological organism. Inside the cell there will be many chemical reactions taking place as the cell performs its role in the organism. Many chemicals are being formed and consumed in the cell, but at a given time, t, it is possible to conceive, at least theoretically, that each chemical has reached a certain concentration in the cell as a result of the reactions taking place. Thus the internal state, $q(t)$, at that time would be a list of all the chemical concentrations in the cell then. Clearly this would be a phenomenally complex piece of information, but the number of chemicals involved would again be finite. In this case the complete set of internal states may not be finite, since each chemical could clearly exist in one of an infinite number of concentrations. We overcome this difficulty by using the idea of a threshold. It is clear that some chemicals can exist in very tiny concentrations without substantially changing the behaviour of the cell, and as it is the behaviour of the cell, and in particular its response to changes in its environment, that interests us, we can often replace the infinite sets of chemical concentrations by finite ones, since the behaviour of the cell may change only after a chemical concentration has crossed a threshold value. (See example 2.8.)

Let us now consider the system as being described by a finite set Q of internal states. Changes in its environment will in many cases force changes in its internal state and we will now make the added assumption that this is the only way that internal states can be changed. So that if there is no change in the environment between times t and t_1 then $q(t)=q(t_1)$.

How do we model the environmental influences? Consider the environment and the system at a given time, t, and note all the relevant environmental factors that can affect the system; this particular environmental profile will be denoted by $\sigma(t)$. The set of all such $\sigma(t)$ that can affect the system is called Σ, the set of *environmental inputs* or the *input alphabet.*

In the examples discussed above the environmental input to the computer system at time t is either a *pulse of electricity* applied to the system or *no electrical charge.* For the biological example the input will be a particular profile involving, perhaps temperature, quantity of light, concentrations of various chemicals etc. and as before it may be considered to be a finite set by applying the threshold principle.

We are then left with two finite sets, Q representing the internal states of the system, and Σ representing the possible environmental influences acting on the system. Since the system will react to different environmental inputs by changing its internal state, the final stage in the modelling of the system is a function that tells us how the system will behave. We agree first that the internal changes and the reception of inputs take place in the context of a suitable discrete time scale based on the length of time that the system takes to react. In this way we will remove problems associated with the influence of time on the inputs and the internal states.

Let the system be in state $q \in Q$ and suppose that the environment changes to $\sigma \in \Sigma$. The change will cause the state to change at the next point on the time scale to a new state $q' \in Q$ and so we have the *resultant* of applying the environmental input σ to the system in state q. If we specify this resultant for some of the possible combinations of internal state and environmental input, we will be specifying a partial function $F: Q \times \Sigma \to Q$ in such a way that $F((q, \sigma))=q'$ where $q \in Q$, $\sigma \in \Sigma$ and q' is the result of applying σ to the system in state q.

A *state machine* or *semiautomaton* is a triple $\mathcal{M} = (Q, \Sigma, F)$ where Q and Σ are finite sets and F is a partial function $F: Q \times \Sigma \to Q$. (We allow the possibilities that either Q or Σ or both are empty.) A state machine $\mathcal{M} = (Q, \Sigma, F)$ is called *complete* if the partial function $F: Q \times \Sigma \to Q$ is in fact a *function*. In this situation we can specify what the resultant $F((q, \sigma))$ is for all possible combinations of $q \in Q$ and $\sigma \in \Sigma$.

Such a system is clearly very general and can be applied to many different situations. It is almost too general, from a mathematical point of view, and the fact that we can investigate such systems successfully using algebraic techniques, is, in my opinion, one of the most remarkable achievements of modern mathematics.

One advantage of such a general definition is that it is easy to find simple examples and their study amply repays the effort involved. We shall look at some now.

Examples 2.1

(i) Some simple cases are where Q and Σ are both singletons. Let $Q = \{0\}$ and $\Sigma = \{\sigma\}$, then we can have $F: Q \times \Sigma \to Q$ defined by $F(0, \sigma) = 0$. This is illustrated with a simple diagram:

σ
0

where the arrow is labelled by the only input, σ.

(ii) Suppose that $|Q| = 1$ and Σ is any finite set, we don't really get anything very different, just $F((0, \sigma)) = 0\ \forall \sigma \in \Sigma$, or in diagrammatic form:

all σ
0

(iii) Letting $|Q| > 1$ does introduce some more interesting examples, thus if $Q = \{0, 1\}$ and $|\Sigma| = 1$ we could have any of the following:

(*a*) σ 0 σ 1

or

(*b*) 0 σ 1 σ

or

(*c*) σ 0 σ 1

or

(*d*) 0 σ σ 1

(iv) These examples are all complete, and in fact incomplete state machines need not have any arrows. For example:

0 could represent $(\{0\}, \{\sigma\}, F)$ where

$F: Q \times \Sigma \to Q$ is not defined for $(0, \sigma) \in Q \times \Sigma$.

Another incomplete state machine is:

$$0 \xrightarrow{\sigma} 1$$

and here $F(1, \sigma)$ is undefined.

(v) We will introduce a *cyclic state machine* as follows. Let p, r be positive integers and put $Q = \{0, 1, 2, \ldots, r+p-1\}$, $\Sigma = \{\sigma\}$. Consider the diagram

$$0 \xrightarrow{\sigma} 1 \xrightarrow{\sigma} 2 \xrightarrow{\sigma} \cdots \xrightarrow{\sigma} r-1 \xrightarrow{\sigma} r \xrightarrow{\sigma} r+1 \xrightarrow{\sigma} \cdots \xrightarrow{\sigma} r+p-1 \xrightarrow{\sigma} r$$

so that $F(0, \sigma) = 1$, $F(1, \sigma) = 2$ etc.

This is called the cyclic state machine with *stem* of length r and *cycle* of length p; we note that this machine is complete.

These diagrams, or directed graphs, are sometimes quite useful tools. In these simple cases they clearly define the state machine precisely, and we will often use them for this purpose.

They can, however, also tell us something about the properties of the state machines. Take a look at the cyclic machine above, the states $0, \ldots, r-1$ have the property that once the machine leaves them it can never return. We could call these 'states of no return'. The cycle of states $r, r+1, \ldots, r+p-1$ is a 'cycle of no escape'.

Another way of specifying a state machine is by writing out the partial function F in tabular form, for example:

(vi) $Q = \{0, 1, 2\}$, $\Sigma = \{\sigma, \tau\}$ and

F	0	1	2
σ	0	0	2
τ	1	$\varnothing$	1

(Here $F((1, \tau))$ is undefined, we write it as $\varnothing$ in the table.)

This represents the same machine as the diagram:

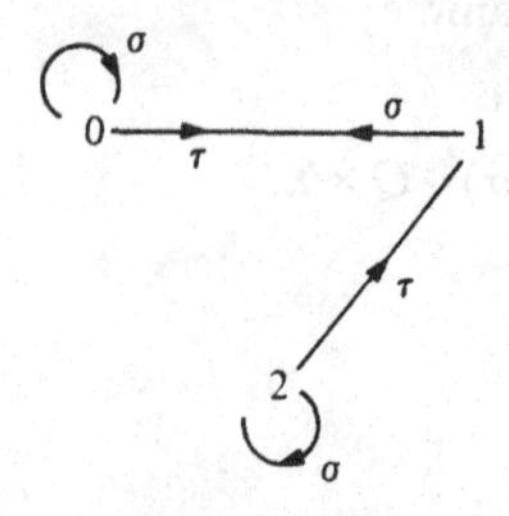

The fact that $F((1, \tau))$ is undefined is indicated on the diagram by the lack of an arrow labelled by τ emanating from the state 1.

An incomplete state machine can be completed by introducing new arrows from states that are lacking them. However we choose a more systematic method.

We introduce a new state to the machine and arrange for all the missing arrows to go to this new state.

For example in (vi) above we introduce the new state z so that $Q = \{0, 1, 2, z\}$, $\Sigma = \{\sigma, \tau\}$ and the graph of the completed machine is:

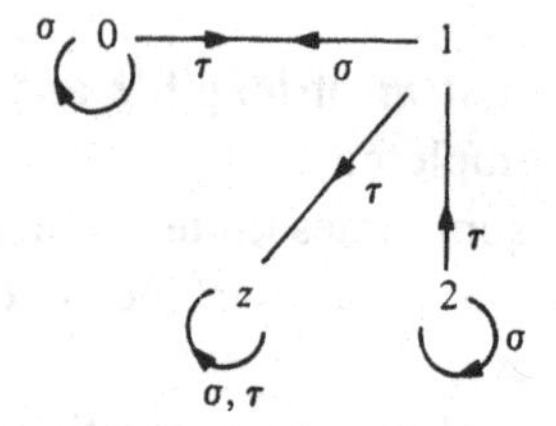

Formally let $\mathcal{M} = (Q, \Sigma, F)$ be an incomplete state machine. Define the *completion* $\mathcal{M}^c = (Q', \Sigma, F')$, of $\mathcal{M}$ by putting

$$Q' = Q \cup \{z\}$$

where $z \notin Q$, and

$$F'((q', \sigma)) = \begin{cases} F(q, \sigma) & \text{if } q \in Q \text{ and } F(q, \sigma) \text{ is defined} \\ z & \text{otherwise.} \end{cases}$$

The new state z is called the *sink state* of $\mathcal{M}^c$.

We will examine some practical examples of such machines at the end of the chapter. Our next task is to look at the way these machines operate.

Generally speaking we will present the machine $\mathcal{M} = (Q, \Sigma, F)$ with a symbol $\sigma \in \Sigma$ while it is in some state, say $q \in Q$. The machine then moves to state $F((q, \sigma)) \in Q$. This notation is a little cumbersome and we will introduce the idea of the state mapping induced by the input.

This concept is defined thus:

let $\sigma \in \Sigma$, define $F_\sigma : Q \to Q$ by

$qF_\sigma = F((q, \sigma))$ for each $q \in Q$.

Since the machine may not be complete, F_σ may only be a partial function of Q to itself. Each input symbol σ from Σ yields a partial function $F_\sigma : Q \to Q$.

Now suppose that σ is applied to the machine in the state q and consequently the machine moves to state qF_σ. (Using the usual convention that $qF_\sigma = \varnothing$ if $F((q, \sigma))$ is undefined.) If, further, another input, say $\sigma' \in \Sigma$, is applied to the machine we get the resultant state $qF_\sigma F_{\sigma'}$. We may extend our notation in the following way. Let $\alpha \in \Sigma^+$ be a word of length at least 1 with symbols from Σ.

Suppose that $\alpha = \sigma_1\sigma_2 \ldots \sigma_k$ then we define

$F_\alpha : Q \to Q$

by

$qF_\alpha = qF_{\sigma_1}F_{\sigma_2} \ldots F_{\sigma_k}$.

Now it is perhaps apparent why we are writing the result of state mappings in the form qF_σ rather than $F_\sigma(q)$, it is caused by our convention of writing words from left to right!

Each word from Σ^+ will therefore correspond to some partial function of Q to itself.

Returning, once more, to example 2.1(vi) we note that some of the partial functions induced by words from Σ^+ are:

	0	1	2
F_σ	0	0	2
F_τ	1	$\varnothing$	1
$F_{\sigma\sigma}$	0	0	2
$F_{\tau\tau}$	$\varnothing$	$\varnothing$	$\varnothing$
$F_{\sigma\tau}$	1	1	1
$F_{\tau\sigma}$	0	$\varnothing$	0

Note that $F_{\tau\tau}$ is the empty function $\boldsymbol{\theta} : Q \to Q$ and so we have a natural example of what one might have originally thought was a rather artificial concept. The function $F_{\tau\sigma}$ is, like F_τ, a partial function.

2.2 The semigroup of a state machine

The state set of a state machine is finite and so the number of partial mappings definable on the state set is also finite. Therefore the number of *distinct* state mappings induced by words from the symbol set is also finite. Consequently some words will yield the same state mappings. We will use this idea to introduce a relation on the free semigroup generated by the symbol set.

Let $\mathcal{M} = (Q, \Sigma, F)$ be a state machine and consider the set Σ^+ of all words of length greater than or equal to 1 in the alphabet Σ. Define a relation $\sim$ on Σ^+ by

$$\alpha \sim \beta \Leftrightarrow F_\alpha = F_\beta \quad \text{where } \alpha, \beta \in \Sigma^+$$

This relation is easily seen to be an equivalence relation. Since Σ^+ has a natural semigroup structure, using concatenation of words as the operation, it is natural to ask whether $\sim$ is a congruence on Σ^+. This is indeed the case, for example if $\alpha, \beta, \gamma \in \Sigma^+$ and $\alpha \sim \beta$ then $F_\alpha = F_\beta$ and for any $q \in Q$, $qF_{\gamma\alpha} = qF_\gamma F_\alpha = (qF_\gamma)F_\alpha = (qF_\gamma)F_\beta = qF_{\gamma\beta}$ and so $F_{\gamma\alpha} = F_{\gamma\beta}$ which yields $\gamma\alpha \sim \gamma\beta$, etc. We now construct the quotient semigroup $\Sigma^+/\sim$ and call it the *semigroup of the state machine* $\mathcal{M}$, the notation used being $\mathbf{S}(\mathcal{M})$. The elements of $\mathbf{S}(\mathcal{M})$ will be equivalence classes $[\alpha]$, $\alpha \in \Sigma^+$.

We have already noted that each $\sigma \in \Sigma$ defines a partial mapping $F_\sigma : Q \to Q$ and so there is a natural function $\mathbf{F} : \Sigma \to \mathbf{PF}(Q)$, given by $\mathbf{F}(\sigma) = F_\sigma$ for $\sigma \in \Sigma$. If we denote by $\langle \mathbf{F}(\mathcal{M}) \rangle$ the subsemigroup of $\mathbf{PF}(Q)$ generated by the set of functions $\{F_\sigma \mid \sigma \in \Sigma\}$ we obtain an isomorphic copy of the semigroup $\mathbf{S}(\mathcal{M})$ of the state machine $\mathcal{M}$. To see this just note that there is a surjection θ from Σ^+ onto $\langle \mathbf{F}(\mathcal{M}) \rangle$ defined by $\theta(\alpha) = F_\alpha$ for $\alpha \in \Sigma^+$, with corresponding congruence defined by the relation $\sim$. The first isomorphism theorem for semigroups yields the result. We thus have:

Proposition 2.2.1

Let $\mathcal{M} = (Q, \Sigma, F)$ be a state machine and $\langle \mathbf{F}(\mathcal{M}) \rangle$ the subsemigroup of $\mathbf{PF}(Q)$ generated by $\{F_\sigma \mid \sigma \in \Sigma\}$, then $\langle \mathbf{F}(\mathcal{M}) \rangle \cong \mathbf{S}(\mathcal{M}) = \Sigma^+/\sim$. Furthermore $\mathbf{S}(\mathcal{M})$ is a finite semigroup.

The last statement follows from the fact that $\mathbf{PF}(Q)$ is finite when Q is finite.

The semigroup $\mathbf{PF}(Q)$ is actually a monoid and while $\mathbf{S}(\mathcal{M})$ may also be a monoid it can happen that $\mathbf{S}(\mathcal{M})$ does not possess an identity. We can easily construct a monoid from the state machine $\mathcal{M}$ by forming the monoid Σ^* of all words in Σ, including the empty word Λ, and extending the relation $\sim$ to Σ^* by putting

$$\alpha \sim \beta \Leftrightarrow F_\alpha = F_\beta \quad \text{for } \alpha, \beta \in \Sigma^*.$$

Again $\sim$ is a congruence, but $\Sigma^*/\sim$ is a finite monoid isomorphic to $\langle F(\mathcal{M})\rangle \cup \{1_Q\}$. We write $\Sigma^*/\sim$ as $\mathbf{M}(\mathcal{M})$, and call it the *monoid of* $\mathcal{M}$.

Note that in both cases the relations $\sim$ defined on Σ^+ and Σ^* depend on the state machine $\mathcal{M}$. However, it is quite possible for different state machines to have the same, or at least isomorphic, semigroups.

Given a state machine $\mathcal{M} = (Q, \Sigma, F)$ we have now associated with it a semigroup $\mathbf{S}(\mathcal{M})$. In many situations it is more convenient to study this semigroup rather than the original machine $\mathcal{M}$. However we don't want to lose sight of the set of states and so we consider the pair $(Q, \mathbf{S}(\mathcal{M}))$ consisting of the set of states Q of $\mathcal{M}$ and the semigroup $\mathbf{S}(\mathcal{M})$ of $\mathcal{M}$. Each element of $\mathbf{S}(\mathcal{M})$ is an equivalence class of Σ^+, which acts on Q as follows: $q[\alpha] = qF_\alpha$ where $q \in Q$, $\alpha \in \Sigma^+$. This is an example of a transformation semigroup and it is these that we will be studying in detail.

A *transformation semigroup* is a pair (Q, S) consisting of a finite set Q, a finite semigroup S and an *action* of S on Q, that is a partial function $\lambda : Q \times S \to Q$ satisfying two conditions:

(i) $\lambda(\lambda(q, s), s_1) = \lambda(q, ss_1)$ for all $q \in Q$; $s, s_1 \in S$.

(ii) $\lambda(q, s) = \lambda(q, s_1)$ for all $q \in Q$ implies $s = s_1$ where $s, s_1 \in S$.

It is usual to write $\lambda(q, s)$ as $q \cdot s$ or qs for $q \in Q$, $s \in S$ and these conditions become

(i)′ $(qs)s_1 = q(ss_1)$ for all $q \in Q$; $s, s_1 \in S$.

(ii)′ $qs = qs_1$ for *all* $q \in Q$ implies $s = s_1$ where $s, s_1 \in S$.

We write the operation of S on Q on the right to preserve the connection with state machines. Notice that there is a natural embedding of the semigroup S into the monoid $\mathbf{PF}(Q)$ obtained by defining $\theta(s): Q \to Q$ to be given by $q\theta(s) = qs$ for each $q \in Q$, and each $s \in S$. Then $\theta : S \to \mathbf{PF}(Q)$ is a semigroup monomorphism. Conversely given any set Q and a subsemigroup $S \subseteq \mathbf{PF}(Q)$ then (Q, S) is a transformation semigroup.

Associated with any state machine $\mathcal{M} = (Q, \Sigma, F)$ there is then a transformation semigroup $(Q, \mathbf{S}(\mathcal{M}))$ which we will denote by $\mathbf{TS}(\mathcal{M})$ and call the *transformation semigroup of* $\mathcal{M}$.

Now each transformation semigroup determines a state machine, for suppose that $\mathcal{A} = (Q, S)$ is a transformation semigroup, we define the state machine $\mathcal{M} = (Q, S, F)$ where

$$F : Q \times S \to Q$$

is given by

$$F(q, s) = qs \quad \text{for all } q \in Q, s \in S.$$

Clearly $\mathcal{M}$ is a state machine, we call it the *state machine of* (Q, S) and denote it by $\mathbf{SM}(\mathcal{A})$. The relationship between state machines and transformation semigroups is very close.

In some situations the semigroup $\mathbf{S}(\mathcal{M})$ of a state machine $\mathcal{M}$ is in fact a monoid and $\mathbf{TS}(\mathcal{M})$ is a *transformation monoid.* Generally we define a transformation monoid as a transformation semigroup (Q, S) where S is a monoid *and* the identity 1 of S satisfies

$$q \cdot 1 = q \quad \text{for all } q \in Q.$$

For a given state machine $\mathcal{M} = (Q, \Sigma, F)$ we may define the *transformation monoid of* $\mathcal{M}$, $\mathbf{TM}(\mathcal{M})$, as being $(Q, \mathbf{M}(\mathcal{M}))$.

Now is the time to look at some examples.

Examples 2.2

(i) The examples of state machines discussed in 2.1(i) and (ii) both yield the transformation monoid $(\{0\}, S)$ where S is the group of order 1.

(ii) The transformation semigroups of the examples in 2.1(iii) are

(*a*) $(\{0, 1\}, S)$, which is a transformation monoid, with $S = \{1_Q\}$;

(*b*) $(\{0, 1\}, \{\sigma\})$, which is not a transformation monoid although $\sigma^2 = \sigma$;

(*c*) also of the form $(\{0, 1\}, \{\sigma\})$ with $\sigma^2 = \sigma$ although the action is not the same;

(*d*) $(\{0, 1\}, \{\sigma, \sigma^2\})$ with $\sigma^2 = 1_Q$.

(iii) Example 2.1(vi) has the transformation semigroup $(\{0, 1, 2\}, \{\boldsymbol{\theta}, \sigma, \tau, \sigma\tau, \tau\sigma, \sigma\tau\sigma\})$ with the semigroup composition given by the following table:

	$\boldsymbol{\theta}$	σ	τ	$\sigma\tau$	$\tau\sigma$	$\sigma\tau\sigma$
$\boldsymbol{\theta}$	$\boldsymbol{\theta}$	$\boldsymbol{\theta}$	$\boldsymbol{\theta}$	$\boldsymbol{\theta}$	$\boldsymbol{\theta}$	$\boldsymbol{\theta}$
σ	$\boldsymbol{\theta}$	σ	$\sigma\tau$	$\sigma\tau$	$\sigma\tau\sigma$	$\sigma\tau\sigma$
τ	$\boldsymbol{\theta}$	$\tau\sigma$	$\boldsymbol{\theta}$	τ	$\boldsymbol{\theta}$	$\tau\sigma$
$\sigma\tau$	$\boldsymbol{\theta}$	$\sigma\tau\sigma$	$\boldsymbol{\theta}$	$\sigma\tau$	$\boldsymbol{\theta}$	$\sigma\tau\sigma$
$\tau\sigma$	$\boldsymbol{\theta}$	$\tau\sigma$	τ	τ	$\tau\sigma$	$\tau\sigma$
$\sigma\tau\sigma$	$\boldsymbol{\theta}$	$\sigma\tau\sigma$	$\sigma\tau$	$\sigma\tau$	$\sigma\tau\sigma$	$\sigma\tau\sigma$

Since we will be repeatedly dealing with transformation semigroups it will be convenient to introduce some notation to help us refer to some of the more common types. First we will consider a general finite set Q. Let $q \in Q$, then there is a mapping $\bar{q} : Q \to Q$ defined by $y\bar{q} = q$ for all $y \in Q$. Thus $\bar{q}$ is the constant mapping defined by the element q. The set of all the constant mappings on Q generates a semigroup $\bar{Q}$ as a subsemigroup of $\mathbf{PF}(Q)$. We can now consider the transformation semigroup $(Q, \bar{Q})$.

Let $\mathcal{A} = (Q, S)$ be any transformation semigroup, define the *closure* $\bar{\mathcal{A}}$ of $\mathcal{A}$ to be the transformation semigroup $(Q, \langle S \cup \bar{Q} \rangle)$. We call $\mathcal{A}$ *closed* if $\bar{\mathcal{A}} = \mathcal{A}$.

Given any transformation semigroup $\mathcal{A} = (Q, S)$ define the transformation monoid $\mathcal{A}^{\cdot} = (Q, S \cup \{1_Q\})$. For any finite set Q we can form a transformation semigroup $\mathcal{Q} = (Q, \varnothing)$. Then if n is a positive integer recall that the set $\mathbf{n} = \{0, 1, \ldots, n-1\}$ and so we have a transformation semigroup, also denoted by $\mathbf{n}$ and given by $\mathbf{n} = (\mathbf{n}, \varnothing)$. We can now specify some of the transformation semigroups in examples 2.2, these are:

2.2(i) $\mathbf{1}^{\cdot}$, 2.2(ii) $\mathbf{2}^{\cdot}$.

The example 2.2(ii)(b) will be denoted by $\mathcal{C}$.

The transformation semigroup generated by the state machine

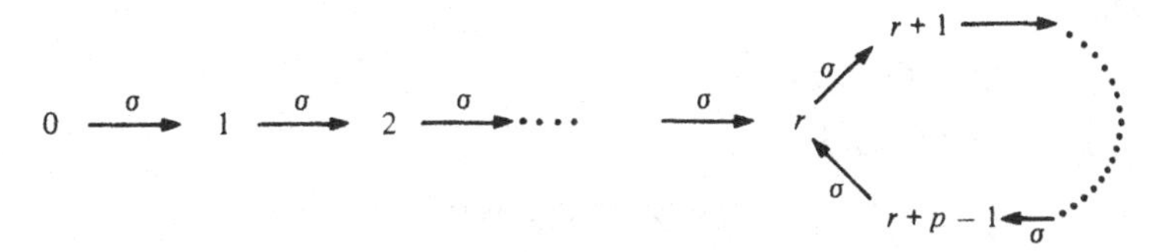

will be written $\mathcal{C}_{(p,r)}$

If G is a group then G may be considered as the transformation monoid (G, G) where the group G acts on the set G by right multiplication, that is $g'g = g' \cdot g$ $(g' \in G, g \in G)$, we will denote this transformation monoid by $\mathcal{G}$, and since the monoid is a group it will be sensible to call it a *transformation group*. Thus example 2.1(iii)(d) generates $(\mathbb{Z}_2, \mathbb{Z}_2)$, a transformation group; we will write this as $\mathbb{Z}_2$.

In the case of the transformation group $\mathcal{G}$ formed from a group G it is clear that the action is faithful. However if S is a semigroup the action of S on the set S defined by right multiplication need not be faithful. This is a special case of a more general situation.

Suppose that Q is a finite set and S is a semigroup, suppose further that an action qs $(q \in Q, s \in S)$ is given. The pair (Q, S) may not be a transformation semigroup even if the action satisfies $(qs)s_1 = q(ss_1)$ for all $q \in Q$, $s, s_1 \in S$. However such a pair may be converted into a transformation semigroup. Let $\sim$ define a relation on S defined by $s \sim s_1 \Leftrightarrow qs = qs_1$ for all $q \in Q$. Then $\sim$ is a congruence and we may form the quotient semigroup $S/\sim$. The pair $(Q, S/\sim)$ now becomes a transformation semigroup with action defined by $q[s] = qs$, $q \in Q$, $[s] \in S/\sim$. We call this the transformation semigroup *represented* by the pair (Q, S).

Now if S is a semigroup then the pair (S, S) represents a transformation semigroup $(S, S/\sim)$.

Another way of defining a transformation semigroup from an arbitrary semigroup is to consider the 'semigroup made into a monoid' by the adjunction of an identity element. So if S is a semigroup which is not a monoid then $S^{\cdot} = S \cup \{e\}$ where $e \notin S$ is suitably chosen and is defined to act as an identity for $S^{\cdot}$. Then we can construct a transformation semigroup $(S^{\cdot}, S)$ with action by right multiplication; this is denoted by $\mathscr{S}$. If S is a monoid then (S, S) is a transformation monoid, also written as $\mathscr{S}$.

A transformation semigroup $\mathscr{A} = (Q, S)$ may not be complete, that is qs may not be defined for some $q \in Q$, $s \in S$. The *completion*, $\mathscr{A}^c$, is defined to be

$$\mathscr{A}^c = (Q', S)$$

where

$$Q' = Q \cup \{z\}$$

for some $z \notin Q$ and

$$q' \cdot s = \begin{cases} qs & \text{if } q \in Q \text{ and } qs \text{ is defined in } \mathscr{A} \\ z & \text{otherwise.} \end{cases}$$

Naturally, if $\mathscr{A}$ is complete we will define $\mathscr{A}^c = \mathscr{A}$.

If $\mathscr{A} = (Q, S)$ is a transformation monoid, $Q \neq \varnothing$ and S is a group, we call $\mathscr{A}$ a transformation group. If $\mathscr{A} = (Q, S)$ is such that either $\mathscr{A}$ is a transformation group or $Q \neq \varnothing$ and $S = \varnothing$ we say that $\mathscr{A}$ is a *generalized transformation group.*

2.3 Homomorphisms and quotients

Let $\mathcal{M} = (Q, \Sigma, F)$ and $\mathcal{M}' = (Q', \Sigma', F')$ be state machines. Let $\alpha : Q \to Q'$, $\beta : \Sigma \to \Sigma'$ be mappings such that

$$\alpha(qF_\sigma) \subseteq (\alpha(q))F'_{\beta(\sigma)}$$

for any $q \in Q$, $\sigma \in \Sigma$. (This means that if qF_σ is undefined we put $\alpha(qF_\sigma) = \varnothing$ and if qF_σ is defined then so is $(\alpha(q))F'_{\beta(\sigma)}$ and $\alpha(qF_\sigma) = (\alpha(q))F'_{\beta(\sigma)}$.)

We call the pair (α, β) a *state machine homomorphism* from $\mathcal{M}$ to $\mathcal{M}'$ and write $(\alpha, \beta) : \mathcal{M} \to \mathcal{M}'$.

If α and β are both one-one mappings then we call (α, β) a *monomorphism* and if α and β are both onto mappings then (α, β) is called an *epimorphism*. An *isomorphism* of state machines is both a monomorphism and an epimorphism, in this case we write $\mathcal{M} \cong \mathcal{M}'$.

Example 2.3
Let $\mathcal{M} = (Q, \Sigma, F)$ be the state machine defined by the diagram

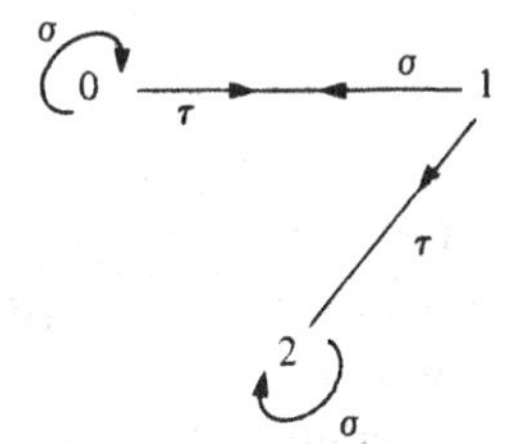

This is example 2.1 (vi).

If $\mathcal{M}' = (Q', \Sigma', F')$ is the state machine defined by the diagram

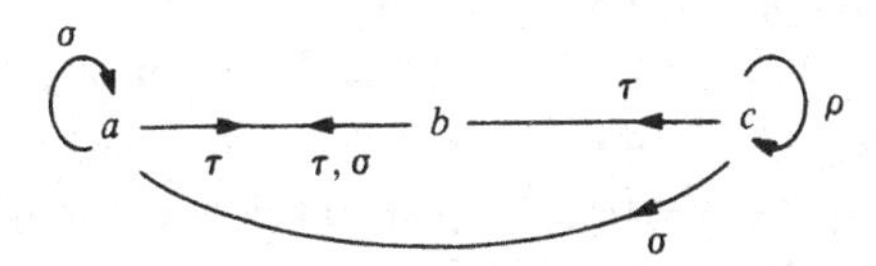

where $Q' = \{a, b, c\}$ and $\Sigma' = \{\sigma, \tau, \rho\}$,
define

$$\alpha : Q \to Q' \quad \text{by } \alpha(0) = \alpha(2) = a, \alpha(1) = b$$
$$\beta : \Sigma \to \Sigma' \quad \text{by } \beta(\sigma) = \sigma, \beta(\tau) = \tau.$$

Then $(\alpha, \beta) : \mathcal{M} \to \mathcal{M}'$ is a homomorphism; note that

$$\alpha(2F_\tau) = \varnothing \subseteq (\alpha(2))F'_{\beta(\tau)} = b$$
$$\alpha(0F_\sigma) = a = (\alpha(0)F'_{\beta(\sigma)})$$

etc.

If $\mathcal{A} = (Q, S)$, $\mathcal{A}' = (Q', S')$ are transformation semigroups, $f : Q \to Q'$ is a mapping and $g : S \to S'$ a semigroup homomorphism, then the pair (f, g) is said to be a *transformation semigroup homomorphism* from $\mathcal{A}$ to $\mathcal{A}'$ if

$$f(qs) \subseteq f(q) \cdot g(s) \quad \text{for all } q \in Q, s \in S.$$

(It should be realized that in incomplete transformation semigroups the left hand side may be undefined and is then by convention the empty set.) As before we write $(f, g) : \mathcal{A} \to \mathcal{A}'$.

(f, g) is a *monomorphism* if f and g are one-one; an *epimorphism* if f and g are onto; and an *isomorphism* if (f, g) is both a monomorphism and an epimorphism, we then write $\mathscr{A} \cong \mathscr{A}'$.

Theorem 2.3.1
Let $\mathcal{M} = (Q, \Sigma, F)$, $\mathcal{M}' = (Q', \Sigma', F')$ be complete state machines and $(\alpha, \beta): \mathcal{M} \to \mathcal{M}'$ a homomorphism with α onto. There exists a homomorphism

$$(f_\alpha, g_\beta): \mathbf{TS}(\mathcal{M}) \to \mathbf{TS}(\mathcal{M}').$$

Proof Define $f_\alpha: Q \to Q'$ by $f_\alpha = \alpha$. Let $S = \mathbf{S}(\mathcal{M})$, $S' = \mathbf{S}(\mathcal{M}')$ and suppose that $s \in S$. Then there exists $a \in \Sigma^+$ such that $s = [a]$, the $\sim$-equivalence class containing a. Suppose that $a = \sigma_1 \ldots \sigma_n$, $\sigma_i \in \Sigma$ define $g_\beta(s) = [\beta(a)]'$ where $[\beta(a)]'$ is the $\sim'$-equivalence class containing $\beta(a) = \beta(\sigma_1) \ldots \beta(\sigma_n) \in (\Sigma')^+$. (Note that $\sim$ is induced by $\mathcal{M}$ and $\sim'$ is induced by $\mathcal{M}'$.)

We must first establish that $g_\beta: S \to S'$ is well-defined. Suppose that $s = [b]$ where $b \in \Sigma^+$, then $b = \tau_1 \ldots \tau_m$ where $\tau_j \in \Sigma$. Now for any $q \in Q$, $qF_a = qF_b$. Let $q' \in Q'$, there exists $q \in Q$ such that $q' = \alpha(q)$. Then $q'F'_{\beta(a)} = (\alpha(q))F'_{\beta(a)}$ and $q'F'_{\beta(b)} = (\alpha(q))F'_{\beta(b)}$. However $\alpha(qF_a) = \alpha(qF_b)$ and then $\alpha(qF_a) = (\alpha(q))F'_{\beta(a)} = (\alpha(q))F'_{\beta(b)}$ so $q'F'_{\beta(a)} = q'F'_{\beta(b)}$. Thus $\beta(a) \sim' \beta(b)$ and g_β is well-defined. Now let $q \in Q$, $s \in S$, then $f_\alpha(qs) = \alpha(qs) = \alpha(qF_a) = (\alpha(q))F'_{\beta(a)} = f_\alpha(q)[\beta(a)]' = f_\alpha(q)g_\beta(s)$, where $s = [a]$ and $a \in \Sigma^+$. □

This result gives us some useful information concerning the relationship between a state machine homomorphism and a homomorphism of the related transformation semigroups. We consider, now, two elementary results that link state machines with transformation semigroups.

Theorem 2.3.2
Let $\mathscr{A} = (Q, S)$ be a transformation semigroup; then

$$\mathbf{TS}(\mathbf{SM}(\mathscr{A})) \cong \mathscr{A}.$$

Proof Let $\mathbf{SM}(\mathscr{A}) = (Q, S, F)$ and consider the semigroup $K = \langle \mathbf{F}(\mathbf{SM}(\mathscr{A})) \rangle$ generated by $\mathbf{SM}(\mathscr{A})$. There is clearly an isomorphism $\theta: S \to K$ defined by $\theta(s) = F_s$, $s \in S$ and this yields the isomorphism $(1_Q, \theta): \mathscr{A} \to \mathbf{TS}(\mathbf{SM}(\mathscr{A}))$. □

Theorem 2.3.3

Let $\mathcal{M}=(Q,\Sigma,F)$ be a state machine; there exists a state machine monomorphism

$$(\alpha,\beta):\mathcal{M}\to \mathbf{SM}(\mathbf{TS}(\mathcal{M})).$$

As with most algebraic systems, homomorphisms are closely linked with 'congruence' relations. Suppose that we have a homomorphism of state machines

$$(\alpha,\beta):\mathcal{M}\to\mathcal{M}' \quad \text{where } \mathcal{M}=(Q,\Sigma,F) \text{ and } \mathcal{M}'=(Q',\Sigma',F').$$

The mapping $\alpha: Q\to Q'$ induces an equivalence relation R_α on the set Q defined by

$$qR_\alpha q_1 \Leftrightarrow \alpha(q)=\alpha(q_1) \quad \text{for } q, q_1\in Q.$$

The relation R_α satisfies the following condition: let $qR_\alpha q_1$ and $\sigma\in\Sigma$ and suppose that qF_σ and q_1F_σ are both defined, then

$$(qF_\sigma)R_\alpha(q_1F_\sigma).$$

This follows because $\alpha(qF_\sigma)=\alpha(q)F'_{\beta(\sigma)}=\alpha(q_1)F'_{\beta(\sigma)}=\alpha(q_1F_\sigma)$.

The relation R_α is an example of an *admissible* relation on Q. Formally if $\mathcal{M}=(Q,\Sigma,F)$ is a state machine then a relation R on Q is *admissible* if:

(i) R is an equivalence relation;

(ii) given q, $q_1\in Q$, $\sigma\in\Sigma$ such that qRq_1 and both qF_σ, q_1F_σ are defined then $(qF_\sigma)R(q_1F_\sigma)$.

An admissible relation R defines a partition on the set Q of the state machine $\mathcal{M}=(Q,\Sigma,F)$. Suppose we denote this partition by $\pi=\{H_i\}_{i\in I}$ where each H_i is an equivalence class of Q for $i\in I$. Then $Q=\bigcup_{i\in I}H_i$ and $H_i\cap H_j=\varnothing$ for $i\neq j$, $i, j\in I$. Furthermore given $H_i\in\pi$ and $\sigma\in\Sigma$ we form the set $H_iF_\sigma=\{qF_\sigma \mid q\in H_i\}$ and then $H_iF_\sigma\subseteq H_j$ for some $j\in I$. (Clearly H_iF_σ may be empty.) Consequently we have a special type of partition π which will be called an admissible partition.

Thus a *partition* $\pi=\{H_i\}_{i\in I}$ of the state set Q of the state machine $\mathcal{M}=(Q,\Sigma,F)$ is called *admissible* if given $i\in I$, $\sigma\in\Sigma$ either there exists $j\in I$ such that

$$H_iF_\sigma\subseteq H_j$$

or

$$H_iF_\sigma=\varnothing.$$

If the machine $\mathcal{M}$ is complete then the choice of j, given i and σ, is unique. For incomplete machines this is not always true.

Turning now to transformation semigroups we make the following parallel definitions.

Let $\mathscr{A} = (Q, S)$ be a transformation semigroup, an *admissible relation on* Q is a relation R such that if $q, q_1 \in Q$, $s \in S$, $qs \neq \varnothing$, $q_1 s \neq \varnothing$ and qRq_1 then $qsRq_1s$.

A *partition* $\pi = \{H_i\}_{i\in I}$ on Q is *admissible* if given $i \in I$, $s \in S$ either there exists $j \in I$ such that

$$H_i s \subseteq H_j$$

or

$$H_i s = \varnothing.$$

The idea of an admissible partition leads to a procedure for constructing quotient systems in the following way.

Let $\mathcal{M} = (Q, \Sigma, F)$ be a state machine and $\pi = \{H_i\}_{i\in I}$ an admissible partition on Q, construct a state machine $\mathcal{M}/\pi = (Y, \Sigma, G)$ by defining $Y = \pi$, the set of π-blocks, and putting $H_i G_\sigma = H_j$ where

$$\left.\begin{aligned} &H_i F_\sigma \subseteq H_j \\ &H_i G_\sigma = \varnothing \quad \text{if } H_i F_\sigma = \varnothing \end{aligned}\right\}(i, j \in I, \sigma \in \Sigma).$$

This definition of G_σ is well-defined since π is a partition and admissible. The state machine $\mathcal{M}/\pi$ is called the *quotient state machine* of $\mathcal{M}$ *with respect to* π.

If we change the scene to that of transformation semigroups a similar construction emerges.

Let $\mathscr{A} = (Q, S)$ be a transformation semigroup and $\pi = \{H_i\}_{i\in I}$ an admissible partition on Q, construct a pair (Y, S) where $Y = \pi$, the set of π-blocks. Now S acts on Y with respect to the operation $*$ defined by:

$$\left.\begin{aligned} &H_i * s = H_j \Leftrightarrow H_i s \subseteq H_j \\ &H_i * s = \varnothing \Leftrightarrow H_i s = \varnothing \end{aligned}\right\}(i, j \in I, s \in S).$$

Clearly $(H_i * s) * s' = H_i * (ss')$ but it may be that $H_i * s = H_i * s'$ for all $H_i \in Y$ and yet $s \neq s'$. To make (Y, S) into a transformation semigroup it is necessary that we remove this possibility. The usual procedure is to define a relation $\sim$, this time on the semigroup S.

Put $s \sim s' \Leftrightarrow H_i * s = H_i * s'$, $i \in I$, where $s, s' \in S$. This relation is clearly a congruence on S and if we form the quotient semigroup $S' = S/\sim$ we now obtain a transformation semigroup

$$\mathscr{A}/\langle\pi\rangle = (Y, S')$$

with the operation $*$ defined by

$$H_i * [s] = H_i * s$$

where $[s]$ denotes the $\sim$-class containing s ($H_i \in Y$, $s \in S$).

Some remarks concerning the relationships between these two concepts of quotients are worth making.

First of all consider the state machine $\mathcal{M} = (Q, \Sigma, F)$ and its transformation semigroup $\mathbf{TS}(\mathcal{M})$. A partition π on Q is admissible with respect to $\mathcal{M}$ if and only if it is admissible with respect to $\mathbf{TS}(\mathcal{M})$.

Secondly the transformation semigroup of $\mathcal{M}/\pi$, $\mathbf{TS}(\mathcal{M}/\pi)$, is isomorphic to

$$(\mathbf{TS}(\mathcal{M}))/\langle\pi\rangle.$$

There are natural epimorphisms defined by quotient state machines and quotient transformation semigroups.

If $\mathcal{M} = (Q, \Sigma, F)$ is a state machine and $\pi = \{H_i\}_{i\in I}$ is an admissible partition on $\mathcal{M}$ then the epimorphism $(\alpha^\pi, 1_\Sigma): \mathcal{M} \to \mathcal{M}/\pi$ defined by $\alpha^\pi(q) = H_i \Leftrightarrow q \in H_i$ $(q \in Q, H_i \in \pi)$, is called the *natural epimorphism defined by* π. If $\mathcal{A} = (Q, S)$ is a transformation semigroup and $\pi = \{H_i\}$ is an admissible partition on $\mathcal{A}$ then the epimorphism $(f^\pi, g^\pi): \mathcal{A} \to \mathcal{A}/\pi$ defined by

$$f^\pi(q) = H_i \Leftrightarrow q \in H_i \quad (q \in Q, H_i \in \pi)$$
$$g^\pi(s) = [s] \quad (s \in S)$$

is called the *natural epimorphism defined by* π.

Suppose that $\mathcal{M} = (Q, \Sigma, F)$ is a state machine, and let $\pi = \{H_i\}_{i\in I}$, $\pi' = \{K_j\}_{j\in J}$ be admissible partitions on $\mathcal{M}$. If $\pi \le \pi'$, that is, if given $i \in I$ there exists $j \in J$ with $H_i \subseteq K_j$, we can construct an epimorphism $(\alpha, 1_\Sigma): \mathcal{M}/\pi \to \mathcal{M}/\pi'$ by $\alpha(H_i) = K_j$. This leads us to a homomorphism theorem for state machines.

Theorem 2.3.4

Let $\mathcal{M} = (Q, \Sigma, F)$ and $\mathcal{M}' = (Q', \Sigma', F')$ be state machines and $(\alpha, \beta): \mathcal{M} \to \mathcal{M}'$ an epimorphism. Suppose that π_α is the admissible partition defined by α on $\mathcal{M}$ (so π_α is the partition of the relation R_α defined by α) and that π is an admissible partition on $\mathcal{M}$ satisfying the condition $\pi \le \pi_\alpha$ then there exists an epimorphism $(\lambda, \mu): \mathcal{M}/\pi \to \mathcal{M}'$ such that the following diagram of homomorphisms is commutative:

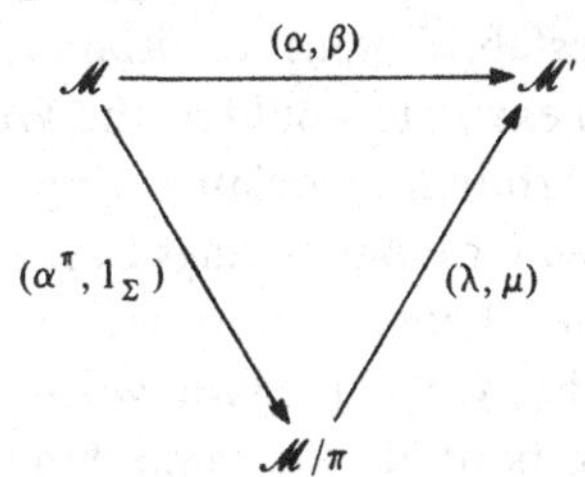

Furthermore if $\pi = \pi_\alpha$ then (λ, μ) is an isomorphism.

Proof Let $\pi=\{H_i\}_{i\in I}$, $\pi_\alpha=\{K_j\}_{j\in J}$. We define $\lambda:\pi\to Q'$ by $\lambda(H_i)=\alpha(q)$ where $q\in H_i$ $(i\in I)$. This is well-defined for if $q_1\in H_i$ then $q, q_1\in H_i\subseteq K_j$ for some $j\in J$ and so $\alpha(q)=\alpha(q_1)$. If we define $\mu:\Sigma\to\Sigma'$ by putting $\mu=\beta$ the result then follows easily. □

Theorem 2.3.5
Let $\mathcal{A}=(Q, S)$ and $\mathcal{A}'=(Q', S')$ be transformation semigroups and $(f, g):\mathcal{A}\to\mathcal{A}'$ an epimorphism. Suppose that π_f is the admissible partition defined on $\mathcal{A}$ by f and that π is an admissible partition on $\mathcal{A}$ satisfying the condition $\pi\leq\pi_f$ then there exists an epimorphism $(l, m):\mathcal{A}/\langle\pi\rangle\to\mathcal{A}'$ such that the following diagram of homomorphisms is commutative,

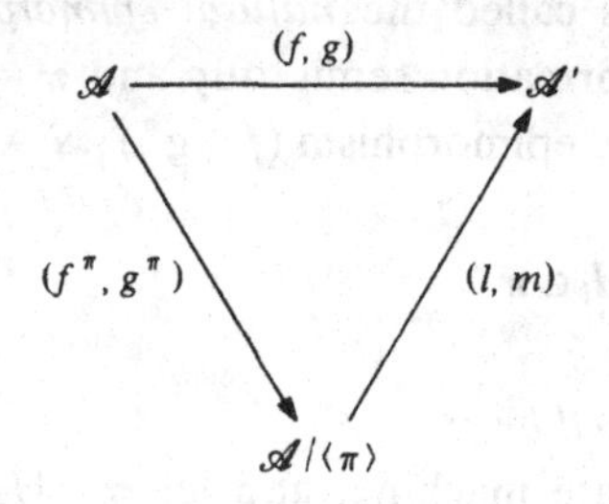

Furthermore if $\pi=\pi_f$ then (l, m) is an isomorphism.

Proof See exercise 2.2.

There are many other results concerned with homomorphisms and quotients of both state machines and transformation semigroups. While these are of independent algebraic interest they have not yet proved particularly useful in the study of automata and related areas. In fact the algebraic theory of machines diverges from the direction taken in other algebraic theories in one important respect. The idea of isomorphism is crucially important in many algebraic theories and many important classification theorems involve the establishment of isomorphisms between particular algebraic objects: an example would be the Wedderburn–Artin theorem for semi-simple Artinian associative rings which are shown to be *isomorphic* to a direct sum of matrix rings over various division rings. The emphasis in automata theory is, however, not what machines 'look like' but what 'they can do'. We will regard two machines as being very closely related if they can both 'do the same thing', they may however not be algebraically isomorphic!

2.4 Coverings

Before we can talk about two state machines doing the same thing we must first examine what the function of a state machine actually is. Let $\mathcal{M} = (Q, \Sigma, F)$ be a complete state machine and choose any $q \in Q$. Each word $\alpha \in \Sigma^*$ defines a partial function $F_\alpha : Q \to Q$ given by

$$qF_\alpha = F(q, \alpha) \text{ for all } q \in Q.$$

Therefore $\mathcal{M}$ is just a collection of partial functions $\{F_\alpha \mid \alpha \in \Sigma^*\}$. Now suppose that $\mathcal{M}' = (Q', \Sigma, F')$ is another state machine that 'performs the same function' as $\mathcal{M}$. Each state in $\mathcal{M}$ must correspond to a state in $\mathcal{M}'$ in such a way that the image under F_α in $\mathcal{M}$ corresponds to the image under F'_α in $\mathcal{M}'$ for each $\alpha \in \Sigma^*$. Formally we require a surjective partial function $\eta : Q' \to Q$, called a *covering*, such that $\eta(q')F_\alpha = \eta(q'F'_\alpha)$ for all $\alpha \in \Sigma^*$ and all q' belonging to the domain of η. To tidy up the notation and also to extend the notion to incomplete state machines we will write $\eta(q') = \varnothing$ if q' does not belong to the domain of η and also $qF_\alpha = \varnothing$ if $F(q, \alpha)$ is undefined. We may then define the covering requirement as

$$\eta(q')F_\alpha \subseteq \eta(q'F'_\alpha),$$

so that if q' is not in the domain of η we have

$$\varnothing \subseteq \eta(q'F'_\alpha),$$

similarly if

$$F(\eta(q'), \alpha)$$

is undefined then again

$$\varnothing \subseteq \eta(q'F'_\alpha).$$

However, if for some reason $\eta(q'F'_\alpha) = \varnothing$, then unless $\eta(q')F_\alpha = \varnothing$ also, the partial function η will not be a covering.

In general the input alphabets of $\mathcal{M}$ and $\mathcal{M}'$ may not be the same and so we must extend our covering concept to include this case.

Let $\mathcal{M} = (Q, \Sigma, F)$, $\mathcal{M}' = (Q', \Sigma', F')$ be state machines. If $\xi : \Sigma \to \Sigma'$ is a function and $\eta : Q' \to Q$ is a surjective partial function such that

$$\eta(q')F_\alpha \subseteq \eta(q'F'_{\xi(\alpha)})$$

for each $q' \in Q'$ and $\alpha \in \Sigma^*$, we say that (η, ξ) is a *covering of* $\mathcal{M}$ *by* $\mathcal{M}'$, written $\mathcal{M} \leq \mathcal{M}'$.

Examples 2.4

(i) Let $\mathcal{M} = (Q, \Sigma, F)$ be a state machine, define a relation $\sim$ on Σ by

$$\sigma \sim \sigma_1 \Leftrightarrow F_\sigma = F_{\sigma_1} \quad \text{for } \sigma, \sigma_1 \in \Sigma.$$

Construct a state machine $\mathcal{M}' = (Q, \Sigma/\sim, \bar{F})$ by defining $\bar{F}(q, [\sigma]) = F(q, \sigma)$ for $q \in Q$ and $[\sigma] \in \Sigma/\sim$. Now form $\xi : \Sigma \to \Sigma/\sim$ by putting

$$\xi(\sigma) = [\sigma] \quad \text{for } \sigma \in \Sigma$$

and

$$\eta : Q \to Q$$

by

$$\eta(q) = q \quad \text{for } q \in Q$$

and we will obtain a covering (η, ξ) of $\mathcal{M}$ by $\mathcal{M}'$. We say that $\mathcal{M}'$ has been constructed from $\mathcal{M}$ by 'coinciding equal inputs'.

(ii) Let $\mathcal{M}$ be defined by the diagram

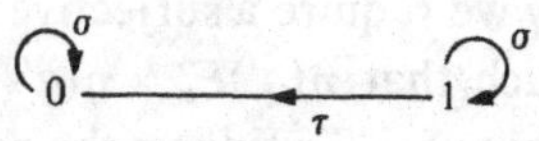

and $\mathcal{M}'$ by

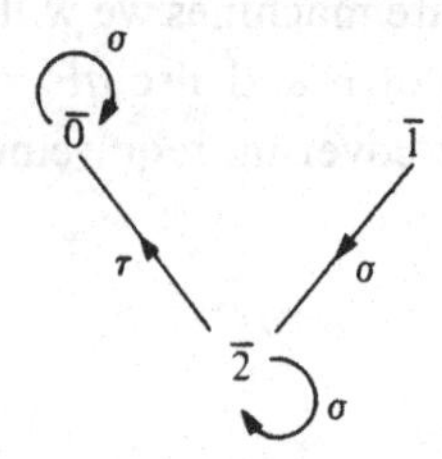

Defining $\eta : \bar{0} \to 0, \bar{1} \to 1$ does not yield a covering $(\eta, 1_\Sigma)$ since

$$\eta(\bar{1})F_\tau = 0 \not\subseteq \eta(\bar{1}F'_\tau) = \varnothing.$$

However by putting $\eta' : \bar{0} \to 0, \bar{2} \to 1$ we may check that $(\eta', 1_\Sigma)$ gives a covering $\mathcal{M} \leq \mathcal{M}'$.

(iii) Let $\mathcal{M} = (Q, \Sigma, F)$, $\mathcal{M}_1 = (Q_1, \Sigma, F_1)$ and $\mathcal{M}' = (Q', \Sigma, F')$ be state machines. Suppose that $(f, 1_\Sigma) : \mathcal{M}' \to \mathcal{M}$ and $(f_1, 1_\Sigma) : \mathcal{M}' \to \mathcal{M}_1$ are homomorphisms with f an injective function and f_1 surjective. Construct a partial mapping $\eta : Q \to Q_1$ by $\eta(q) = f_1(f^{-1}(q))$ for $q \in f(Q')$. For $q \in f(Q')$, $\alpha \in \Sigma^*$,

$$\begin{aligned}
\eta(q)F_{1\alpha} &= [f_1(f^{-1}(q))]F_{1\alpha} \\
&= f_1((f^{-1}(q))F'_\alpha) \\
&= f_1((f^{-1}f(q'))F'_\alpha) \quad \text{if } q = f(q'), q' \in Q' \\
&= f_1(q'F'_\alpha) \\
&= f_1(f^{-1}f(q'F'_\alpha)) \\
&= f_1 f^{-1}(f(q')F_\alpha) \\
&= \eta(qF_\alpha)
\end{aligned}$$

For $q \in Q \backslash f(Q')$, $\eta(q)F_{1\alpha} = \varnothing \subseteq \eta(qF_\alpha)$. Thus $\mathcal{M}_1 \leq \mathcal{M}$.

The concept of covering also has an important role to play in transformation semigroups.

Let $\mathcal{A} = (Q, S)$, $\mathcal{B} = (P, T)$ be transformation semigroups and suppose that $\eta : P \to Q$ is a surjective partial function and that for each $s \in S$ there exists a $t_s \in T$ such that

$$\eta(p) \cdot s \subseteq \eta(p \cdot t_s) \quad \text{for } p \in P. \qquad (*)$$

We say that $\mathcal{B}$ *covers* $\mathcal{A}$, written $\mathcal{A} \leq \mathcal{B}$ and that η is *a covering of* $\mathcal{A}$ *by* $\mathcal{B}$. Furthermore we will say that t_s is a *covering element* for s.

If $s, s' \in S$ then

$$\eta(p) \cdot ss' = (\eta(p) \cdot s) \cdot s' \subseteq \eta(p \cdot t_s) \cdot s' \subseteq \eta(p \cdot t_s \cdot t_{s'})$$

and so by defining $t_{ss'} = t_s \cdot t_{s'}$ the relationship $(*)$ is satisfied.

This is a slightly more general concept for transformation semigroups than might seem necessary from the analogy with state machines. We could have asked for a semigroup homomorphism $\xi : S \to T$ such that $\eta(p) \cdot s \subseteq \eta(p \cdot \xi(s))$ for $p \in P$, however this possibility for a definition of covering is too restrictive for our purposes. See exercise 2.33. If we define $\xi(s)$ to be some element $t_s \in T$ that covers s we will have to show that $\xi(s \cdot s') = \xi(s) \cdot \xi(s')$, for $s, s' \in S$ and the element chosen as $\xi(s \cdot s')$ may differ from $\xi(s) \cdot \xi(s')$.

Theorem 2.4.1

Let $\mathcal{M}$, $\mathcal{M}'$ be state machines such that $\mathcal{M} \leq \mathcal{M}'$, then

$$\mathbf{TS}(\mathcal{M}) \leq \mathbf{TS}(\mathcal{M}').$$

Proof Let $\mathcal{M} = (Q, \Sigma, F)$, $\mathcal{M}' = (Q', \Sigma', F')$, let $\eta : Q' \to Q$ be a surjective partial covering function and $\xi : \Sigma \to \Sigma'$ a function, then

$$\eta(q')F_\alpha \subseteq \eta(q'F'_{\xi(\alpha)})$$

for $q' \in Q'$, $\alpha \in \Sigma^*$.

Suppose that $\mathbf{TS}(\mathcal{M}) = (Q, S)$ and $\mathbf{TS}(\mathcal{M}') = (Q', S')$. Let $s \in S$, then $a \in \Sigma^*$ such that $s = [a]$. Put $t_s = [\xi(a)] \in S'$. Now if $q' \in Q'$, $\eta(q') \cdot s = \eta(q') \cdot F_a \subseteq \eta(q'F'_{\xi(a)}) = \eta(q't_s)$ and so η defines a covering

$$\mathbf{TS}(\mathcal{M}) \leq \mathbf{TS}(\mathcal{M}'). \qquad \square$$

Recall the definition of the transformation semigroup of a finite semigroup $S^{\cdot}$, it is the pair $(S^{\cdot}, S)$. Suppose that T is also a semigroup and form the transformation semigroup $(T^{\cdot}, T)$; if $(S^{\cdot}, S)$ is covered by $(T^{\cdot}, T)$ what can be said about the relationship between S and T?

Theorem 2.4.2

Let S, T be finite semigroups, then $(S^{\cdot}, S) \leq (T^{\cdot}, T)$ if and only if there exists a subsemigroup $\bar{T}$ of T such that S is a homomorphic image of $\bar{T}$.

Proof Let $(S^{\cdot}, S) \leq (T^{\cdot}, T)$, then there exists a surjective partial function $\eta : T^{\cdot} \to S^{\cdot}$, and for each $s \in S$ there exists $t_s \in T$ such that

$$\eta(y) \cdot s \subseteq \eta(y \cdot t_s) \quad \text{for all } y \in T^{\cdot}.$$

If η is a surjective partial function there exists a right inverse $\eta^{-1} : S^{\cdot} \to T^{\cdot}$ defined by choosing a y such that $\eta(y) = x$ and putting $\eta^{-1}(x) = y$. Then $\eta(\eta^{-1}(x)) = x$ for each $x \in S^{\cdot}$. Now $\eta(\eta^{-1}(x)) \cdot s \subseteq \eta(\eta^{-1}(x) \cdot t_s)$ and so

$$x \cdot s \subseteq \eta(\eta^{-1}(x) \cdot t_s).$$

However, $x \cdot s \neq \varnothing$ in this case, thus $x \cdot s = \eta(\eta^{-1}(x) \cdot t_s)$ for $x \in S^{\cdot}$. Now suppose that there exists $s' \in S$ such that

$$\eta(y) \cdot s' \subseteq \eta(y \cdot t_s) \quad \text{for all } y \in T^{\cdot},$$

so that when $x \in S^{\cdot}$,

$$x \cdot s' = \eta(\eta^{-1}(x) \cdot t_s) = x \cdot s.$$

However this implies that $s' = s$ because of the faithfulness of the semigroup action. We may now define a partial function $f : T \to S$ by

$$f(t) = s \Leftrightarrow t = t_s, \ t \in T.$$

Suppose that $\bar{T}$ is the domain of f, and $t_1, t_2 \in \bar{T}$, then $t_1 = t_{s_1}$, $t_2 = t_{s_2}$ for $s_1, s_2 \in S$. Since $t_{s_1} \cdot t_{s_2} = t_{s_1 \cdot s_2} = t_1 \cdot t_2$ we see that $t_1 \cdot t_2 \in \bar{T}$ and so $\bar{T}$ is a subsemigroup of T. Finally, $f(t_1 \cdot t_2) = f(t_{s_1 \cdot s_2}) = s_1 \cdot s_2 = f(t_1) \cdot f(t_2)$ and so f restricted to $\bar{T}$ is a *semigroup* homomorphism onto S.

Conversely let $g : \bar{T} \to S$ be a *semigroup* homomorphism from a subsemigroup $\bar{T}$ of T onto S. Consider the partial function $g^{\cdot} : T^{\cdot} \to S^{\cdot}$ defined by

$$g^{\cdot}(t) = g(t) \quad \text{if } t \in \bar{T}$$
$$g^{\cdot}(1) = 1 \quad \text{if } 1 \in T^{\cdot} \backslash T.$$

Then $g^{\cdot}$ is a surjective partial function from $T^{\cdot}$ onto $S^{\cdot}$. Let $s \in S$, there exists a $t \in T$ such that $s = g(t)$ and we write $t_s = t$.

Now for $y \in T^{\cdot}$, $g^{\cdot}(y) \cdot s = g(y) \cdot s = g(y) \cdot g(t) = g(y \cdot t) = g^{\cdot}(y \cdot t_s)$ if $y \in T$, and $g^{\cdot}(1) \cdot s = 1 \cdot s = g(t) = g(1 \cdot t_s) = g^{\cdot}(1 \cdot t_s)$ and so $g^{\cdot}$ is a covering map and

$$(S^{\cdot}, S) \leq (T^{\cdot}, T). \qquad \square$$

We say that S *divides* T in this situation and write $S|T$.

2.5 Mealy machines

So far we have examined state machines without any formal output, and we will now digress for a short while to look at machines with outputs. The reason for this is to motivate the next section on state machine products. Throughout this section all state machines are assumed to be complete.

Let $\mathcal{M} = (Q, \Sigma, F)$ be a state machine, and suppose that Θ is a non-empty finite set and $G: Q \times \Sigma \to \Theta$ is a function. The quintuple $\hat{\mathcal{M}} = (Q, \Sigma, \Theta, F, G)$ will be called a *Mealy machine* (after G. Mealy, 1955), Σ is the input alphabet, F the state transition function, Θ the output alphabet and G the output function. The machine works as follows.

Suppose that the input word $\sigma \in \Sigma$ is applied to the machine in state q, the machine then moves to state qF_σ and produces an output $G(q, \sigma) \in \Theta$ at the same instant. We will have, for each $\sigma \in \Sigma$, a mapping

$$G_\sigma : Q \to \Theta \quad \text{defined by} \quad qG_\sigma = G(q, \sigma),\ q \in Q.$$

To see what the machine does when we apply an input word $\alpha = \sigma_1 \ldots \sigma_k \in \Sigma^*$ to the machine in state q it is best to imagine the symbols printed on a tape and treat the machine as a black box that changes state and at the same time prints symbols from Θ on an output tape. The input tape will be fed into the machine on the right hand side and will move from right to left. The output tape also moves from right to left. See figure 2.1.

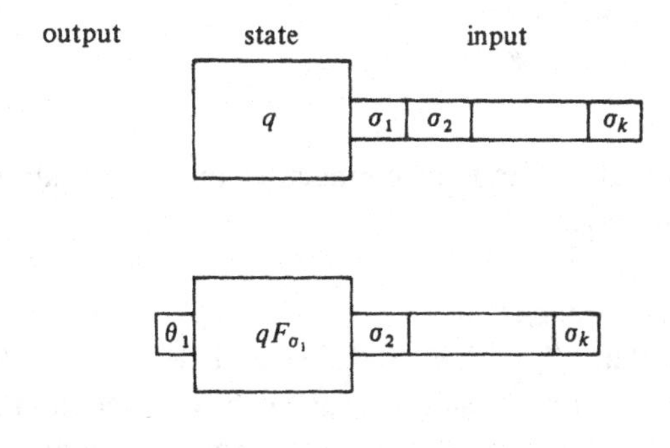

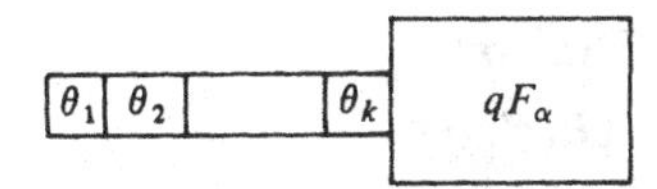

Figure 2.1. The action of a Mealy machine.

The final state will be qF_α and the output word is $\beta = \theta_1\theta_2 \ldots \theta_k \in \Theta^*$ where

$$\theta_1 = qG_{\sigma_1}, \theta_2 = qF_{\sigma_1}G_{\sigma_2}, \ldots$$
$$\theta_k = qF_{\sigma_1} \ldots F_{\sigma_{k-1}}G_{\sigma_k}.$$

We will study the theory of Mealy machines in more detail later, it is sufficient to remark that a suitable notion of covering of Mealy machines can be formulated and this concept is closely related to the covering of the underlying state machines, for inside every Mealy machine there is a state machine.

A Mealy machine is just a set of translators, one for each internal state, which translates words of length k in Σ^* into words of length k in Θ^*, in fact each translator is nothing more than a rather special semigroup homomorphism.

These Mealy machines have been introduced here for the sole purpose of examining how they may be connected together to produce other Mealy machines. Each machine will be regarded as a black box with an input channel and an output channel.

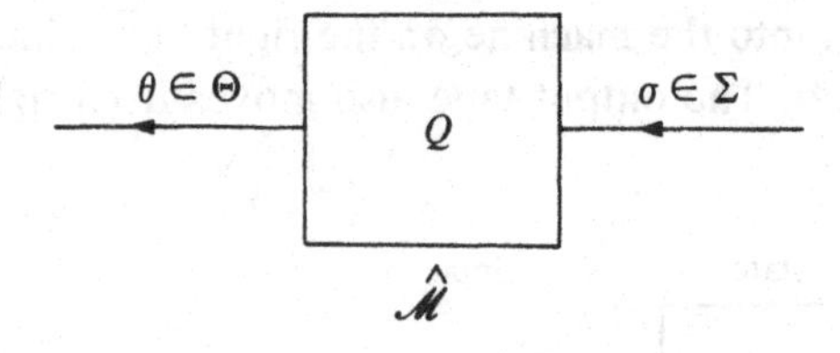

There are two major methods of connecting up two Mealy machines, by parallel and by series.

Parallel connections

Suppose that $\hat{\mathcal{M}} = (Q, \Sigma, \Theta, F, G)$ and $\hat{\mathcal{M}}' = (Q', \Sigma, \Theta', F', G')$ are Mealy machines with the same input set Σ. Connecting them up in parallel as in figure 2.2 will produce a new Mealy machine:

$$\hat{\mathcal{M}} \wedge \hat{\mathcal{M}}' = (Q \times Q', \Sigma, \Theta \times \Theta', F \wedge F', G \wedge G')$$

where

$$(F \wedge F')((q, q'), \sigma) = (F(q, \sigma), F'(q', \sigma));$$
$$(G \wedge G')((q, q'), \sigma) = (G(q, \sigma), G'(q', \sigma))$$

for each $\sigma \in \Sigma$, $(q, q') \in Q \times Q'$.

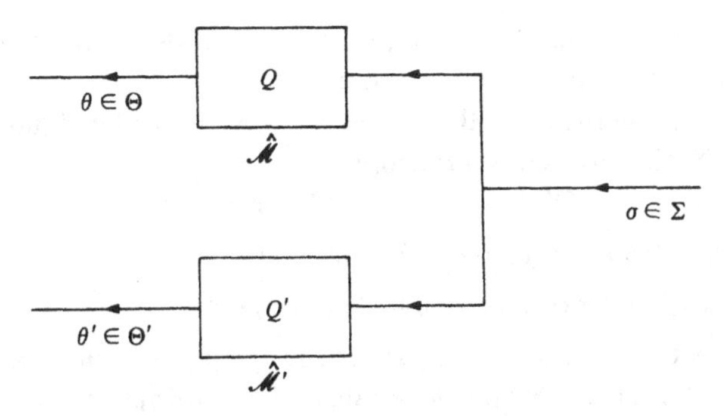

Figure 2.2. A restricted parallel connection.

We call this machine the *restricted direct product* of $\hat{\mathcal{M}}$ and $\hat{\mathcal{M}}'$ and it clearly produces words in $(\Theta \times \Theta')^*$ as outputs in response to input words from Σ^*.

Another type of parallel connection can be made, even when the input alphabets are different.

Let $\hat{\mathcal{M}} = (Q, \Sigma, \Theta, F, G)$, $\hat{\mathcal{M}}' = (Q', \Sigma', \Theta', F', G')$ be Mealy machines and define

$$\hat{\mathcal{M}} \times \hat{\mathcal{M}}' = (Q \times Q', \Sigma \times \Sigma', \Theta \times \Theta', F \times F', G \times G')$$

where

$$(F \times F')((q, q'), (\sigma, \sigma')) = (F(q, \sigma), F'(q', \sigma'))$$
$$(G \times G')((q, q'), (\sigma, \sigma')) = (G(q, \sigma), G'(q', \sigma'))$$

for each $(\sigma, \sigma') \in \Sigma \times \Sigma'$, $(q, q') \in Q \times Q'$.

This Mealy machine is called the (full) *direct product* of $\hat{\mathcal{M}}$ and $\hat{\mathcal{M}}'$. It converts words from $(\Sigma \times \Sigma')^*$ into words from $(\Theta \times \Theta')^*$. See figure 2.3. Note that each input word $(\sigma_1, \sigma'_1)(\sigma_2, \sigma'_2) \ldots (\sigma_k, \sigma'_k) \in (\Sigma \times \Sigma')^*$ can be written as $(\sigma_1\sigma_2 \ldots \sigma_k, \sigma'_1\sigma'_2 \ldots \sigma'_k) \in \Sigma^* \times (\Sigma')^*$ but not any word

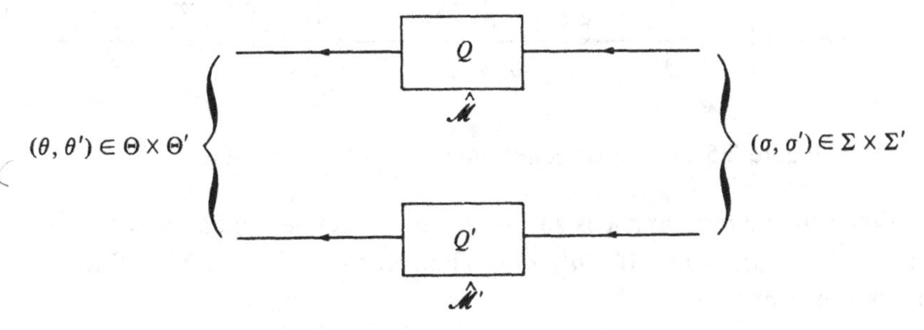

Figure 2.3. A full parallel connection.

from $\Sigma^* \times (\Sigma')^*$ can be used as an input unless it is of the form (α, α') where the length of α equals the length of α'.

This can be generalized in the following way. Consider figure 2.4, where $\lambda : \bar{\Sigma} \to \Sigma \times \Sigma'$ represents a mapping.

The machine is $(Q \times Q', \bar{\Sigma}, \Theta \times \Theta', F^\lambda, G^\lambda)$ where

$$F^\lambda((q, q'), \bar{\sigma}) = (F(q, p_1\lambda(\bar{\sigma})), F'(q', p_2\lambda(\bar{\sigma})))$$
$$G^\lambda((q, q'), \bar{\sigma}) = (G(q, p_1\lambda(\bar{\sigma})), G'(q', p_2\lambda(\bar{\sigma})))$$

for $(q, q') \in Q \times Q'$, $\bar{\sigma} \in \bar{\Sigma}$ and p_1, p_2 are the projection mappings associated with $\Sigma \times \Sigma'$. This machine generalizes both forms of the direct product and will be denoted by $\hat{\mathcal{M}} * \hat{\mathcal{M}}'$ and called the *general direct product.*

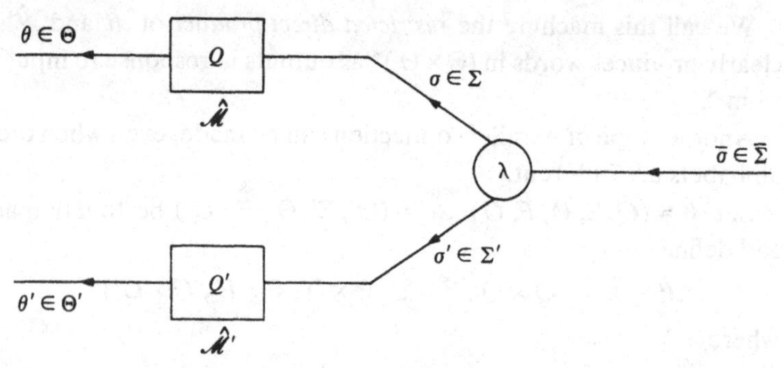

Figure 2.4. A general parallel connection.

Series connections

If we wish to connect two Mealy machines up in series we must ensure that we can 'hook up' the output from the first machine to the input of the second. See figure 2.5. One way of doing this is to define a function $\lambda : \Theta' \to \Sigma$ and so convert each output word $\beta' \in (\Theta')^*$ into an input word $\lambda(\beta') \in \Sigma^*$ before applying it to the machine $\hat{\mathcal{M}}$.

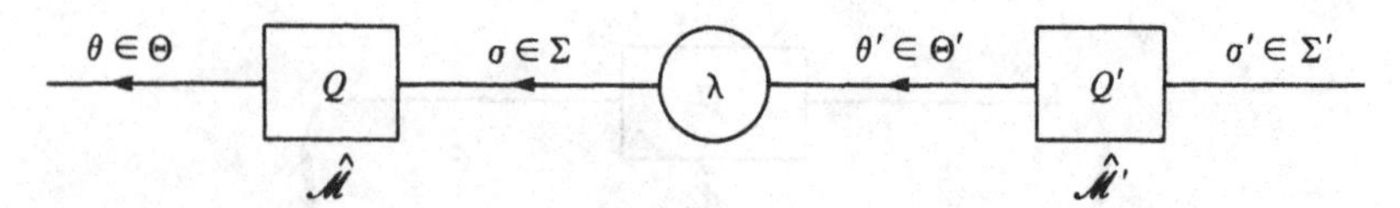

Figure 2.5. A cascade connection.

Once such a mapping λ is specified we can define a mapping $\omega : Q' \times \Sigma' \to \Sigma$ by $\omega(q', \sigma') = \lambda(G'(q', \sigma'))$. Then each input $\sigma' \in \Sigma'$ defines a mapping

$$\omega_{\sigma'} : Q' \to \Sigma$$

by

$$\omega_{\sigma'}(q') = \omega(q', \sigma')$$

for $q' \in Q'$.

The Mealy machine we have formed is:

$$\hat{\mathcal{M}}\omega\hat{\mathcal{M}}' = (Q \times Q', \Sigma', \Theta, F^\omega, G^\omega)$$

where

$$F^\omega((q, q'), \sigma') = (F(q, \omega_{\sigma'}(q')), F'(q', \sigma'))$$
$$G^\omega((q, q'), \sigma') = G(q, \omega_{\sigma'}(q'))$$

for $\sigma' \in \Sigma'$, $(q, q') \in Q \times Q'$.

Such a machine is called the *cascade product* of $\hat{\mathcal{M}}$ and $\hat{\mathcal{M}}'$ *induced* by ω.

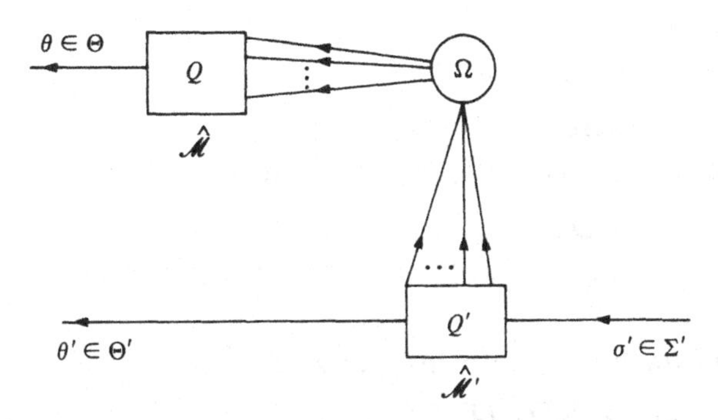

Figure 2.6. An alternative interpretation of a cascade connection.

Since ω defines a set of mappings $\Omega = \{\omega_{\sigma'} : Q' \to \Sigma\}_{\sigma' \in \Sigma'} \subseteq \Sigma^{Q'}$ we can visualize the connections as depicted in figure 2.6. This is but a short step from the useful generalization of figure 2.7 where $\Omega = \Sigma^{Q'}$. The

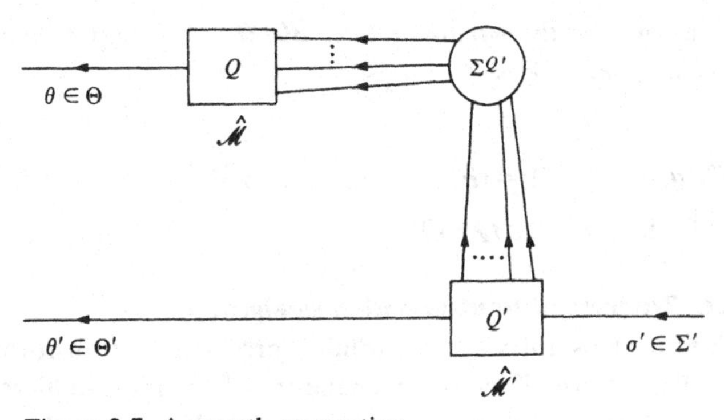

Figure 2.7. A wreath connection.

wreath product of $\hat{\mathcal{M}}$ and $\hat{\mathcal{M}}'$ is

$$\hat{\mathcal{M}} \circ \hat{\mathcal{M}}' = (Q \times Q', \Sigma^{Q'} \times \Sigma', \Theta \times \Theta', F^\circ, G^\circ)$$

where

$$F^\circ((q, q'), (f, \sigma')) = (F(q, f(q')), F'(q', \sigma'))$$
$$G^\circ((q, q'), (f, \sigma')) = (G(q, f(q')), G'(q', \sigma'))$$

for $\sigma' \in \Sigma', f \in \Sigma^{Q'}, (q, q') \in Q \times Q'$.

There are further types of connection, most notably the feedback connections, but we will not be requiring them here.

The various products of Mealy machines show us how to define products of state machines; we merely remove the output sets and functions.

Let $\mathcal{M} = (Q, \Sigma, F)$, $\mathcal{M}' = (Q', \Sigma', F')$ be state machines. Define their *restricted direct product*:

$$\mathcal{M} \wedge \mathcal{M}' = (Q \times Q', \Sigma, F \wedge F'),$$

in the special case where $\Sigma = \Sigma'$ only, by:

$$(F \wedge F')((q, q'), \sigma) = (F(q, \sigma), F'(q', \sigma))$$

for $\sigma \in \Sigma$, $(q, q') \in Q \times Q'$.

Let $\mathcal{M} \times \mathcal{M}' = (Q \times Q', \Sigma \times \Sigma', F \times F')$ be the (full) *direct product* of $\mathcal{M}$ and $\mathcal{M}'$ where

$$(F \times F')((q, q'), (\sigma, \sigma')) = (F(q, \sigma), F'(q', \sigma'))$$

for $\sigma \in \Sigma$, $\sigma' \in \Sigma'$, $(q, q') \in Q \times Q'$.

Define the *cascade product* of $\mathcal{M}$ and $\mathcal{M}'$ with *respect to* $\omega : Q' \times \Sigma' \to \Sigma$ by

$$\mathcal{M} \omega \mathcal{M}' = (Q \times Q', \Sigma', F^\omega)$$

where $F^\omega((q, q'), \sigma') = (F(q, \omega(q', \sigma')), F'(q', \sigma'))$, for $\sigma' \in \Sigma'$, $(q, q') \in Q \times Q'$.

Finally we consider the *wreath product*, $\mathcal{M} \circ \mathcal{M}'$, of $\mathcal{M}$ and $\mathcal{M}'$ where

$$\mathcal{M} \circ \mathcal{M}' = (Q \times Q', \Sigma^{Q'} \times \Sigma', F^\circ)$$

and

$$F^\circ((q, q'), (f, \sigma')) = (F(q, f(q')), F'(q', \sigma'))$$

for $\sigma' \in \Sigma', f \in \Sigma^{Q'}, (q, q') \in Q \times Q'$.

2.6 Products of transformation semigroups

The most useful ways of forming products of transformation semigroups will emerge if we consider the transformation semigroup of a product of state machines and compare it with the transformation semigroups of the original state machines. Again all state machines and transformation semigroups will be assumed to be complete in this section. We examine the restricted direct product $\mathcal{M} \wedge \mathcal{M}'$, where $\mathcal{M} = (Q, \Sigma, F)$, and $\mathcal{M}' = (Q', \Sigma, F')$, and find its transformation semigroup.

Let $\alpha \in \Sigma^+$, then α defines a class $[\alpha]_\wedge$ with respect to the state machine $\mathcal{M} \wedge \mathcal{M}'$. Now let $\beta \in \Sigma^+$, then $\beta \in [\alpha]_\wedge$ if and only if

$$(F \wedge F')((q, q'), \beta) = (F \wedge F')((q, q'), \alpha)$$

for all $(q, q') \in Q \times Q'$ i.e.

$$(qF_\beta, q'F'_\beta) = (qF_\alpha, q'F'_\alpha)$$

i.e.

$$qF_\beta = qF_\alpha \quad \text{for all } q \in Q$$

and

$$q'F'_\beta = q'F'_\alpha \quad \text{for all } q' \in Q'.$$

Thus $\beta \in [\alpha]_\wedge$ if and only if $\beta \in [\alpha] \cap [\alpha]'$ where $[\alpha]$ and $[\alpha]'$ are the equivalence classes containing α with respect to the state machines $\mathcal{M}$ and $\mathcal{M}'$ respectively. Therefore the semigroup of $\mathcal{M} \wedge \mathcal{M}'$ is isomorphic to the quotient semigroup

$$\Sigma^+/\sim \cap \sim'$$

where $\sim$ and $\sim'$ are the equivalence relations defined by $\mathcal{M}$ and $\mathcal{M}'$ respectively.

It is clear that we will not be able to form a restricted direct product between two arbitrary transformation semigroups. If $\mathcal{A} = (Q, S)$, $\mathcal{A}' = (Q', S')$ are transformation semigroups and there exists a free semigroup Σ^+ with epimorphisms $\theta : \Sigma^+ \to S$, $\theta' : \Sigma^+ \to S'$ then we can form the transformation semigroup

$$\mathcal{A} \wedge \mathcal{A}' = (Q \times Q', T)$$

where $T = \Sigma^+/(R_\theta \cap R_{\theta'})$ (here R_θ corresponds to $\sim$ and $R_{\theta'}$ to $\sim'$) and the action is given by:

$$(q, q')[\alpha]_\wedge = (q\theta(\alpha), q'\theta'(\alpha))$$

for $(q, q') \in Q \times Q'$ and $[\alpha]_\wedge \in T$. (It is an easy matter to check that T acts faithfully on $Q \times Q'$.)

The definition of $\mathcal{A} \wedge \mathcal{A}'$ will depend on the choice of θ and θ', so we call $\mathcal{A} \wedge \mathcal{A}'$ the *restricted direct product* of $\mathcal{A}$ and $\mathcal{A}'$ (with respect to θ and θ'). We can now state:

Theorem 2.6.1
Let $\mathcal{M} = (Q, \Sigma, F)$ and $\mathcal{M}' = (Q', \Sigma, F')$ then

$$\mathbf{TS}(\mathcal{M} \wedge \mathcal{M}') = \mathbf{TS}(\mathcal{M}) \wedge \mathbf{TS}(\mathcal{M}')$$

(for suitable epimorphisms $\theta : \Sigma^+ \to \mathbf{S}(\mathcal{M})$, $\theta' : \Sigma^+ \to \mathbf{S}(\mathcal{M}')$).

Turning our attention to the full direct product we immediately see that the situation is more straightforward. In chapter 1 the direct product of two semigroups was introduced. We can extend this concept easily to the (full) direct product of two transformation semigroups.

Let $\mathcal{A} = (Q, S)$, $\mathcal{A}' = (Q', S')$ be transformation semigroups, define the (full) *direct product*

$$\mathcal{A} \times \mathcal{A}' = (Q \times Q', S \times S')$$

where the action is given by:

$$(q, q')(s, s') = (qs, q's')$$

for $(q, q') \in Q \times Q'$, $(s, s') \in S \times S'$.

Clearly the action is faithful.

We write $\overset{r}{\Pi} \mathcal{A}$ to denote $\mathcal{A} \times \mathcal{A} \times \ldots \times \mathcal{A}$ (r times).

Theorem 2.6.2
Let $\mathcal{M} = (Q, \Sigma, F)$, $\mathcal{M}' = (Q', \Sigma', F')$ be state machines.
$\mathbf{TS}(\mathcal{M} \times \mathcal{M}') \leqslant \mathbf{TS}(\mathcal{M}) \times \mathbf{TS}(\mathcal{M}')$.

Proof Now $\mathcal{M} \times \mathcal{M}' = (Q \times Q', \Sigma \times \Sigma', F \times F')$, and so $\mathbf{S}(\mathcal{M} \times \mathcal{M}')$ will be a quotient semigroup of the free semigroup $(\Sigma \times \Sigma')^+$. Let $(\alpha, \beta) \in (\Sigma \times \Sigma')^+$, then

$$(\alpha, \beta) = (\sigma_1, \sigma'_1) \ldots (\sigma_n, \sigma'_n)$$

for some $\sigma_i \in \Sigma$, $\sigma'_i \in \Sigma'$, $i = 1, \ldots, n$. The elements of $\mathbf{S}(\mathcal{M} \times \mathcal{M}')$ will be equivalence classes of the form $[(\alpha, \beta)]_\times$ where

$$(\alpha, \beta) \sim_\times (\alpha_1, \beta_1) \Leftrightarrow (F \times F')_{(\alpha,\beta)} = (F \times F')_{(\alpha_1,\beta_1)}$$
$$\Leftrightarrow F_\alpha = F_{\alpha_1} \text{ and } F'_\beta = F'_{\beta_1} \Leftrightarrow \alpha \sim \alpha_1 \text{ and } \beta \sim' \beta_1.$$

Define a function $g : \mathbf{S}(\mathcal{M} \times \mathcal{M}') \to \mathbf{S}(\mathcal{M}) \times \mathbf{S}(\mathcal{M}')$ by

$$g([(\alpha, \beta)]_\times) = ([\alpha], [\beta]')$$

for each $[(\alpha, \beta)]_\times \in \mathbf{S}(\mathcal{M} \times \mathcal{M}')$. It is a routine matter to establish that g is a semigroup monomorphism and finally $(1_{Q \times Q'}, g) : \mathbf{TS}(\mathcal{M} \times \mathcal{M}') \to \mathbf{TS}(\mathcal{M}) \times \mathbf{TS}(\mathcal{M})'$ is a transformation semigroup covering. □

Theorem 2.6.3

Let $\mathscr{A}=(Q,S)$, $\mathscr{A}'=(Q',S')$ be transformation semigroups, Σ a finite non-empty set and $\theta:\Sigma^+\to S$, $\theta':\Sigma^+\to S'$ semigroup epimorphisms, then

$$\mathbf{TS}(\mathscr{A}\wedge\mathscr{A}')\le\mathbf{TS}(\mathscr{A}\times\mathscr{A}').$$

Proof Using the notation of 2.6.1 we will take the identity map $1_{Q\times Q'}$ as the covering map. Now let $[\alpha]_\wedge\in T$, the semigroup of $\mathscr{A}\wedge\mathscr{A}'$, so that $\alpha\in\Sigma^+$. Consider $\theta(\alpha)\in S$ and $\theta'(\alpha)\in S'$ and define $(\theta(\alpha),\theta'(\alpha))$ to be a covering element for $[\alpha]_\wedge$. Then, for $(q,q')\in Q\times Q'$, we have

$$\begin{aligned}(q,q')[\alpha]_\wedge&=(q\theta(\alpha),q'\theta'(\alpha))\\&=(q,q')(\theta(\alpha),\theta'(\alpha))\end{aligned}$$

and so the covering exists. □

Our next topic is the examination of the cascade and wreath products and their implications for transformation semigroup theory. Suppose that $\mathscr{M}=(Q,\Sigma,F)$, $\mathscr{M}'=(Q',\Sigma',F')$ are state machines and $\omega:Q'\times\Sigma'\to\Sigma$ is a mapping. Let $\mathscr{A}=(Q,S)$, $\mathscr{A}'=(Q',S')$ be the transformation semigroups of $\mathscr{M}$ and $\mathscr{M}'$ respectively. If $\mathscr{B}=(Q\times Q',T)$ is the transformation semigroup of $\mathscr{M}\omega\mathscr{M}'$, we wish to find a relationship between $\mathscr{A}$, $\mathscr{A}'$ and $\mathscr{B}$. Unfortunately there is no simple straightforward construction that yields the transformation semigroup $\mathscr{B}$ from a suitable combination of $\mathscr{A}$ and $\mathscr{A}'$. What we will do here is to show that $\mathscr{B}$ can be covered by the wreath product of the transformation semigroups $\mathscr{A}$ and $\mathscr{A}'$. This will now be defined.

Let $\mathscr{A}=(Q,S)$, $\mathscr{A}'=(Q',S')$ be transformation semigroups. Define

$$\mathscr{A}\circ\mathscr{A}'=(Q\times Q',S^{Q'}\times S')$$

where $S^{Q'}$ is the set of all mappings from Q' to S. The set $S^{Q'}\times S'$ is a semigroup, for if $f:Q'\to S$, $f_1:Q'\to S$, s', $s_1'\in S'$ then we may define a mapping $f*f_1:Q'\to S$ by

$$f*f_1(q')=f(q')\cdot f_1(q's')$$

for all $q'\in Q'$ and put $(f,s)\cdot(f_1,s')=(f*f_1,ss')$. Then the action of $S^{Q'}\times S'$ on $Q\times Q'$ is defined by

$$(q,q')(f,s')=(q(f(q')),q's')$$

for $(q,q')\in Q\times Q'$, $(f,s')\in S^{Q'}\times S'$.

The faithfulness of this action is easily checked and thus $\mathscr{A}\circ\mathscr{A}'$ is a transformation semigroup, called the *wreath product* of $\mathscr{A}$ and $\mathscr{A}'$. If

$\mathscr{S} = (S^{\cdot}, S)$ and $\mathscr{T} = (T^{\cdot}, T)$ are transformation semigroups with S and T semigroups then $\mathscr{S} \circ \mathscr{T} = (S^{\cdot} \times T^{\cdot}, S \circ T)$ where $S \circ T$ is defined in section 1.3. We have the following result.

Theorem 2.6.4

Let $\mathcal{M} = (Q, \Sigma, F)$, $\mathcal{M}' = (Q', \Sigma', F')$ be state machines and $\omega : Q' \times \Sigma' \to \Sigma$ a mapping.

(i) $\mathbf{TS}(\mathcal{M}\omega\mathcal{M}') \leq \mathbf{TS}(\mathcal{M}) \circ \mathbf{TS}(\mathcal{M}')$

(ii) $\mathbf{TS}(\mathcal{M} \circ \mathcal{M}') \leq \mathbf{TS}(\mathcal{M}) \circ \mathbf{TS}(\mathcal{M}')$.

Proof (i) Let $\alpha' \in (\Sigma')^+$ and consider the element $[\alpha']_\omega$ defined by α' in the semigroup $\mathbf{S}(\mathcal{M}\omega\mathcal{M}')$. We must find an element of the semigroup $S^{Q'} \times S'$ which will cover this element $[\alpha']_\omega$ where $S = \mathbf{S}(\mathcal{M})$ and $S' = \mathbf{S}(\mathcal{M}')$. The word α' will clearly define an element $[\alpha']'$ of S' but we must also find a suitable mapping $f_{\alpha'} : Q' \to S$. First of all we have to examine the mapping $\omega : Q' \times \Sigma' \to \Sigma$. Each $\sigma' \in \Sigma'$ defines a mapping $\omega_{\sigma'} : Q' \to \Sigma$ by

$$\omega_{\sigma'}(q') = \omega(q', \sigma')$$

for all $q' \in Q'$. The mapping ω describes the link-up between the two machines in the cascade connection, so we must investigate what happens when we input a word from $(\Sigma')^+$ into the leading machine. Suppose that we apply a word of length 2 from $(\Sigma')^+$.

If $(q, q') \in Q \times Q'$, $\sigma', \sigma_1' \in \Sigma'$ then

$$\begin{aligned}(q, q')F^\omega_{\sigma'\sigma_1'} &= (qF_{\omega(q',\sigma')}, q'F'_{\sigma'})F^\omega_{\sigma_1'} \\ &= (qF_{\omega(q',\sigma')}F_{\omega(q'F'_{\sigma'},\sigma_1')}, q'F'_{\sigma'}F'_{\sigma_1'}) \\ &= (qF_{\omega(q',\sigma')\omega(q'F'_{\sigma'},\sigma_1')}, q'F'_{\sigma'\sigma_1'}). \qquad (*)\end{aligned}$$

It makes sense to define a generalization of the map ω to cover the cases of words in $(\Sigma')^+$ of length greater than 1. To do this we will define $\omega^+ : Q' \times (\Sigma')^+ \to \Sigma^+$ in such a way that

$$(q, q')F^\omega_{\sigma'\sigma_1'} = (qF_{\omega^+(q',\sigma'\sigma_1')}, q'F'_{\sigma'\sigma_1'})$$

and so we need

$$\omega^+(q', \sigma'\sigma_1) = \omega(q', \sigma')\omega(q'F'_{\sigma'}, \sigma_1)$$

by analogy with (*). Generalizing further we define $\omega^+ : Q' \times (\Sigma')^+ \to \Sigma^+$ inductively by $\omega^+(q', \sigma'\alpha') = \omega(q', \sigma') \cdot \omega^+(q'F'_{\sigma'}, \alpha')$ if $\alpha' \in (\Sigma')^+$ and $\omega^+(q', \sigma') = \omega(q', \sigma')$ where $\sigma' \in \Sigma'$, $\alpha' \in (\Sigma')^+$, $q' \in Q'$. Then $(q, q')F^\omega_{\alpha'} = (qF_{\omega^+(q',\alpha')}, q'F'_{\alpha'})$ for $\alpha' \in (\Sigma')^+$. Now, for each $\alpha' \in (\Sigma')^+$ we have a mapping $\omega^+_{\alpha'} : Q' \to (\Sigma)^+$ defined by $\omega^+_{\alpha'}(q') = \omega^+(q', \alpha')$, $q' \in Q'$. Returning to our problem we can now define

$$f_{\alpha'} : Q' \to S \quad \text{by} \quad f_{\alpha'}(Q') = [\omega^{+}_{\alpha'}(q')]$$

for $q' \in Q'$. Then to each $[\alpha']_\omega$ we will associate the pair $(f_{\alpha'}, [\alpha']')$. The first thing to check is that this definition is well-defined. Suppose that $\beta' \in (\Sigma')^+$ and that $[\alpha']_\omega = [\beta']_\omega$, then for $(q, q') \in Q \times Q'$, $(q, q')F^{\omega}_{\alpha'} = (q, q')F^{\omega}_{\beta'}$ that is

$$qF_{\omega^+(q', \alpha')} = qF_{\omega^+(q', \beta')}$$

for each $q \in Q$ and

$$q'F'_{\alpha'} = q'F'_{\beta'}$$

for each $q' \in Q'$.

Therefore

$$\begin{aligned} f_{\beta'}(q') &= [\omega^{+}_{\beta'}(q')] \\ &= [\omega^{+}(q', \beta')] \\ &= [\omega^{+}(q', \alpha')] \\ &= f_{\alpha'}(q') \end{aligned}$$

for each $q' \in Q'$ and $[\alpha']'_\omega = [\beta']'_\omega$. Hence $[\alpha']_\omega = [\beta']_\omega$ implies $(f_{\alpha'}, [\alpha']') = (f_{\beta'}, [\beta']')$. Our final task is the verification of the covering relationship using the identity mapping on $Q \times Q'$. That is

$$(q, q') \cdot [\alpha'] \subseteq (q, q') \cdot (f_{\alpha'}, [\alpha']')$$

or in other words,

$$\begin{aligned} (qF_{\omega^+(q', \alpha')}, q'F'_{\alpha'}) &\subseteq (q \cdot [\omega^{+}(q', \alpha')], q' \cdot [\alpha']') \\ &= (q \cdot f_{\alpha'}(q'), q' \cdot [\alpha']') \\ &= (q, q') \cdot (f_{\alpha'}, [\alpha']'). \end{aligned}$$

The covering is thus established.

(ii) We will leave the proof of the covering $\mathbf{TS}(\mathcal{M} \circ \mathcal{M}') \le \mathbf{TS}(\mathcal{M}) \circ \mathbf{TS}(\mathcal{M}')$ as an exercise. □

The product $\mathcal{A} \circ \mathcal{A} \circ \ldots \circ \mathcal{A}$ with r factors is written $\mathcal{A}^r$.

Example 2.5

Consider the state machine $\mathcal{M}$ defined by:

0 ——σ—— 1 (σ: 0 → 1, σ: 1 → 0)

with transformation semigroup $\mathbb{Z}_2$ and the state machine $\mathcal{M}'$ defined by:

0 ——σ—— 1 ↻ σ

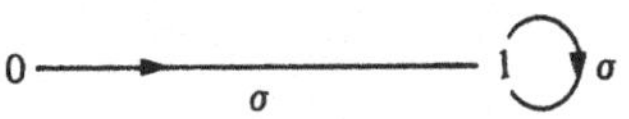

with transformation semigroup $\mathscr{C}$. We may form the restricted direct product $\mathcal{M} \wedge \mathcal{M}'$ which can be described by the diagram:

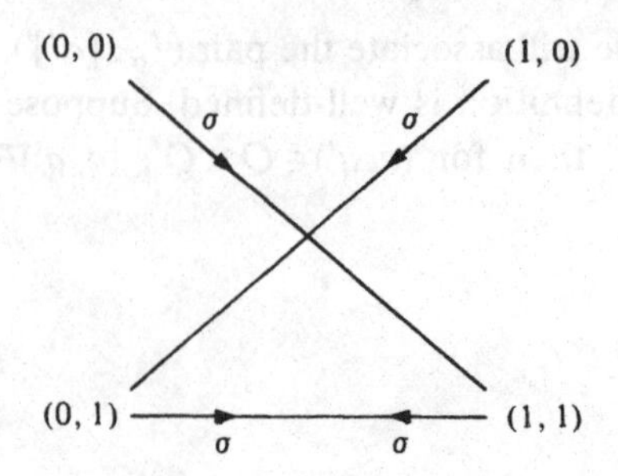

This has semigroup $\{\sigma, \sigma^2\}$ with the identity $\sigma^3 = \sigma$. The full direct product $\mathcal{M} \times \mathcal{M}'$ has the same description: in this case because the input alphabets to $\mathcal{M}$ and $\mathcal{M}'$ are both singletons. However if $\mathcal{M}''$ is given by

with transformation semigroup $\bar{\mathbf{2}}$ then $\mathcal{M} \wedge \mathcal{M}''$ cannot be defined and $\mathcal{M} \times \mathcal{M}''$ is given by

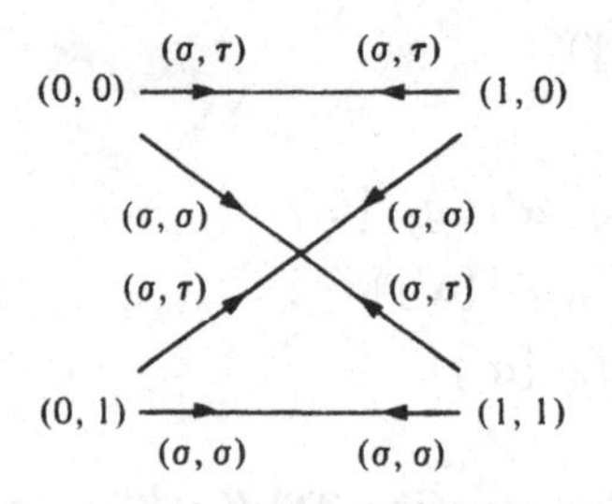

The semigroup of $\mathcal{M} \times \mathcal{M}''$ has four elements and $\mathbf{TS}(\mathcal{M} \times \mathcal{M}'') \cong \mathbb{Z}_2 \times \bar{\mathbf{2}}$.

Example 2.6
Let $\mathcal{M}$ be given by

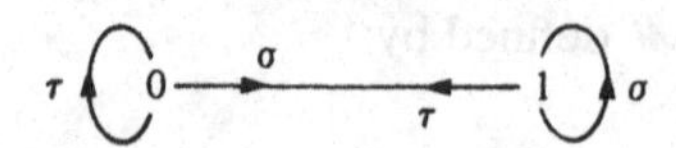

and $\mathcal{M}'$ be given by

Define a mapping $\omega : Q' \times \Sigma' \to \Sigma$ where $Q' = \{0, 1\}$, $\Sigma' = \{\sigma\}$, $\Sigma = \{\sigma, \tau\}$ by

$$\omega(0, \sigma) = \sigma, \quad \omega(1, \sigma) = \tau.$$

It is now possible to define the cascade product $\mathcal{M}\omega\mathcal{M}'$ which has diagram

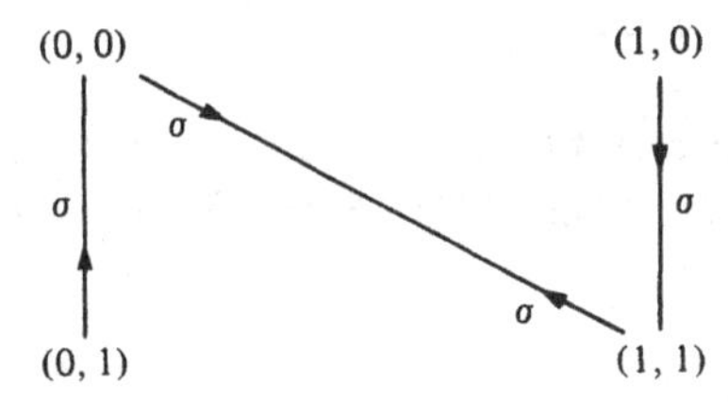

and semigroup $\{\sigma, \sigma^2\}$ subject to the relation $\sigma^3 = \sigma$.

Finally we examine the wreath product of $\mathcal{M}$ and $\mathcal{M}'$, this has input alphabet $\Sigma^{Q'} \times \Sigma'$. Denote the four elements of $\Sigma^{Q'}$ by $\alpha, \beta, \gamma, \delta$ where

$$\alpha(0) = \alpha(1) = \sigma, \ \beta(0) = \sigma, \ \beta(1) = \tau,$$

$$\gamma(0) = \tau, \ \gamma(1) = \sigma, \ \delta(0) = \delta(1) = \tau.$$

Then the state machine $\mathcal{M} \circ \mathcal{M}'$ has the table

	(0, 0)	(1, 0)	(0, 1)	(1, 1)
(α, σ)	(1, 1)	(1, 1)	(1, 0)	(1, 0)
(β, σ)	(1, 1)	(1, 1)	(0, 0)	(0, 0)
(γ, σ)	(0, 1)	(0, 1)	(1, 0)	(1, 0)
(δ, σ)	(0, 1)	(0, 1)	(0, 0)	(0, 0)

The semigroup of this machine has eight elements and the transformation semigroup is isomorphic to the wreath product $\bar{\mathbf{2}} \circ \mathbb{Z}_2$.

However the state table of $\mathcal{M}' \circ \mathcal{M}$ is given by

	(0, 0)	(1, 0)	(0, 1)	(1, 1)
(α, σ)	(1, 1)	(0, 1)	(1, 1)	(0, 1)
(α, τ)	(1, 0)	(0, 0)	(1, 0)	(0, 0)

where $\alpha : Q \to \Sigma'$. This has a semigroup with four elements whereas the semigroup of $\mathbb{Z}_2 \circ \bar{\mathbf{2}}$ has eight elements. Hence the covering in 2.6.4(ii) cannot in general be replaced by an isomorphism.

Our final task in this section is to examine the associativity of the two main product constructions.

Theorem 2.6.5
Let (Q_i, S_i) be transformation semigroups for $i = 1, 2, 3$, then $(Q_1, S_1) \times ((Q_2, S_2) \times (Q_3, S_3)) \cong ((Q_1, S_1) \times (Q_2, S_2)) \times (Q_3, S_3)$.

Proof This is left to the reader. □

Before embarking on a similar result for the wreath product it is best to examine in detail what the wreath product $(Q_1, S_1) \circ ((Q_2, S_2) \circ (Q_3, S_3))$ looks like.

Now

$$(Q_2, S_2) \circ (Q_3, S_3) = (Q_2 \times Q_3, S_2^{Q_3} \times S_3)$$

and

$$(Q_1, S_1) \circ ((Q_2, S_2) \circ (Q_3, S_3)) = (Q_1 \times (Q_2 \times Q_3), S_1^{Q_2 \times Q_3} \times (S_2^{Q_3} \times S_3)).$$

Similarly

$$\begin{aligned}((Q_1, S_1) \circ (Q_2, S_2)) \circ (Q_3, S_3) &= (Q_1 \times Q_2, S_1^{Q_2} \times S_2) \circ (Q_3, S_3) \\ &= ((Q_1 \times Q_2) \times Q_3, (S_1^{Q_2} \times S_2)^{Q_3} \times S_3).\end{aligned}$$

However $(S_1^{Q_2} \times S_2)^{Q_3}$ is the set of maps from Q_3 to $S_1^{Q_2} \times S_2$ and if $f: Q_3 \to S_1^{Q_2} \times S_2$ we can consider the mappings $f_1 = p_1 \circ f$, $f_2 = p_2 \circ f$ obtained by projecting f onto the factors. Then

$$f_1: Q_3 \to S_1^{Q_2} \quad \text{and} \quad f_2: Q_3 \to S_2.$$

Let $q_3 \in Q_3$, then $f_1(q_3) = g$ say, where $g: Q_2 \to S_1$. We can construct a mapping $\bar{f}_1: Q_2 \times Q_3 \to S_1$ by

$$\bar{f}_1(q_2, q_3) = g(q_2) = (f_1(q_3))(q_2).$$

It is a routine matter to check that the function

$$\Theta: (S_1^{Q_2} \times S_2)^{Q_3} \times S_3 \to S_1^{Q_2 \times Q_3} \times (S_2^{Q_3} \times S_3)$$

defined by $\Theta(f, s_3) = (\bar{f}_1, (f_2, s_3))$ for $f \in (S_1^{Q_2} \times S_2)^{Q_3}$, $s_3 \in S_3$, is an isomorphism of semigroups. Thus we may establish:

Theorem 2.6.6
If (Q_i, S_i) are transformation semigroups for $i = 1, 2, 3$ then

$$(Q_1, S_1) \circ ((Q_2, S_2) \circ (Q_3, S_3)) \cong ((Q_1, S_1) \circ (Q_2, S_2)) \circ (Q_3, S_3).$$

Associativity relations for the other products defined in this chapter will be examined in the exercises. As for the direct product and wreath product of any finite number of transformation semigroups, we shall usually rearrange brackets or remove them altogether when this serves to clarify the notation.

There are some 'distributive laws' connecting the direct product and the wreath product but we will postpone discussion of these until the next section.

2.7 More on products

We now extend our definitions of products of state machines and transformation semigroups to include the incomplete cases.

First let $\mathcal{M} = (Q, \Sigma, F)$ be a state machine and suppose that $P \subseteq Q$. Define the *restriction of* $\mathcal{M}$ *to* P to be the state machine

$$\mathcal{M}|_P = (P, \Sigma, F')$$

where

$$F' : P \times \Sigma \to P$$

is defined by

$$F'(p, \sigma) = \begin{cases} F(p, \sigma) & \text{if } F(p, \sigma) \in P, \\ \varnothing & \text{otherwise.} \end{cases}$$

What this amounts to is that all the states in $Q \backslash P$ have been removed together with all the arrows leading to or from these states.

From example 2.1(vi) we have the state machine $\mathcal{M}$ given by

$\mathcal{M}$	0	1	2
F_σ	0	0	2
F_τ	1	$\varnothing$	1

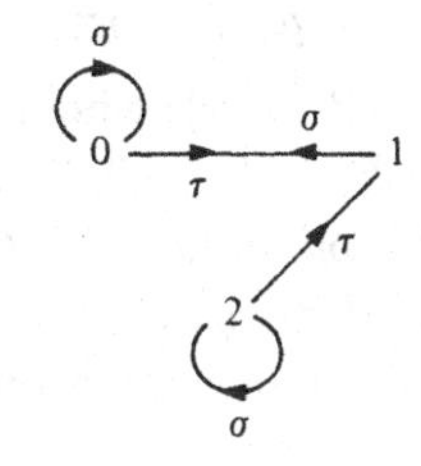

and if $P = \{0, 2\}$ then $\mathcal{M}|_P$ is given by:

$\mathcal{M}\|_P$	0	2
F_σ	0	2
F_τ	$\varnothing$	$\varnothing$

Now suppose that $\mathcal{A} = (Q, S)$ is a transformation semigroup and $P \subseteq Q$. We define the restriction $\mathcal{A}|_P$ to be the transformation semigroup $\mathbf{TS}((\mathbf{SM}(\mathcal{A}))|_P)$. Now $\mathbf{SM}(\mathcal{A}) = (Q, S, F)$ where $F : Q \times S \to Q$ is defined by $qF_s = qs$ for all $q \in Q$, $s \in S$. The restriction $(\mathbf{SM}(\mathcal{A}))|_P$ is (P, S, F')

where $F':P\times S\to P$ is defined by

$$pF'_s=\begin{cases} ps & \text{if } ps\in P\\ \varnothing & \text{otherwise}\end{cases}$$

Now $\mathbf{TS}((\mathbf{SM}(\mathcal{A}))|_P)=(P,T)$ where $T=\mathbf{S}((P,S,F'))=S^*/\sim$ and $\sim$ is defined by $\alpha\sim\beta\Leftrightarrow pF'_\alpha=pF'_\beta$ for all $p\in P$, and $\alpha,\beta\in S^*$. Note that if α is the word $s_1\ldots s_n\in S^*$ then $[\alpha]=[s]$ where s is the product in S of $s_1\ldots s_n$, and $[s]$ is the equivalence class containing s.

Thus

$$\mathcal{A}|_P=(P,S/\sim).$$

Another way of looking at $\mathcal{A}|_P$ is to consider the pair (P,S) with the operation $*$ defined by

$$p*s=\begin{cases} ps & \text{if } ps\in P\\ \varnothing & \text{otherwise.}\end{cases}$$

Under this operation S may not be faithful on P and so we have to define the relation ρ on S by

$$s\rho s'\Leftrightarrow p*s=p*s' \quad \text{for all } p\in P, \text{ where } s,s'\in S.$$

Thus $\mathcal{A}|_P$ may be regarded as the transformation semigroup $(P,S/\rho)$ under the operation induced by $*$, namely

$$p\langle s\rangle=p*s \quad \text{for } p\in P,\ \langle s\rangle\in S/\rho.$$

In our example $\mathcal{A}|_P=(P,\{\sigma,\boldsymbol{\theta}\})$ where $\sigma^2=\sigma=1_P$.

Now let $\mathcal{M}=(Q,\Sigma,F)$, $\mathcal{M}'=(Q',\Sigma',F')$ be two state machines which are not necessarily complete. Define $\mathcal{M}\times\mathcal{M}'=(\mathcal{M}^c\times(\mathcal{M}')^c)|_{Q\times Q'}$. So we complete $\mathcal{M}$ and $\mathcal{M}'$ if necessary, form the direct product with the complete machines and then restrict the resultant to the product of the original state sets.

Similarly $\mathcal{M}\wedge\mathcal{M}'=(\mathcal{M}^c\wedge(\mathcal{M}')^c)|_{Q\times Q'}$ if $\Sigma=\Sigma'$, and $\mathcal{M}\circ\mathcal{M}'=(\mathcal{M}^c\circ(\mathcal{M}')^c)|_{Q\times Q'}$. Now let $\omega:Q'\times\Sigma'\to\Sigma$ be a function, and suppose that $\mathcal{M}'$ is incomplete. If $\bar{q}'$ is the new sink state for $\mathcal{M}'$ define

$$\omega^c:(Q'\cup\{\bar{q}'\})\times\Sigma'\to\Sigma$$

by

$$\omega^c(q',\sigma')=\omega(q',\sigma')$$

for $q'\in Q'$, $\sigma'\in\Sigma'$

$$\omega^c(\bar{q}',\sigma')=\text{arbitrary}$$

for $\sigma'\in\Sigma'$. Now put $\mathcal{M}\omega\mathcal{M}'=(\mathcal{M}^c\omega^c(\mathcal{M}')^c)|_{Q\times Q'}$. We make similar definitions for transformation semigroups. If $\mathcal{A}=(Q,S)$, $\mathcal{A}'=(Q',S')$ then

$$\mathcal{A}\times\mathcal{A}'=(\mathcal{A}^c\times(\mathcal{A}')^c)|_{Q\times Q'}$$

and

$$\mathcal{A} \circ \mathcal{A}' = (\mathcal{A}^c \circ (\mathcal{A}')^c)|_{Q \times Q'}.$$

We now prove some straightfoward identities. As usual we will assume that all things are complete but the necessary adjustments for the incomplete cases should be regarded as exercises.

Theorem 2.7.1
Let $\mathcal{A} = (Q, S)$, $\mathcal{B} = (P, T)$ be transformation semigroups. Then $(\mathcal{A} \circ \mathcal{B})^{\cdot} \leq \mathcal{A}^{\cdot} \circ \mathcal{B}^{\cdot}$.

Proof We will only consider the case where $\mathcal{A}$ and $\mathcal{B}$ are both *not* transformation monoids. Then $\mathcal{A}^{\cdot} = (Q, S \cup \{1_Q\})$ and $\mathcal{B}^{\cdot} = (P, T \cup \{1_P\})$. We have $(\mathcal{A} \circ \mathcal{B})^{\cdot} = (Q \times P, (S^P \times T) \cup \{1_{Q \times P}\})$. Define $1_{Q \times P}$ as the covering function, for $(f, t) \in S^P \times T$ we will use $(\bar{f}, t)$ where $\bar{f} : P \to S \cup \{1_Q\}$ is defined by

$$\bar{f}(p) = f(p)$$

for $p \in P$; and for $(1_Q, t)$ we will use (g, t) where

$$g : P \to S \cup \{1_Q\}$$

is defined by

$$g(p) = 1_Q$$

for $p \in P$.
Now

$$(q, p)(f, t) = (qf(p), pt) \subseteq (q, p)(\bar{f}, t)$$

and

$$(q, p)(1_Q, t) = (q, pt) \subseteq (q, p)(g, t)$$

and the covering is established. □

Theorem 2.7.2
Let $\mathcal{A}_1$, $\mathcal{A}_2$, $\mathcal{B}_1$, $\mathcal{B}_2$ be transformation semigroups, then $(\mathcal{A}_1 \circ \mathcal{A}_2) \times (\mathcal{B}_1 \circ \mathcal{B}_2) \leq (\mathcal{A}_1 \times \mathcal{B}_1) \circ (\mathcal{A}_2 \times \mathcal{B}_2)$.

Proof Let $\mathcal{A}_i = (Q_i, S_i)$, $\mathcal{B}_i = (P_i, T_i)$ for $i = 1, 2$.

$$(\mathcal{A}_1 \circ \mathcal{A}_2) \times (\mathcal{B}_1 \circ \mathcal{B}_2) = (Q_1 \times Q_2 \times P_1 \times P_2, S_1^{Q_2} \times S_2 \times T_1^{P_2} \times T_2)$$

$$(\mathcal{A}_1 \times \mathcal{B}_1) \circ (\mathcal{A}_2 \times \mathcal{B}_2) = (Q_1 \times P_1 \times Q_2 \times P_2, (S_1 \times T_1)^{Q_2 \times P_2} \times S_2 \times T_2).$$

Define $\phi : Q_1 \times P_1 \times Q_2 \times P_2 \to Q_1 \times Q_2 \times P_1 \times P_2$ by

$$\phi(q_1, p_1, q_2, p_2) = (q_1, q_2, p_1, p_2)$$

for $q_i \in Q_i$, $p_i \in P_i$, $i = 1, 2$. Then (f_1, s_2, g_1, t_2) where $f_1 : Q_2 \to S_1$, $g_1 : P_2 \to T_1$, $s_2 \in S_2$, $t_2 \in T_2$ is covered by $(f_1 \times g_1, s_2, t_2)$ where

$$f_1 \times g_1 : Q_2 \times P_2 \to S_1 \times T_1$$

is defined by

$$(f_1 \times g_1)(q_2, p_2) = (f_1(q_2), g_1(p_2)). \qquad \square$$

Theorem 2.7.3

Let $\mathcal{A}$, $\mathcal{A}'$, $\mathcal{B}$ be transformation semigroups such that $\mathcal{A} \leq \mathcal{B}$. Then

$$\mathcal{A} \circ \mathcal{A}' \leq \mathcal{B} \circ \mathcal{A}'.$$

Proof Suppose that $\mathcal{A} = (Q, S)$, $\mathcal{B} = (P, T)$, $\mathcal{A}' = (Q', S')$ and $\phi : P \to Q$ is a surjective partial covering function, and given $s \in S$ there is an element $t_s \in T$ covering s. Define $\phi : P \times Q' \to Q \times Q'$ by $\phi(p, q') = (\phi(p), q')$ for $(p, q') \in P \times Q'$.

Now let $s' \in S'$ and $f : Q' \to S$, define the pair (g, s') where $g : Q' \to T$ is given by

$$g(q') = t_{f(q')} \quad \text{for } q' \in Q',$$

($t_{f(q')}$ is an element of T that covers the element $f(q')$).

Now

$$\begin{aligned} \phi(p, q')(f, s') &= (\phi(p)f(q'), q's') \\ &\subseteq (\phi(pt_{f(q')}), q's') \\ &= \phi((p, q')(g, s')) \end{aligned}$$

and so (g, s') covers (f, s') with respect to ϕ. $\square$

2.8 Examples and applications

Having patiently worked through some of the abstract theory of state machines we can now come to a brief survey of situations that give rise to such objects.

Example 2.7 Transistor components

The NAND_2 component consists of two transistors connected up in a simple circuit as in figure 2.8. The input at terminal I_1 is either *a current applied* denoted by 1; or *no current applied*, denoted by 0. The same choice of inputs are applied to I_2. The complete input description is an ordered pair of the form (a, b) where $a, b \in \{0, 1\}$. The transistors T_1 and T_2 are either 'off', when a current is flowing through CE; or 'on', when no current can flow through CE. The internal states of the

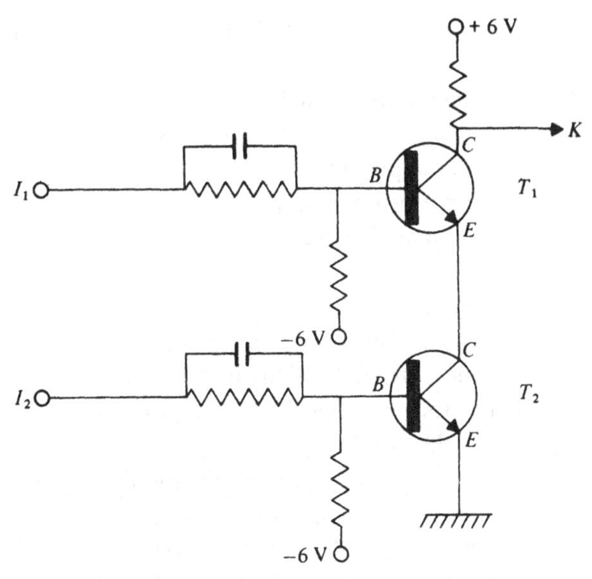

Figure 2.8. $NAND_2$ circuit.

circuit are thus described by the ordered pair $\langle\alpha, \beta\rangle$ where $\alpha, \beta \in$ {OFF, ON}. There is an output at K, either a current (1) or no current (0). We can now describe this circuit by means of a Mealy machine $\hat{\mathcal{M}} = (Q, \Sigma, \Theta, F, G)$ where

$$Q = \{\langle\text{off, off}\rangle, \langle\text{off, on}\rangle, \langle\text{on, off}\rangle, \langle\text{on, on}\rangle\}$$

$$\Sigma = \{0, 1\} \times \{0, 1\}, \Theta = \{0, 1\},$$

F is given by:

F	⟨off, off⟩	⟨off, on⟩	⟨on, off⟩	⟨on, on⟩
(0, 0)	⟨off, off⟩	⟨off, off⟩	⟨off, off⟩	⟨off, off⟩
(0, 1)	⟨off, on⟩	⟨off, on⟩	⟨off, on⟩	⟨off, on⟩
(1, 0)	⟨on, off⟩	⟨on, off⟩	⟨on, off⟩	⟨on, off⟩
(1, 1)	⟨on, on⟩	⟨on, on⟩	⟨on, on⟩	⟨on, on⟩

G is given by:

G	⟨off, off⟩	⟨off, on⟩	⟨on, off⟩	⟨on, on⟩
(0, 0)	1	1	1	1
(0, 1)	1	1	1	1
(1, 0)	1	1	1	1
(1, 1)	0	0	0	0

Many types of electronic components, from simple cases like this to complete computers can be analysed in terms of Mealy machines. Naturally the Mealy machine associated with a computer will have an enormous set of internal states, but since this set is finite the theory of finite state machines is still applicable.

As well as computer hardware it is possible to consider computer software as a type of state machine. See Chittenden [1978].

Example 2.8 Biological cells

The many complex chemical reactions that take place within various biological organisms provide a difficult problem for us to model. These reactions involve the input of various types of environmental stimulus, ranging from light of varying intensity and wavelength, temperature, different chemicals of varying concentrations, etc., both from the external environment of the organism and the more immediate environment of the surrounding cells. In many cases the chemical reactions that take place inside the cell are controlled by the *genetic* component of the *nucleus*, by the synthesis of various enzymes, etc. The net result of all this *metabolic activity* in the cell is the *synthesis* of certain chemicals necessary for growth and the operation of the organism, the storing of energy, heat, etc. So we may regard the activity of the cell as a kind of biological machine with inputs and outputs. It is not immediately apparent that the cell behaves like a Mealy machine: for example are there a finite number of internal states and a finite number of inputs and outputs? To answer this we will examine an argument which we could call the *threshold principle.*

Most of the parameters involved in the description of the state of the cell at a particular moment will be concerned with the concentrations of various chemicals, temperature, etc. and these are measured on a continuous scale. However in many situations minute changes in the concentration of a chemical do not affect the behaviour of a cell and it is only when the cell concentrations pass a *threshold value* on the measurement scale that the cell enters a different phase of behaviour. The same is true of environmental inputs: small changes in these may have no influence but larger changes, exceeding certain threshold values, will cause the cell to react in some way. The next step is to assume that there are only a finite number of these threshold values for each parameter involved in the internal description of the cell and a finite number are also assumed to exist for each input and output parameter. Let $A_1, \ldots, A_n$ be the sets of input parameters and let $\bar{A}_i$ be the set of threshold values of the parameter set A_i. Now we form the set $\Sigma = \prod_{i=1}^{n} \bar{A}_i$, which will be the finite input set. Similarly if $B_1, \ldots, B_m$ are the sets of output parameters then we form the sets $\bar{B}_j$ $(j = 1, \ldots, m)$ of associated threshold values and put $\Theta = \prod_{j=1}^{m} \bar{B}_j$. Finally let $C_1, \ldots, C_l$ be the sets of internal parameters and $\bar{C}_k$ $(k = 1, \ldots, l)$ the associated sets of threshold values. Let $S = \prod_{k=1}^{l} \bar{C}_k$, this will be the set of internal states; Σ is the input set and Θ is the output set. The whole system can now be represented by the Mealy machine $(S, \Sigma, \Theta, F, G)$ where the next state and output functions F and G are defined appropriately. One benefit of this view is that we can consider groups of biological cells also as a Mealy machine, we just have to extend our sets S, Σ, Θ and the mappings F, G suitably. (See, e.g. Rosen [1972].)

Example 2.9 Neural networks

A model of the brain can be constructed using a simple model of the brain cell or *neuron*, see figure 2.9. We will not concern ourselves with the neurological and anatomical details of a typical brain cell here. We concentrate, instead, on the basic function of such a cell.

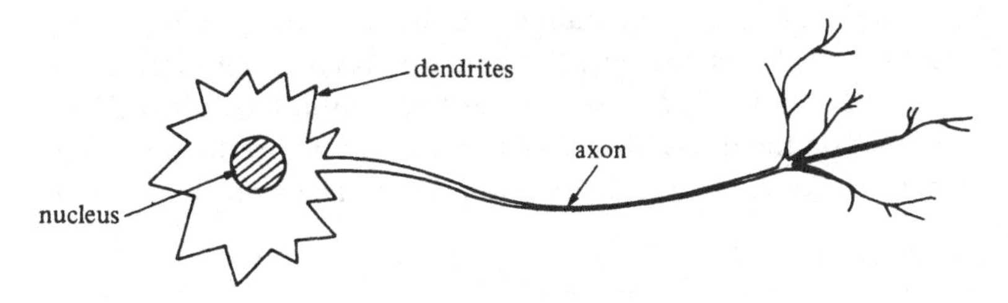

Figure 2.9. An example of a neuron.

Small electrical impulses arrive at the dendrites of the neuron, and over a short period of time they are 'summed up' by the nucleus. If the total exceeds a particular threshold value, the cell reacts by sending an impulse down the axon, the end of which branches into a number of small filaments which are in electrical contact with the dendrites of other neurons. In this way the neuron receives and propagates electrical impulses in the network of interconnected neurons which is the brain. The details of this will be found in Arbib [1964].

We can represent a neuron by a diagram such as figure 2.10, where $w_1, \ldots, w_n$ are the weights associated with the neuron's dendrites, some will be positive real numbers, indicating *excitatory* dendrites and others will be negative and mark the *inhibitory* dendrites. The threshold value

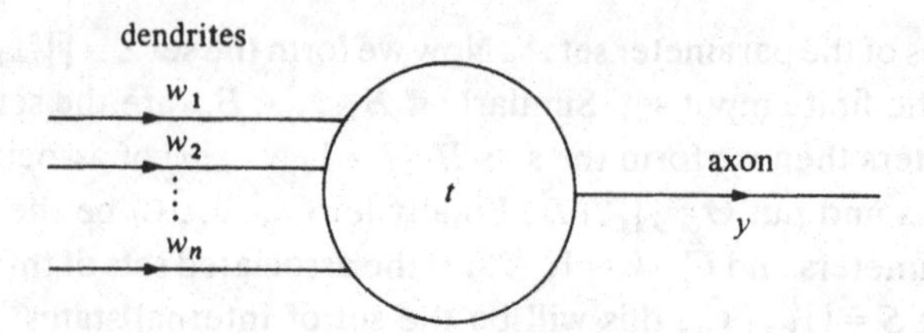

Figure 2.10. A general neuron.

is t and y is the weight of the output axon, indicating the strength of an output impulse.

Consider the simple example shown in figure 2.11. We interpret this as a Mealy machine in the following way. The set of states

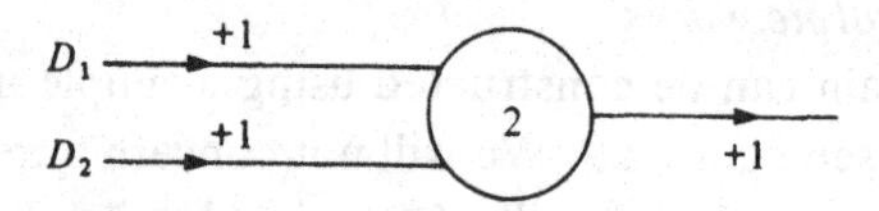

Figure 2.11. A simple neuron.

$Q = \{\text{'}on\text{'}, \text{'}off\text{'}\}$. The input and output sets are $\Sigma = \{(0,0), (1,0), (0,1), (1,1)\}$ and $\Theta = \{0, 1\}$ respectively. If the neuron is *on* and it receives an input of 1 at D_1 and 0 at D_2 we see that the neuron then turns *off* at the next point on the discrete time scale because the threshold has not been reached. However there is an output of 1. The state and output tables are:

	on	off	on	off
(0, 0)	off	off	1	0
(1, 0)	off	off	1	0
(0, 1)	off	off	1	0
(1, 1)	on	on	1	0

By constructing a model of the neural network, using simple mathematical models of the neurons connected together in certain ways we can investigate the way in which information, in the form of electrical impulses, is conveyed around the nervous system and the brain. There are, however, certain limitations to the use of this model as there are several basic assumptions that have to be made in order that the mathematics can be handled. These include the synchronization of the neurons in a convenient manner. Despite these drawbacks the model, known as the *neural network*, has proved useful. It can be shown that such a neural network is equivalent, in a natural way, to a Mealy machine. See Arbib [1964], Minsky [1967].

Other systems which model cell behaviour, for example metabolism repair systems, are also known to be equivalent to Mealy machines (see Rosen [1972]).

Example 2.10 Metabolic pathways

The following metabolic pathway illustrates the complex series of chemical reactions that make up the Krebs cycle in mammals, this is the process by which carbohydrates are converted into energy. Figure 2.12 illustrates the progress of one molecule of oxalacetic acid through the cycle.

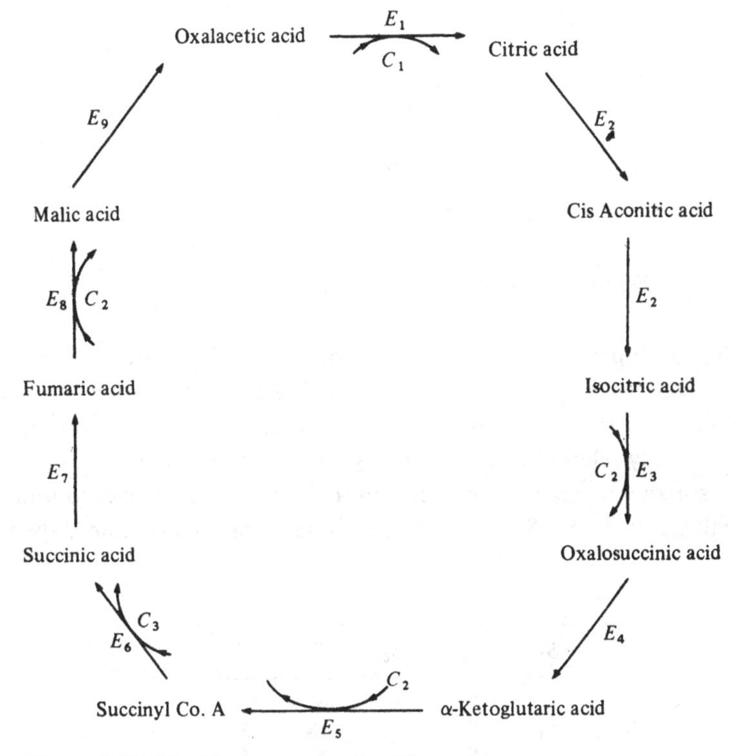

Figure 2.12. The Krebs or tricarboxylic cycle.

The letters $E_1, \ldots, E_9$ represent enzymes that are assumed to be present in sufficient concentrations during the cycle to permit the individual reactions to occur. The letters C_1, C_2, C_3 represent coenzymes.

These are also necessary for some of the reactions to take place, they combine with atoms from the substrates and are then involved in further reactions in other metabolic pathways and are eventually released for further use back in the cycle. If we make certain assumptions about the rates at which the reactions proceed and combine reactions that do not involve coenzymes we can simplify the diagram to figure 2.13. This new diagram involves the substrates:

S_1 – Oxalacetic acid, S_4 – Isocitric acid,
S_6 – α-Ketoglutaric acid, S_7 – Succinyl Co. A,
S_9 – Fumaric acid

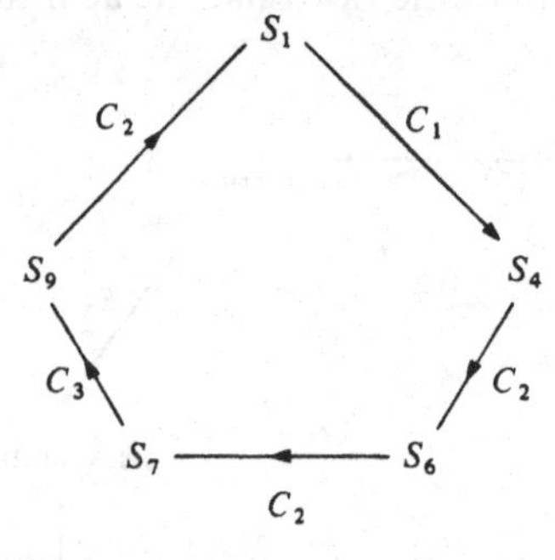

Figure 2.13. Reduced Krebs cycle.

Now an input of C_2 applied to a molecule of S_4 produces eventually a molecule of S_6 plus a molecule involving C_2 and two hydrogen atoms which are then 'passed on' in a separate series of reactions which eventually release the molecule C_2 for further use in the cycle. It is then reasonable to regard the coenzymes C_1, C_2, C_3 as inputs to a machine with states S_1, S_4, S_6, S_7, S_9. The state table of the machine is then given by:

	S_1	S_4	S_6	S_7	S_9
C_1	S_4	S_4	S_6	S_7	S_9
C_2	S_1	S_6	S_7	S_7	S_1
C_3	S_1	S_4	S_6	S_9	S_9

Note that coenzyme C_1 does not play any role in the reactions involving the states S_4, S_6, S_7, S_9 and so we regard C_1 as acting as an identity on these states. The semigroup of this state machine can now be calculated, however the size of this semigroup is rather large. Clearly it cannot exceed $5^5 = 3125$ and in this case it is only practical to use a computer for this task. The number of elements is in fact 183. A program (in Pascal suitable for use on an Apple or ITT 2020 microcomputer) is in the appendix. This program evaluates the semigroup of a complete state machine with 5 states or less and 9 inputs or less. See Krohn, Langer and Rhodes [1967].

State machines have been used in various investigations in psychology. The work of Chomsky in psycholinguistics caused a considerable amount of interest in the use of machines for modelling the learning of language etc. For some time it appeared that the stimulus–response (S–R) theory of learning was incapable of dealing with the acquisition of machine-like behaviour, but a paper by P. Suppes [1969] claimed to indicate the ways in which S–R techniques would enable subjects to behave like simple machines. This paper resulted in some controversy (Arbib [1969], Nelson [1975]) which was then re-examined by Kieras [1976] who pointed out some ambiguities in Suppes' original work.

Another aspect of psychology that has been influenced by machines is the problem of systems that can answer questions, see Fiksel and Bower [1976]. In this paper the authors consider a network with a type of finite automaton at each node. This network models human memory and the process of question-answering is then considered to be the process of determining paths in the network bearing given sequences of labels. The automata at the nodes are only in direct communication with their immediate neighbours but this is enough for the system to determine a shortest such path.

Other areas where automata have been used as models are in economics. W. Roedding [1975] has examined the use of 'indeterminate' Mealy machines in the modelling of various economic processes.

It is likely that automata theory will feature in many modelling processes in many different subjects. This seems justification for a continued detailed study of the theory of automata and in our next chapter we will take the first steps in the procedures for simplifying and decomposing finite state machines.

2.9 Exercises

2.1 If (Q, S) is a transformation semigroup and $\mathcal{M} = (Q, S, F)$ is the state machine of (Q, S) show that $\mathbf{TS}(\mathcal{M}) = (Q, S)$. If $\mathcal{M}$ is a state machine and $(Q, S) = \mathbf{TS}(\mathcal{M})$ show that $\mathcal{M}$ may be embedded isomorphically inside the state machine of (Q, S).

2.2 Prove theorems 2.3.3 and 2.3.5.

2.3 Prove that if π is an admissible partition on $\mathcal{M} = (Q, \Sigma, F)$ then $\mathbf{TS}(\mathcal{M}/\pi) = (\mathbf{TS}(\mathcal{M}))/\pi$. Prove that $(\alpha, 1): \mathcal{M} \to \mathcal{M}/\pi$ induces the homomorphism

$$(f, g): \mathbf{TS}(\mathcal{M}) \to \mathbf{TS}(\mathcal{M}/\pi).$$

2.4 Find an example of an incomplete transformation semigroup $\mathcal{A}$ and a proper admissible partition π such that $\mathcal{A}/\langle\pi\rangle$ is complete.

2.5 Prove the homomorphism theorem for transformation semigroups.

2.6 Let $\mathcal{M}$ and $\mathcal{M}'$ be given by the tables:

$\mathcal{M}$	q_0	q_1	q_2
a	q_1	q_2	q_0
b	q_0	q_1	q_2

$\mathcal{M}'$	p_0	p_1	p_2
a	p_2	p_2	p_0
b	p_0	p_1	p_2

Calculate $\mathbf{S}(\mathcal{M})$ and $\mathbf{S}(\mathcal{M}')$ and verify theorem 2.6.3.

2.7 A partition π on the state set Q of the machine $\mathcal{M} = (Q, \Sigma, F)$ is called *elementary* if π is admissible and if given any admissible partition π' with $\pi' < \pi$ then π' is the identity partition. Let $(f, 1): (Q, \Sigma, F) \to (Q', \Sigma, F')$ be a homomorphism, show that there exists a sequence of state machines and homomorphisms

$$(Q, \Sigma, F) \to (Q_1, \Sigma, F_1) \to \ldots \to (Q_r, \Sigma, F_r) \cong (Q', \Sigma, F')$$

such that each epimorphism $(Q_i, \Sigma, F_i) \to (Q_{i+1}, \Sigma, F_{i+1})$ is induced by an elementary partition.

2.8 $\mathcal{M} = (Q, \Sigma, F)$ is called *transitive* if, for any q, $q_1 \in Q$, there exists $\alpha \in \Sigma^*$ such that

$$q_1 = qF_\alpha.$$

Let C be the group of all state machine isomorphisms $(f, 1): \mathcal{M} \to \mathcal{M}$. Prove that if $(f, 1) \in C$ and $f \neq 1_Q$ then $f(q) \neq q$ for each $q \in Q$. Define a relation $\sim$ on Q by

$$q \sim q' \Leftrightarrow q' = f(q) \quad \text{for some } (f, 1) \in C.$$

Prove that $\sim$ defines an admissible partition on $\mathcal{M}$.

2.9 Find the semigroup of the state machine $\mathcal{M}$ given by

$\mathcal{M}$	a	b	c
0	a	a	a
1	b	a	c

Construct a semigroup homomorphism between this semigroup and a proper subsemigroup of it.

2.10 Let $\mathcal{M}=(Q,\Sigma,F)$. Prove that the set $G=\{\alpha\in\mathbf{S}(\mathcal{M})\mid Q\alpha=Q\}$ is a group.

2.11 If $\mathcal{M}$ and $\mathcal{M}'$ are state machines with $\mathcal{M}\le\mathcal{M}'$ then there exists a subsemigroup $A\subseteq\mathbf{S}(\mathcal{M}')$ and a semigroup epimorphism $f:A\to\mathbf{S}(\mathcal{M})$.

2.12 Establish the following identities where $\mathcal{M}$, $\mathcal{M}'$ and $\mathcal{M}''$ are state machines:

$$(\mathcal{M}\times\mathcal{M}')\times\mathcal{M}''\cong\mathcal{M}\times(\mathcal{M}'\times\mathcal{M}'')$$
$$(\mathcal{M}\wedge\mathcal{M}')\wedge\mathcal{M}''\cong\mathcal{M}\wedge(\mathcal{M}'\wedge\mathcal{M}'')$$

(when the restricted direct product is defined).
Investigate the situation for the cascade product.

2.13 A transformation semigroup (Q,S) is called *irreducible* if

(i) $|Q|>1$

an (ii) the only admissible partitions are the trivial partitions, i.e. the identity partition consisting of singleton blocks and the partition $\{Q\}$.
Prove that given any $q\in Q$ either $|qS|=1$ or $qS=Q$.

2.14 If $\mathcal{M}=(Q,\Sigma,F)$ is an incomplete state machine show that

$$(\mathbf{TS}(\mathcal{M}))^c\le\mathbf{TS}(\mathcal{M}^c).$$

2.15 Let $\mathcal{M}=(Q,\Sigma,F)$ be a state machine and suppose that $P\subseteq Q$. Calculate $\mathbf{TS}(\mathcal{M}|_P)$.

2.16 Prove theorem 2.6.4(ii).

2.17 Prove theorem 2.6.5.

2.18 If $\mathcal{M}$ and $\mathcal{N}$ are suitably defined state machines show that $\mathcal{M}\wedge\mathcal{N}\le\mathcal{M}\times\mathcal{N}$ and $\mathcal{M}\omega\mathcal{N}\le\mathcal{M}\circ\mathcal{N}$.

2.19 If $\mathcal{M}\le\mathcal{M}_1$ and $\mathcal{M}_1\le\mathcal{M}_2$ show that $\mathcal{M}\le\mathcal{M}_2$ where $\mathcal{M}$, $\mathcal{M}_1$, $\mathcal{M}_2$ are state machines. Establish a similar result for transformation semigroups.

2.20 Let $(\alpha, \beta): \mathcal{M} \to \mathcal{M}_1$ be a state machine homomorphism. If (α, β) is an epimorphism, prove that $\mathcal{M}_1 \leq \mathcal{M}$ and if (α, β) is a monomorphism prove that $\mathcal{M} \leq \mathcal{M}_1$.

2.21 Prove that if $\mathcal{A}$ and $\mathcal{B}$ are transformation semigroups such that

$$\mathcal{A} \leq \mathcal{B}$$

then $\mathcal{A}^{\cdot} \leq \mathcal{B}^{\cdot}$, $\bar{\mathcal{A}} \leq \bar{\mathcal{B}}$, $\mathcal{A}^c \leq \mathcal{B}^c$.

2.22 Prove that for state machines $\mathcal{M}$, $\mathcal{N}$, $\overline{\mathcal{M} \times \mathcal{N}} \leq \bar{\mathcal{M}} \times \bar{\mathcal{N}}$, $\overline{\mathcal{M} \wedge \mathcal{N}} \leq \bar{\mathcal{M}} \wedge \bar{\mathcal{N}}$, $\overline{\mathcal{M} \circ \mathcal{N}} \leq \bar{\mathcal{M}} \circ \bar{\mathcal{N}}$, and $\overline{\mathcal{M} \omega \mathcal{N}} \leq \bar{\mathcal{M}} \bar{\omega} \bar{\mathcal{N}}$ for suitable ω, $\bar{\omega}$.

2.23 Let $\mathcal{M} = (Q, \Sigma, F)$ be a state machine. Put $\mathcal{M}^{\cdot} = (Q, \Sigma \cup \{1_Q\}, F^{\cdot})$ where $qF^{\cdot}_\sigma = qF_\sigma$, $qF^{\cdot}_{1_Q} = q$, $q \in Q$. Prove that $\mathbf{TS}(\mathcal{M}^{\cdot}) \leq (\mathbf{TS}(\mathcal{M}))^{\cdot}$.

2.24 *If* $\mathcal{M}$, $\mathcal{N}$, $\mathcal{P}$, $\mathcal{R}$ are state machines and $\mathcal{M} \leq \mathcal{N}$ prove that $\mathcal{P} \omega \mathcal{M} \leq \mathcal{P} \omega \mathcal{N}$, and

$$\mathcal{P} \leq \mathcal{M} \omega \mathcal{R} \Rightarrow \mathcal{P} \leq \mathcal{N} \omega \mathcal{R}$$

$$\mathcal{P} \leq \mathcal{R} \omega \mathcal{M} \Rightarrow \mathcal{P} \leq \mathcal{R} \omega \mathcal{N}.$$

2.25 Show that if $\mathcal{A}$ is a transformation semigroup then

$$\mathcal{A} \circ \mathbf{1}^{\cdot} \cong \mathcal{A} \cong \mathbf{1}^{\cdot} \circ \mathcal{A}.$$

2.26 Let $\mathcal{M} = (Q, \Sigma, F)$ and consider $\bar{\mathcal{M}} = (Q, \Sigma \cup Q, \bar{F})$ where $\bar{F}_\sigma(q) = F_\sigma(q)$, $\bar{F}_{q'}(q) = q'$ for $q, q' \in Q$, $\sigma \in \Sigma$. Show that $\mathbf{TS}(\bar{\mathcal{M}}) = \overline{\mathbf{TS}(\mathcal{M})}$.

2.27 *Let* $\mathcal{M}$ be a state machine. Prove that

$$\mathbf{TS}(\mathcal{M})^{\cdot} = \mathbf{TM}(\mathcal{M}).$$

2.28 $\mathcal{A}$, $\mathcal{B}$ are transformation semigroups, show that

$$\overline{\mathcal{A} \circ \mathcal{B}} \leq \bar{\mathcal{A}} \circ \bar{\mathcal{B}}, \quad (\mathcal{A} \circ \mathcal{B})^c \leq \mathcal{A}^c \circ \mathcal{B}^c.$$

2.29 Let $\mathcal{A} = (Q, S)$, $\mathcal{B} = (P, T)$ be transformation semigroups such that $Q \cap P = \varnothing$. We define the *join* of $\mathcal{A}$ and $\mathcal{B}$ to be $\mathcal{A} \vee \mathcal{B} = (Q \cup P, S \vee T)$ where the action $*$ is defined by:

$$q * s = qs \quad (q \in Q, s \in S)$$
$$q * t = \varnothing \quad (q \in Q, t \in T)$$
$$p * s = \varnothing \quad (p \in P, s \in S)$$
$$p * t = pt \quad (p \in P, t \in T)$$
$$q * 0 = \varnothing \quad (q \in Q)$$
$$p * 0 = \varnothing \quad (p \in P)$$

Prove that $\mathcal{A} \vee \mathcal{B}$ is a transformation semigroup and that the join is an associative product.

2.30 If $\mathscr{A}$, $\mathscr{A}_1$, $\mathscr{B}$, $\mathscr{B}_1$ are transformation semigroups show that

$$\bar{\mathscr{A}} \vee \bar{\mathscr{B}} \leq \overline{\mathscr{A} \vee \mathscr{B}}$$

$$\mathscr{A}^{\cdot} \vee \mathscr{B}^{\cdot} \leq (\mathscr{A} \vee \mathscr{B})^{\cdot}$$

$$\mathscr{A} \circ (\mathscr{B}_1 \vee \mathscr{B}_2) \leq (\mathscr{A} \circ \mathscr{B}_1) \vee (\mathscr{A} \circ \mathscr{B}_2)$$

$$\mathscr{A} \leq \mathscr{B} \text{ implies } \mathscr{A} \vee \mathscr{A}_1 \leq \mathscr{B} \vee \mathscr{A}_1.$$

2.31 *Let* $\mathscr{A} = (Q, S)$, $\mathscr{B} = (P, T)$ be transformation semigroups with $S \neq \emptyset$, $T \neq \emptyset$. Define the *sum* $\mathscr{A} + \mathscr{B} = (Q \cup P, S \times T)$ with the operation $*$ given by

$$q(s, t) = qs \quad (q \in Q)$$

$$p(s, t) = pt \quad (p \in P).$$

In the case where S is empty we put

$$\mathscr{A} + \mathscr{B} = (Q \cup P, T)$$

and if T is empty define

$$\mathscr{A} + \mathscr{B} = (Q \cup P, S).$$

Prove that $\mathscr{A} + \mathscr{B}$ is a transformation semigroup.
Show that $\mathscr{A} \vee \mathscr{B} \leq \mathscr{A} + \mathscr{B}$.

2.32 Find algebraic descriptions for the transformation semigroups $\mathscr{A} \times \mathscr{B}$ and $\mathscr{A} \circ \mathscr{B}$ where $\mathscr{A}$ and $\mathscr{B}$ are incomplete.

2.33 Let $\mathscr{F}$ be given by

$$1 \xrightarrow{\sigma} 0, \quad 0 \xrightarrow{\tau} 1$$

then $\mathscr{F} \leq \bar{\mathbf{2}}$. Show that if the semigroup of $\mathscr{F}$ is $S = \{\theta, \sigma, \tau, \sigma\tau, \tau\sigma\}$ and the semigroup of $\bar{\mathbf{2}}$ is $T = \{\alpha, \beta\}$ then the function $\xi : S \to T$ given by $\xi(\sigma) = \xi(\tau\sigma) = \xi(\theta) = \alpha$, $\xi(\tau) = \xi(\sigma\tau) = \beta$, is not a semigroup homomorphism.

3

Decompositions

The previous chapter established that finite state machines and transformation semigroups are natural subjects for study and our next task is to initiate the algebraic theory of these objects. As with other algebraic theories one approach is to replace a general state machine (or transformation semigroup)by a collection of 'algebraically simpler' machines (or transformation semigroups) connected up in suitable ways. We have already remarked that one distinguishing feature of this algebraic theory compared to others is that we are more interested in what the machines do than what they look like. Consequently we will be using the concept of a covering rather than the concept of an isomorphism. The extra flexibility allowed by this approach will be of great use to us.

Our main aim is the development of decomposition theorems. To take the case of the finite state machines first, we will construct coverings of a given state machine in such a way that the covering machine is a product, either direct or cascade, of 'simpler' machines. So we will expect statements of the form

$$\mathcal{M} \leq \mathcal{N}_1 \omega_1 \mathcal{N}_2 \omega_2 \ldots \omega_{n-1} \mathcal{N}_n$$

where $\mathcal{M}$, $\mathcal{N}_1, \ldots, \mathcal{N}_n$ are state machines and the connecting mappings $\omega_1, \ldots, \omega_n$ are defined suitably. Recall that the cascade product is a generalization of the restricted direct product so that this type of decomposition will be the most general. However if we can replace some of the cascade products by restricted direct products we will do so because this will yield a much more efficient covering. (The semigroup of the covering machine will be smaller.)

Similarly we will attempt to derive coverings of transformation semigroups of the form $\mathscr{A} \leq \mathscr{B}_1 \circ \mathscr{B}_2 \circ \ldots \circ \mathscr{B}_n$ where $\mathscr{A}$, $\mathscr{B}_1, \ldots, \mathscr{B}_n$ are transformation semigroups. To be valuable the transformation semigroups $\mathcal{B}_1, \ldots, \mathcal{B}_n$ should have some desirable properties lacking in $\mathcal{A}$.

There will be a possibility of transferring from decompositions of state machines to decompositions of transformation semigroups and vice versa. Sometimes this process can be carried out without losing much in the way of efficiency, but generally there will be a slight loss.

There are two basic ways of obtaining decompositions. One way examines properties of the state set and for these decompositions it is usually easiest to deal with state machines and state machine decompositions. The other approach is based on properties of the input set and as these properties are usually expressed in terms of the semigroup of the machine, it is the theory of transformation semigroups that will be the most useful here. In the next chapter we will consider a decomposition theory that uses both approaches and this is best examined using transformation semigroups.

First, however, we must find a relationship between the two approaches to decomposition theories.

3.1 Decompositions

Let $\mathcal{M}$ be a state machine. A *cascade decomposition* for $\mathcal{M}$ is a covering

$$\mathcal{M} \leq \mathcal{N}_1 \omega_1 \mathcal{N}_2 \omega_2 \ldots \omega_{n-1} \mathcal{N}_n$$

where $\mathcal{N}_1, \ldots, \mathcal{N}_n$ are state machines. Naturally it is easy to construct trivial' examples of such coverings, but we will only be interested in those cases where the machines $\mathcal{N}_1, \ldots, \mathcal{N}_n$ are in some sense simpler than $\mathcal{M}$; usually this means that the state sets of $\mathcal{N}_1, \ldots, \mathcal{N}_n$ are all 'smaller' than the state set of $\mathcal{M}$ or the semigroups of the $\mathcal{N}_1, \ldots, \mathcal{N}_n$ are 'simpler' than the semigroup of $\mathcal{M}$.

A decomposition of the form

$$\mathcal{M} \leq \mathcal{N}_1 \circ \mathcal{N}_2 \circ \ldots \circ \mathcal{N}_n,$$

where $\mathcal{M}, \mathcal{N}_1, \ldots, \mathcal{N}_n$ are state machines is called a *wreath decomposition of* $\mathcal{M}$. Clearly $\mathcal{M} \leq \mathcal{N}_1 \omega_1 \mathcal{N}_2 \omega_2 \ldots \omega_{n-1} \mathcal{N}_n$ implies $\mathcal{M} \leq \mathcal{N}_1 \circ \mathcal{N}_2 \circ \ldots \circ \mathcal{N}_n$ by exercise 2.18.

Now let $\mathcal{A}$ be a transformation semigroup, a *wreath decomposition* for $\mathcal{A}$ is a covering

$$\mathcal{A} \leq \mathcal{B}_1 \circ \mathcal{B}_2 \circ \ldots \circ \mathcal{B}_n$$

where $\mathcal{B}_1, \ldots, \mathcal{B}_n$ are transformation semigroups. In many cases the semigroups of $\mathcal{B}_1, \ldots, \mathcal{B}_n$ are smaller than the semigroup of $\mathcal{A}$.

To compare the two concepts we need the following results.

Theorem 3.1.1

(i) Let $\mathcal{M}$ and $\mathcal{N}$ be state machines with $\mathcal{M} \leq \mathcal{N}$, then $\mathbf{TS}(\mathcal{M}) \leq \mathbf{TS}(\mathcal{N})$.

(ii) Let $\mathcal{A}$ and $\mathcal{B}$ be transformation semigroups with $\mathcal{A} \leq \mathcal{B}$, then $\mathbf{SM}(\mathcal{A}) \leq \mathbf{SM}(\mathcal{B})$.

Theorem 3.1.2

(i) Let $\mathcal{M} \leq \mathcal{N}_1\omega_1\mathcal{N}_2\omega_2 \ldots \omega_{n-1}\mathcal{N}_{n-1}$ be a cascade covering of state machines, then

$$\mathbf{TS}(\mathcal{M}) \leq \mathbf{TS}(\mathcal{N}_1) \circ \mathbf{TS}(\mathcal{N}_2) \circ \ldots \circ \mathbf{TS}(\mathcal{N}_n)$$

is a wreath decomposition of transformation semigroups.

(ii) Let $\mathcal{A} \leq \mathcal{B}_1 \circ \mathcal{B}_2 \circ \ldots \circ \mathcal{B}_n$ be a wreath decomposition of transformation semigroups, then

$$\mathbf{SM}(\mathcal{A}) \leq \mathbf{SM}(\mathcal{B}_1) \circ \mathbf{SM}(\mathcal{B}_2) \circ \ldots \circ \mathbf{SM}(\mathcal{B}_n)$$

is a wreath decomposition of state machines.

Theorem 3.1.3

(i) Let $\mathcal{M} \leq \mathcal{N}_1 \wedge \mathcal{N}_2 \wedge \ldots \wedge \mathcal{N}_n$ be a covering of state machines, then

$$\mathbf{TS}(\mathcal{M}) \leq \mathbf{TS}(\mathcal{N}_1) \times \ldots \times \mathbf{TS}(\mathcal{N}_n)$$

is a covering of transformation semigroups.

(ii) Let $\mathcal{A} \leq \mathcal{B}_1 \times \mathcal{B}_2 \times \ldots \times \mathcal{B}_n$ be a covering of transformation semigroups, then

$$\mathbf{SM}(\mathcal{A}) \leq \mathbf{SM}(\mathcal{B}_1) \times \mathbf{SM}(\mathcal{B}_2) \times \ldots \times \mathbf{SM}(\mathcal{B}_n)$$

is a covering of state machines.

We will prove theorem 3.1.1; the other two results follow from it and results in chapter 2 using induction.

Proof of theorem 3.1.1

(i) Let $\mathcal{M} = (Q, \Sigma, F)$, $\mathcal{N} = (Q', \Sigma', F')$ and $\phi : Q' \to Q$ a partial surjective function and $\xi : \Sigma \to \Sigma'$ a function. Let $s \in \mathbf{S}(\mathcal{M})$, then $s = [\alpha]$ for some $\alpha \in \Sigma^+$. Put $\beta = \xi(\alpha)$ and consider the element $t_s = [\beta]' \in \mathbf{S}(\mathcal{N})$. We will establish that t_s covers s, for if $q' \in Q$ then

$$\begin{aligned}
\phi(q') \cdot s = \phi(q') \cdot [\alpha] &= (\phi(q'))F_\alpha \\
&\subseteq \phi(q'F'_{\xi(\alpha)}) \\
&= \phi(q'[\beta]') \\
&= \phi(q' \cdot t_s).
\end{aligned}$$

Therefore $\mathbf{TS}(\mathcal{M}) \leq \mathbf{TS}(\mathcal{N})$.

(ii) Now let $\mathscr{A} = (Q, S)$, $\mathscr{B} = (Q', T)$ and consider

$$\mathbf{SM}(\mathscr{A}) = (Q, S, F), \mathbf{SM}(\mathscr{B}) = (Q', T, F')$$

where

$$qF_s = qs \quad \text{for } q \in Q, s \in S$$

and

$$q'F'_t = q't \text{ for } q' \in Q', t \in T.$$

Define a function $\phi : S \to T$ by

$$\xi(s) = t_s$$

where t_s is a suitably chosen element of T that covers $s \in S$.

Then, for $q' \in Q'$, $s \in S$ we have

$$\phi(q')F_s = \phi(q')s \subseteq \phi(q' \cdot t_s) = \phi(q'F'_{\xi(s)})$$

and thus $\mathbf{SM}(\mathscr{M}) \leq \mathbf{SM}(\mathscr{N})$. □

We start our decomposition with some useful results involving state machines.

3.2 Orthogonal partitions

Let $\mathscr{M} = (Q, \Sigma, F)$ be a state machine and suppose that $\pi = \{H_i\}_{i \in I}$ is a non-trivial admissible partition on Q. We call π *orthogonal* if there exists a non-trivial admissible partition τ on Q such that $\pi \cap \tau = 1_Q$.

If $\tau = \{K_j\}_{j \in J}$ then we have $|H_i \cap K_j| \leq 1$ for any $i \in I$, $j \in J$.

Given an admissible partition π on Q it is easy to construct a partition τ on Q such that $\pi \cap \tau = 1_Q$, for example if H_1 is a π-block of maximal

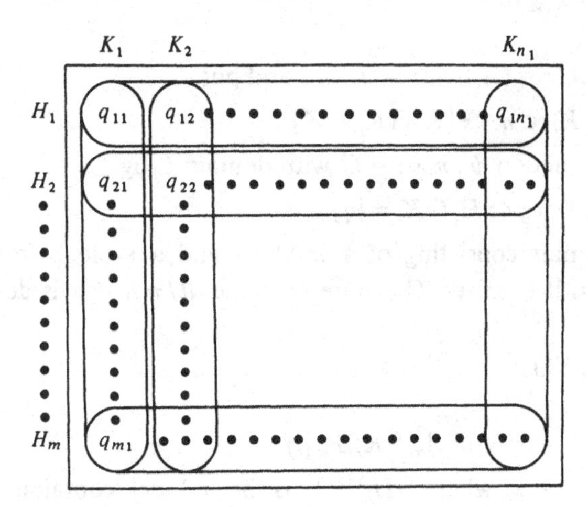

Figure 3.1. A general construction for τ.

order and $H_1 = \{q_{11}, \ldots, q_{1n_1}\}$ we construct K_1 by selecting q_{11} and one element from each other π-block. The block K_2 is constructed from q_{12} plus a new element from every other π-block, and so on. We eventually obtain a partition $\tau = \{K_1, K_2, \ldots, K_{n_1}\}$ satisfying $\pi \cap \tau = 1_Q$ with τ non-trivial. However τ may not be admissible. Note that τ contains n_1 distinct blocks. See figure 3.1.

Example 3.1

Consider the cyclic state machine $\mathcal{M} = (Q, \Sigma, F)$ defined by

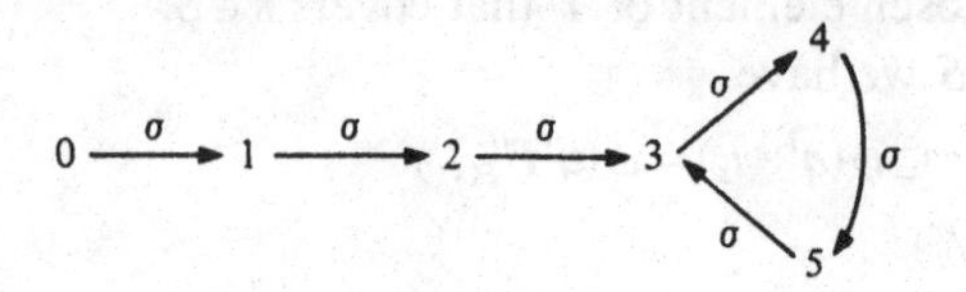

and let $\pi = \{\{0\}, \{1\}, \{2\}, \{3, 4, 5\}\}$, then π is an admissible partition. Let $\tau' = \{\{0, 1, 2, 3\}, \{4\}, \{5\}\}$ then $\pi \cap \tau' = 1_Q$ but τ' is not admissible. However $\tau = \{\{0, 3\}, \{1, 4\}, \{2, 5\}\}$ is an admissible partition, $\pi \cap \tau = 1_Q$ and π is thus orthogonal (of course so is τ!).

The existence of an orthogonal admissible partition is very valuable.

Theorem 3.2.1

Let $\mathcal{M} = (Q, \Sigma, F)$ be a state machine and suppose that π is an orthogonal admissible partition on Q. If τ is a non-trivial admissible partition on Q such that $\pi \cap \tau = 1_Q$ then

$$\mathcal{M} \leq \mathcal{M}/\pi \wedge \mathcal{M}/\tau.$$

Proof Let $\pi = \{H_i\}_{i \in I}$, $\tau = \{K_j\}_{j \in J}$ and put

$$L = \{(H_i, K_j) \in \pi \times \tau \mid H_i \cap K_j \neq \varnothing\}.$$

Define a partial function $\phi : \pi \times \tau \to Q$ with domain L by

$$\phi((H_i, K_j)) = q \Leftrightarrow H_i \cap K_j = \{q\}.$$

Then ϕ maps a pair consisting of a π-block and a τ-block to their common element, if it exists. The state machine $\mathcal{M}/\pi \wedge \mathcal{M}/\tau$ is defined to be

$$(\pi \times \tau, \Sigma, F^{\wedge})$$

where

$$(H_i, K_j)F^{\wedge}_\sigma = ([(H_i)F_\sigma]_\pi, [(K_j)F_\sigma]_\tau)$$

for $H_i \in \pi$, $K_j \in \tau$, $\sigma \in \Sigma$; where $[(H_i)F_\sigma]_\pi$ is the π-block containing the states $(H_i)F_\sigma = \{qF_\sigma \mid q \in H_i\}$ etc.

Now for $(H_i, K_j) \in L$ and $\sigma \in \Sigma$

$$\begin{aligned}(\phi((H_i, K_j)))F_\sigma &= qF_\sigma \qquad \text{where } H_i \cap K_j = \{q\} \\ &\in (H_i)F_\sigma \cap (K_j)F_\sigma \\ &\subseteq [(H_i)F_\sigma]_\pi \cap [(K_j)F_\sigma]_\tau\end{aligned}$$

so

$$\begin{aligned}qF_\sigma &= \phi([(H_i)F_\sigma]_\pi \cap [(K_j)F_\sigma)]_\tau) \\ &= \phi((H_i, K_j)F_\sigma^\wedge) \qquad \text{if } qF_\sigma \neq \varnothing\end{aligned}$$

and generally

$$qF_\sigma \subseteq \phi((H_i, K_j)F_\sigma^\wedge)$$

□

Corollary 3.2.2

$$\begin{aligned}\mathbf{TS}(\mathcal{M}) &\leq \mathbf{TS}(\mathcal{M}/\pi) \times \mathbf{TS}(\mathcal{M}/\tau) \\ &= (\mathbf{TS}(\mathcal{M}))/\langle\pi\rangle \times (\mathbf{TS}(\mathcal{M}))/\langle\tau\rangle\end{aligned}$$

The concept of an orthogonal admissible partition on a transformation semigroup can be defined in an analogous fashion and it is then a straightforward matter to deduce that if π is orthogonal on $\mathscr{A} = (Q, S)$ then

$$\mathscr{A} \leq \mathscr{A}/\langle\pi\rangle \times \mathscr{A}/\langle\tau\rangle.$$

Example 3.2

Returning to our previous example (3.1) we see that

$$\mathcal{M} \leq \mathcal{M}/\pi \times \mathcal{M}/\tau$$

where $\mathcal{M}/\pi$ is given by

$$\{0\} \xrightarrow{\sigma} \{1\} \xrightarrow{\sigma} \{2\} \xrightarrow{\sigma} \{3, 4, 5\} \circlearrowleft \sigma$$

and $\mathcal{M}/\tau$ is given by

$$\{0, 3\} \xrightarrow{\sigma} \{1, 4\} \xrightarrow{\sigma} \{2, 5\} \xrightarrow{\sigma} \{0, 3\}$$

In the notation of transformation semigroups

$$\mathbf{TS}(\mathcal{M}) = \mathscr{C}_{(3,3)}$$

$$\mathbf{TS}(\mathcal{M}/\pi) = \mathscr{C}_{(1,3)}$$

and

$$\mathbf{TS}(\mathcal{M}/\tau) = \mathbb{Z}_3$$

and so

$$\mathscr{C}_{(3,3)} \leq \mathscr{C}_{(1,3)} \times \mathbb{Z}_3.$$

Example 3.3

Let $\mathcal{M} = (Q, \Sigma, F)$ be a state machine with the property that $|(Q)F_\sigma| \leq 1$ for all $\sigma \in \Sigma$. Such machines are called *reset machines.* The transformation semigroup of $\mathcal{M}$ is covered by $\overline{(Q, \varnothing)}$ which is closed. Now let $\pi = \{H_i\}_{i \in I}$ be any partition of Q, then π is admissible, for if $\sigma \in \Sigma$, $i \in I$, then $|(H_i)F_\sigma| \leq 1$ and so $(H_i)F_\sigma \subseteq H_j$ for some $j \in I$. Consequently all partitions of Q are admissible. If $|Q| > 1$ let $q_1, q_2 \in Q$ and consider the partition $\pi = \{\{q_1, q_2\}, Q \backslash \{q_1, q_2\}\}$ which has two blocks. It is admissible and orthogonal; choose any partition τ such that $\pi \cap \tau = 1_Q$. Then

$$\mathcal{M} \leq \mathcal{M}/\pi \times \mathcal{M}/\tau.$$

Now $\mathcal{M}/\pi$ has two states and is a reset machine. The state machine $\mathcal{M}/\tau$ is also a reset machine and has fewer states than $\mathcal{M}$. We can therefore apply the process again to the state machine $\mathcal{M}/\tau$ by choosing a partition with two blocks and continuing as before. Eventually this process will cease since $|Q|$ is finite. We will then have established that $\mathcal{M} \leq \mathcal{N}_1 \times \mathcal{N}_2 \times \ldots \times \mathcal{N}_n$ where each $\mathcal{N}_i$ is a two-state reset machine. For the transformation semigroup case we have

$$\mathbf{TS}(\mathcal{M}) \leq \overline{(Q, \varnothing)} \leq \mathbf{TS}(\mathcal{N}_1) \times \mathbf{TS}(\mathcal{N}_2) \times \ldots \times \mathbf{TS}(\mathcal{N}_n).$$

Each $\mathbf{TS}(\mathcal{N}_i)$ can be covered by $\bar{\mathbf{2}}$ and so

$$\mathbf{TS}(\mathcal{M}) \leq \overline{(Q, \varnothing)} \leq \bar{\mathbf{2}} \times \bar{\mathbf{2}} \times \ldots \times \bar{\mathbf{2}} = \prod^{k} \bar{\mathbf{2}}$$

where $k = |Q| - 1$ and $\prod^k \bar{\mathbf{2}}$ means the direct product of k copies of the transformation semigroup $\bar{\mathbf{2}}$. In fact better decompositions exist for this type of state machine. (See example 3.2.)

The results in theorems 3.1.1, 3.1.2, 3.1.3 have the obvious extensions to transformation monoids and so we have

$$\mathbf{TM}(\mathcal{M}) \leq \overline{(Q, \varnothing)}^{\cdot} \leq \prod^{k} \bar{\mathbf{2}}^{\cdot}$$

where $k = |Q| - 1$.

3.3 General admissible partitions

The next step is to examine the situation where π is an admissible partition which is not necessarily orthogonal. First let $\mathbf{max}(\pi)$ indicate the maximum size of a π-block.

Theorem 3.3.1

Let $\mathcal{M} = (Q, \Sigma, F)$ be a state machine and $\pi = \{H_i\}_{i \in I}$ a non-trivial admissible partition on Q. There exists a state machine $\mathcal{N} = (Q', \Sigma', F')$ such that

$$\mathcal{M} \leq \mathcal{N}\omega\mathcal{M}/\pi$$

for some $\omega : \pi \times \Sigma \to \Sigma'$ and $|Q'| = \mathbf{max}(\pi)$.

Proof Again let $\tau = \{K_j\}_{j \in J}$ be a partition of Q satisfying $\pi \cap \tau = 1_Q$. Construct a state machine $\mathcal{N} = (Q', \Sigma', F')$ by putting

$$Q' = \tau$$

$$\Sigma' = \pi \times \Sigma$$

and defining

$$(K_j)F'_{(H_i, \sigma)} = [(H_i \cap K_j)F_\sigma]_\tau$$

for each $\sigma \in \Sigma$, $H_i \in \pi$, $K_j \in \tau$. Define $\omega = 1_{\Sigma'}$ and consider the state machine

$$\mathcal{N}\omega\mathcal{M}/\pi = (\tau \times \pi, \Sigma, \bar{F})$$

where

$$(K_j, H_i)\bar{F}_\sigma = ((K_j)F'_{(H_i, \sigma)}, [(H_i)F_\sigma]_\pi)$$

for $\sigma \in \Sigma$, $K_j \in \tau$ and $H_i \in \pi$. Put $L = \{(K_j, H_i) \mid K_j \in \tau, H_i \in \pi, K_j \cap H_i \neq \varnothing\}$. Define $\phi : \tau \times \pi \to Q$ to be the surjective partial function with domain L given by

$$\phi(K_j, H_i) = q \Leftrightarrow K_j \cap H_i = \{q\}.$$

Now consider $(K_j, H_i) \in L$, $\sigma \in \Sigma$ and note that

$$(\phi(K_j, H_i))F_\sigma = qF_\sigma$$

where $K_j \cap H_i = \{q\}$.

Now $(K_j \cap H_i)F_\sigma = \{qF_\sigma\} \subseteq [(H_i)F_\sigma]_\pi$ and since $\{qF_\sigma\} \subseteq [(K_j \cap H_i)F_\sigma]_\tau$ we have

$$[(K_j \cap H_i)F_\sigma]_\tau \cap [(H_i)F_\sigma]_\pi = \{qF_\sigma\}.$$

Thus $(\phi(K_j, H_i))F_\sigma \subseteq \phi((K_j, H_i)\bar{F}_\sigma)$ in general. □

Corollary 3.3.2

$\mathbf{TS}(\mathcal{M}) \leq \mathbf{TS}(\mathcal{N}) \circ (\mathbf{TS}(\mathcal{M}))/\langle\pi\rangle$.

Example 3.4

In a previous example (3.2) we obtained the state machine

$$0 \xrightarrow{\sigma} 1 \xrightarrow{\sigma} 2 \xrightarrow{\sigma} 3 \circlearrowleft \sigma$$

which we will now call $\mathcal{M}$.

This has an admissible partition

$$\pi = \{\{0\}, \{1, 2, 3\}\}.$$

Putting $\tau = \{\{0, 1\}, \{2\}, \{3\}\}$ we see that $\mathcal{M}/\pi$ is given by

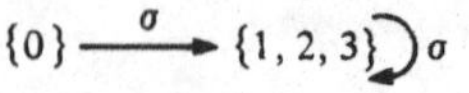

$$\{0\} \xrightarrow{\sigma} \{1, 2, 3\} \circlearrowright \sigma$$

and $\mathcal{N}$ is given by

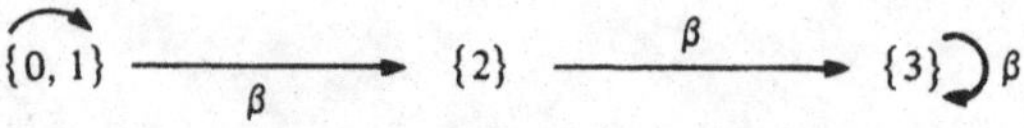

$$\overset{\alpha}{\curvearrowright}\{0, 1\} \xrightarrow[\beta]{} \{2\} \xrightarrow{\beta} \{3\} \circlearrowright \beta$$

where $\alpha = (\sigma, \{0\})$, $\beta = (\sigma, \{1, 2, 3\})$. For the transformation semigroup situation $\mathbf{TS}(\mathcal{M}/\pi) \cong \mathscr{C}$ and $\mathbf{TS}(\mathcal{N}) \leq (\mathscr{C}_{(1,2)})^{\cdot}$. Now $\mathscr{C}_{(1,3)} \leq (\mathscr{C}_{(1,2)})^{\cdot} \circ \mathscr{C}$ and $\mathscr{C}_{(1,2)} \leq \bar{\mathbf{2}} \circ \mathscr{C}$, so $\mathscr{C}_{(1,3)} \leq \bar{\mathbf{2}}^{\cdot} \circ \mathscr{C} \circ \mathscr{C}$.

The result in 3.3.2 is an indication both of the possibilities of finding useful decompositions using an admissible partition π and of the limitations of this approach, because of the difficulties of determining the semigroup of $\mathbf{TS}(\mathcal{N})$. The choice of the partition τ will have a major influence on the ease of determining the semigroup of $\mathbf{TS}(\mathcal{N})$. In some of our later results we will, in effect, make a suitable choice of τ so that $\mathbf{TS}(\mathcal{N})$ has a particularly desirable form. It is then often easier to construct the covering of 3.3.2 directly rather than to calculate $\mathbf{TS}(\mathcal{N})$ and this is the approach that we will usually take.

Let $\mathcal{M} = (Q, \Sigma, F)$ be a state machine, an admissible partition $\pi = \{H_i\}_{i \in I}$ is called *maximal* if π is non-trivial and if τ is any admissible partition with $\pi \leq \tau \leq \{Q\}$ then either $\tau = \pi$ or $\tau = \{Q\}$. So a maximal partition is an admissible partition such that no strictly larger non-trivial admissible partitions exist.

A state machine $\mathcal{N} = (Q', \Sigma', F')$ is called *irreducible* if $|Q'| > 1$ and the only admissible partitions on Q' are trivial (i.e. $1_{Q'}$ and $\{Q'\}$). (See exercise 2.13.)

Theorem 3.3.3

Let $\mathcal{M} = (Q, \Sigma, F)$ be a state machine. π is a maximal admissible partition on Q if and only if $\mathcal{M}/\pi$ is irreducible.

Proof See exercise 3.3.

Theorem 3.3.1 can now be applied to produce the *irreducible decomposition.*

Theorem 3.3.4

Let $\mathcal{M} = (Q, \Sigma, F)$ be any state machine, $|Q| = m \geq 2$, then

$$\mathcal{M} \leq \mathcal{N}_1\omega_1\mathcal{N}_2\omega_2 \ldots \omega_{n-1}\mathcal{N}_n$$

where $\mathcal{N}_1, \ldots, \mathcal{N}_n$ are irreducible state machines each with state sets of order less than m.

Proof Choose a maximal admissible partition π of Q, since $|Q| \neq 1$, and apply theorem 3.3.1, then $\mathcal{M} \leq \mathcal{N}\omega\mathcal{M}/\pi$ for suitable $\mathcal{N}$ and ω. The state set of $\mathcal{N}$ is by construction of order equal to the size of the largest π-block, which is less than m. Similarly $\mathcal{M}/\pi$ has state set equal to the number of distinct π-blocks, which is also less than m. Furthermore $\mathcal{M}/\pi$ is irreducible by 3.3.3. Now apply 3.3.1 to the state machine $\mathcal{N}$, having first found a maximal admissible partition π' for $\mathcal{N}$. Then

$$\mathcal{N} \leq \mathcal{N}'\omega'\mathcal{N}/\pi'$$

and so

$$\mathcal{M} \leq \mathcal{N}'\omega'\mathcal{N}/\pi'\omega\mathcal{M}/\pi.$$

Continuing in this way the finiteness of m forces a halt and clearly all the state machines in the decomposition are irreducible. If $|Q| = 2$ then $\mathcal{M}$ is already irreducible. □

Corollary 3.3.5

Let $\mathcal{A} = (Q, S)$ be any transformation semigroup with $|Q| = m \geq 2$, then

$$\mathcal{A} \leq \mathcal{B}_1 \circ \mathcal{B}_2 \circ \ldots \circ \mathcal{B}_n$$

where $\mathcal{B}_i = (Q_i, S_i)$ are transformation semigroups satisfying

(i) $|Q_i| < m$

(ii) $\mathcal{B}_i$ have no non-trivial admissible partitions.

Proof Apply the transformation semigroup process to the decomposition of 3.3.4 noting that irreducible state machines give rise to transformation semigroups satisfying condition (ii). □

This decomposition is not as useful as it may appear, principally because we do not have a clear idea of what irreducible state machines look like. This problem will be examined in section 3.7 and the exercises. However the study of irreducible state machines may well be of some importance since many seem to arise naturally in applications. The example in chapter 2 of a state machine arising from a metabolic pathway is irreducible (and it may be that biological examples are generally irreducible for reasons of stability).

Example 3.5
Consider the state machine $\mathcal{M} = (Q, \Sigma, F)$ given by

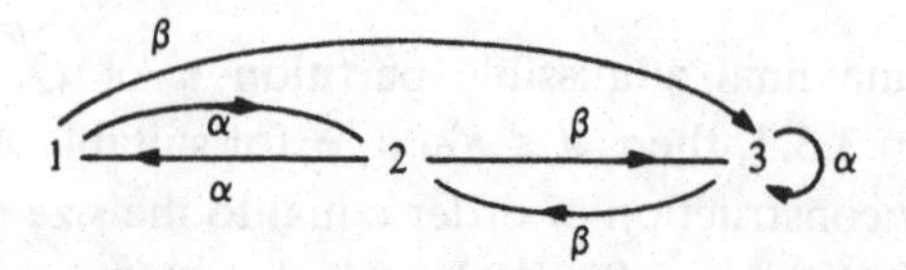

The partition $\pi = \{\{1, 2\}, \{3\}\}$ is admissible and maximal. Choose $\tau = \{\{2, 3\}, \{1\}\}$, then $\pi \cap \tau = 1_Q$ but τ is not admissible. Applying 3.3.1 we obtain

$$\mathcal{M} \leq \mathcal{N} \omega \mathcal{M}/\pi$$

where $\mathcal{M}/\pi$ is given by

α
{1, 2} β {3} α
β

and $\mathcal{N}$ is given by

{2, 3} σ, τ {1}
σ
τ, ρ, δ

where $\sigma = (\alpha, \{1, 2\})$, $\tau = (\beta, \{1, 2\})$, $\rho = (\alpha, \{3\})$, $\delta = (\beta, \{3\})$. Both $\mathcal{N}$ and $\mathcal{M}/\pi$ are irreducible. Converting to transformation semigroups we have

$$\mathbf{TS}(\mathcal{M}) \leq \mathbf{TS}(\mathcal{N}) \circ \mathbb{Z}_2$$

and since

$$\mathbf{TS}(\mathcal{N}) \leq \bar{\mathbb{Z}}_2$$

we obtain

$$\mathbf{TS}(\mathcal{M}) \leq \bar{\mathbb{Z}}_2 \circ \mathbb{Z}_2.$$

(Compare this with example 4.8.)

3.4 Permutation–reset machines

An important class of state machines are the permutation–reset state machines. Let $\mathcal{M} = (Q, \Sigma, F)$ be a state machine with $|Q| > 1$ and suppose that for each $\sigma \in \Sigma$ either $(Q)F_\sigma = Q$ or $|(Q)F_\sigma| = 1$, then we call $\mathcal{M}$ a *permutation–reset machine*. (Each input either defines a permutation of Q or a reset.) We shall see shortly how these arise naturally, but in the meantime we will examine a method of decomposing them.

First we call a state machine $\bar{\mathcal{M}} = (\bar{Q}, \bar{\Sigma}, \bar{F})$ a *permutation machine* if $(\bar{Q})\bar{F}_{\bar{\sigma}} = \bar{Q}$ for all $\bar{\sigma} \in \bar{\Sigma}$. Thus each input gives a permutation of the state machine.

Theorem 3.4.1

Let $\mathcal{M}$ be a permutation–reset machine then

$$\mathcal{M} \leq \mathcal{N}^{\cdot}\omega\mathcal{P}$$

where $\mathcal{N}$ is a reset machine and $\mathcal{P}$ is a permutation machine.

Proof Let $\mathcal{M} = (Q, \Sigma, F)$ and put

$$\Theta = \{\sigma \in \Sigma \mid (Q)F_\sigma = Q\}$$

and

$$\Xi = \{\sigma \in \Sigma \| (Q)F_\sigma| = 1\}.$$

Define G to be the subgroup of $\mathbf{S}(\mathcal{M})$ generated by Θ and put $\mathcal{P} = (G, \Sigma, \bar{F})$ where

$$[\alpha]\bar{F}_\theta = [\alpha\theta] \quad \text{for } \theta \in \Theta, \alpha \in \Theta^*$$

$$[\alpha]\bar{F}_\xi = [\alpha] \quad \text{for } \xi \in \Xi, \alpha \in \Theta^*.$$

Let $\mathcal{N} = (Q, G \times \Sigma, F^*)$ where $qF^*_{(g,\xi)} = qF_\alpha F_\xi (F_\alpha)^{-1}$ if $g = [\alpha] \in G$, $\alpha \in \Theta^*$, $\xi \in \Xi$, $q \in Q$. Now F_α is a permutation of Q and so $(F_\alpha)^{-1}$ is defined, furthermore $|(Q)F^*_{(g,\xi)}| = |(Q)F_\alpha F_\xi (F_\alpha)^{-1}| = 1$ since $(Q)F_\alpha = Q$ and $|(Q)F_\xi| = 1$, and thus $\mathcal{N}$ is a reset machine. The state machine $\mathcal{N}^{\cdot}$ consists of the state machine $\mathcal{N}$ with the identity map 1_Q adjoined. We thus adjoin a *new* symbol Λ to the set $G \times \Sigma$ and $\mathcal{N}^{\cdot} = (Q, (G \times \Sigma) \cup \{\Lambda\}, F^{**})$ where

$$qF^{**}_{(g,\xi)} = qF^*_{(g,\xi)}$$

for $q \in Q$, $g \in G$, $\xi \in \Xi$ and

$$qF^{**}_\Lambda = q$$

for $q \in Q$. Now define

$$\omega : G \times \Sigma \to G \times \Sigma \cup \{\Lambda\}$$

by

$$\omega(g, \sigma) = \begin{cases} \Lambda & \text{if } \sigma \in \Theta \\ (g, \sigma) & \text{if } \sigma \in \Xi. \end{cases}$$

We may now form the cascade product $\mathcal{N}^{\cdot}\omega\mathcal{P}$; the state mapping of this machine will be denoted by F^ω. The covering map $\phi : Q \times G \to Q$

is defined by

$$\phi(q, g) = qF_\alpha$$

where $g = [\alpha] \in G$, $q \in Q$.

We must now establish the covering properties for ϕ. First ϕ is clearly surjective as $G \neq \varnothing$ and F_α is a permutation of Q. Now let $\sigma \in \Theta$ and $(q, g) \in Q \times G$. If $g = [\alpha]$ where $\alpha \in \Theta^*$, then

$$(\phi(q, [\alpha]))F_\sigma = (qF_\alpha)F_\sigma = qF_{\alpha\sigma} = \phi(q, [\alpha\sigma])$$

since $\alpha\sigma \in \Theta^*$. Hence

$$\begin{aligned}\phi((q, [\alpha])F_\sigma^\omega) &= \phi((qF^*_{([\alpha],\sigma)}, [\alpha]\bar{F}_\sigma) \\ &= \phi((q, [\alpha\sigma])).\end{aligned}$$

If $\sigma \in \Xi$ and $(q, g) \in Q \times G$ with $g = [\alpha]$ for $\alpha \in \Theta^*$ then

$$(\phi(q, [\alpha]))F_\sigma = (qF_\alpha)F_\sigma = qF_\sigma.$$

Also

$$\begin{aligned}\phi((q, [\alpha])F_\sigma^\omega) &= \phi(qF^{**}_{([\alpha],\sigma)}, [\alpha]\bar{F}_\sigma) \\ &= \phi(qF^*_{([\alpha],\sigma)}, [\alpha]) \\ &= \phi(qF_\alpha F_\sigma(F_\alpha)^{-1}, [\alpha]) \\ &= qF_\alpha F_\sigma(F_\alpha)^{-1}F_\alpha \\ &= qF_\alpha F_\sigma = qF_\sigma.\end{aligned}$$

Hence in all cases $(\phi(q, [\alpha]))F_\sigma \subseteq \phi((q, [\alpha])F_\sigma^\omega)$. □

This result can be interpreted in the language of transformation semigroups and it is then possible to generalize it slightly. First we have:

Corollary 3.4.2

$\mathbf{TS}(\mathcal{M}) \leq \mathbf{TS}(\mathcal{N})^\cdot \circ \mathcal{G}$

where

$G = \mathbf{S}(\mathcal{M})$.

Proof This follows from 3.4.1 using the fact that $\mathcal{P} \leq \mathbf{SM}(\mathcal{G})$. (*See* 3.4.4.) □

Theorem 3.4.3

Let $\mathcal{A} = (Q, S)$ be a transformation monoid and suppose that G is the maximal subgroup of S. Then

$\mathcal{A} \leq (Q, S \backslash G)^\cdot \circ \mathcal{G}$.

Proof We use the same covering map, namely $\phi : Q \times G \to Q$ defined in 3.4.1, so that $\phi(q, g) = qg$ $(q \in Q, g \in G)$. Now let $s \in S$ and we have two cases. If $s \in G$ define $f_s : G \to (S\backslash G) \cup 1_Q$ by $f_s(g) = 1_Q$ for $g \in G$. If $s \in S\backslash G$ define $f_s : G \to (S\backslash G) \cup 1_Q$ by $f_s(g) = gsg^{-1}$ for $g \in G$. We note that $gsg^{-1} \in S\backslash G$ if $s \in S\backslash G$, for letting $gsg^{-1} = h \in G$ gives $s = g^{-1}hg \in G$.

Next we form the element $k_s \in G$ defined by

$$k_s = s \quad \text{if } s \in G$$

$$k_s = 1 \quad \text{if } s \in S\backslash G.$$

Now we show that the pair of elements (f_s, k_s) will cover s with respect to ϕ. Choose any $q \in Q$, $g \in G$; then

$$\phi(q, g) \cdot s = qgs$$

and

$$\phi((q, g) \cdot (f_s, k_s)) = \begin{cases} \phi(q, gs) & \text{if } s \in G \\ \phi(qgsg^{-1}, g) & \text{if } s \in S\backslash G \end{cases}$$

$$= qgs \quad \text{in both cases.}$$

Thus $\phi(q, g) \cdot s \subseteq \phi((q, g) \cdot ((f_s, k_s))$ for all $q \in Q$, $g \in G$, $s \in S$. □

If we now turn our attention to permutation machines we have the following result.

Theorem 3.4.4

Let $\mathcal{M} = (Q, \Sigma, F)$ be a permutation machine, then $\mathcal{M} \leq \mathbf{SM}(\mathcal{G})$ where G is a finite group.

Proof Let G be the group of all permutations on the set Q. Consider the state machine $\mathbf{SM}(\mathcal{G}) = (G, G, F')$ where

$$g_1 F'_g = g_1 g \quad \text{for } g, g_1 \in G.$$

Define a covering function $\phi : G \to Q$ as follows, let $q_0 \in Q$ be a fixed element of Q, put $\phi(g) = q_0 F_\alpha$ where $g = [\alpha] \in G$. Now ϕ is surjective. Let $\rho : \Sigma \to G$ be defined by $\rho(\sigma) = [\sigma]$ for $\sigma \in \Sigma$.

Given

$$\begin{aligned} \sigma \in \Sigma, g \in G, \phi(g)F_\sigma &= q_0 F_\alpha F_\sigma \quad \text{where } g = [\alpha] \\ &= q_0 F_{\alpha\sigma} \\ &\subseteq \phi([\alpha\sigma]) \\ &= \phi([\alpha]F'_{[\sigma]}) \\ &= \sigma(gF'_{\rho(\sigma)}) \end{aligned}$$

as required. □

Since the group G may be rather larger than $\mathbf{S}(\mathcal{M})$ we may find the next result more useful.

Theorem 3.4.5
Let $\mathcal{M}$ be a permutation machine, then
$\mathbf{TS}(\mathcal{M}) \leq \mathcal{Q}^{\cdot} \circ \mathcal{G}$ where $\mathcal{M} = (Q, \Sigma, F)$ and $G = \mathbf{S}(\mathcal{M})$.

Proof Apply corollary 3.4.2; then $\mathbf{TS}(\mathcal{N}) = (Q, \varnothing)$ and $\mathbf{TS}(\mathcal{N})^{\cdot} = \mathcal{Q}^{\cdot}$. □

Example 3.6
Consider the state machine $\mathcal{M} = (Q, \Sigma, F)$ defined by

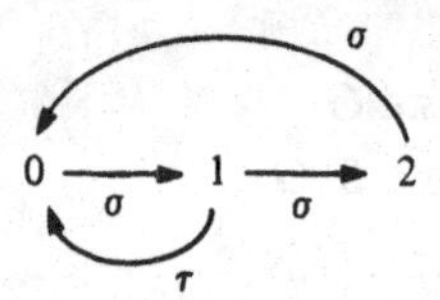

where $\Sigma = \{\sigma, \tau\}$, $Q = \{0, 1, 2\}$. This is a permutation–reset machine. In the notation of 3.4.1 $\mathcal{P} = (\mathbb{Z}_3, \Sigma, \bar{F})$ is given by

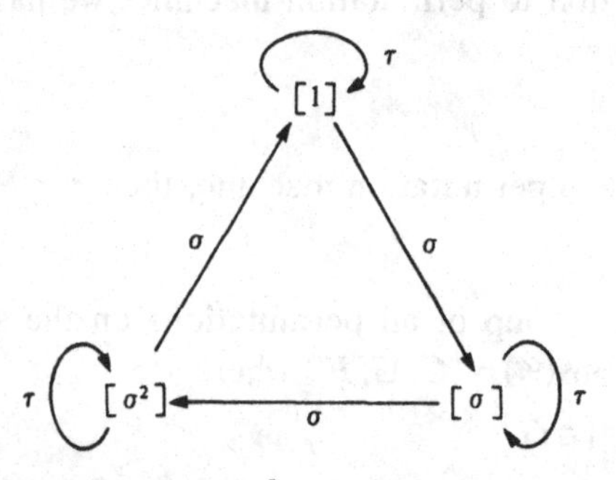

where $\mathbb{Z}_3 = \{[1], [\sigma], [\sigma^2]\}$.

$\mathcal{N}^{\cdot} = (Q, (\mathbb{Z}_3 \times \Sigma) \cup \{\Lambda\}, F^{**})$ is a reset machine given by the table

	0	1	2
Λ	0	1	2
$([1], \sigma)$	∅	∅	∅
$([1], \tau)$	∅	0	∅
$([\sigma], \sigma)$	∅	∅	∅
$([\sigma], \tau)$	∅	2	∅
$([\sigma^2], \sigma)$	∅	∅	∅
$([\sigma^2], \tau)$	∅	1	∅

In transformation semigroups

$$\mathbf{TS}(\mathcal{M}) \le \mathbf{TS}(\mathcal{N})^{\cdot} \circ \mathbf{TS}(\mathcal{P})$$
$$\le (\bar{2} \times \bar{2})^{\cdot} \circ \mathbb{Z}_3$$

since $\mathbf{TS}(\mathcal{N}) \le \bar{2} \times \bar{2}$ and $\mathbf{TS}(\mathcal{P}) \le \mathbb{Z}_3$.

3.5 Group machines

The last result brings us into the world of group machines. As we have seen, given any finite group G we can construct a transformation semigroup $\mathcal{G} = (G, G)$ and a state machine $\mathbf{SM}(\mathcal{G}) = (G, G, F')$ where $g_1 F'_g = g_1 g$ for $g, g_1 \in G$. This state machine is called the *state machine defined by G.*

For the moment it is more natural to use the transformation group terminology. Suppose that H is a subgroup of G, we define a partition on G by using the set of distinct right cosets $\{Hg \mid g \in G\}$. It is immediate that this is an admissible partition since $Hg \cdot g_1 = Hgg_1 \in \pi$ for $Hg \in \pi$ and $g_1 \in G$. Consequently we can construct the quotient transformation semigroup $\mathcal{G}/\pi$. This has state set equal to π or in another terminology G/H (although we should not assume that H is a *normal* subgroup of G: we are just regarding G/H as a set, the set of right cosets of H in G, and not as a group). The semigroup of G/π consists of all the distinct mappings of G/H into G/H induced by elements of G. Define a relation $\sim$ on G by

$$g_1 \sim g_2 \Leftrightarrow Hgg_1 = Hgg_2 \quad \text{for all } Hg \in G/H.$$

So

$$g_1 \sim g_2 \Rightarrow g_1 \in g^{-1}Hgg_2 \quad \text{for all } g \in G$$
$$\in \bigcap_{g \in G} g^{-1}Hg \cdot g_2.$$

Put $H^G = \bigcap_{g \in G} g^{-1}Hg$ then H^G is a subgroup of G and is clearly normal in G.

Lemma 3.5.1

If $H \subseteq G$ and $\sim$ is defined on G by

$$g_1 \sim g_2 \Leftrightarrow Hgg_1 = Hgg_2 \quad \text{for all } Hg \in G/H$$

then $\sim$ is an equivalence relation and the partition of $\sim$ equals the partition consisting of the right cosets of $H^G = \bigcap_{g \in G} g^{-1}Hg$ in G.

Proof Clearly if $g_1, g_2 \in G$ and $g_1 \sim g_2$ then $g_1 \in H^G g_2$ and so $H^G g_1 = H^G g_2$. Now let $H^G g_1 = H^G g_2$, then $g_1 \in H^G g_2$, so $g_1 \in g^{-1}Hgg_2$

for all $g \in G$ and thus $gg_1 \in Hgg_2$. Therefore $gg_1 = hgg_2$ for some $h \in H$ and $Hgg_1 = Hhgg_2 = Hgg_2$, giving $g_1 \sim g_2$. □

We can now write $\mathscr{G}/\pi$ as the transformation group $(G/H, G/H^G)$ with the action of the *group* G/H^G on the *set* G/H given by

$$Hg * H^G g_1 = Hgg_1 \quad \text{for } g, g_1 \in G.$$

It is clear that this action is well-defined and faithful. We now proceed to the central result of this section, namely:

Theorem 3.5.2

Let H be a subgroup of the finite group G, then

$$\mathscr{G} \leq \mathscr{H} \circ (G/H, G/H^G).$$

Proof Let us fix the coset representatives so that

$$G = Hg_1 \cup Hg_2 \cup \ldots \cup Hg_n \quad \text{(say)}.$$

Define a function $\phi : H \times G/H \to G$ by $\phi(h, Hg_i) = hg_i$ for $h \in H$, $Hg_i \in G/H$; this is clearly surjective. Given $g \in G$ we must find a pair $(f_g, s_g) \in H^{G/H} \times G/H^G$, that is a map $f_g : G/H \to H$ and an element $s_g \in G/H^G$, which covers g.

First we put $s_g = H^G g$.

If $Hg_i \in G/H$ then $Hg_ig = Hg_k$ for some $1 \leq k \leq n$ and $g_ig = h'g_k$ for some $h' \in H$. Define $f_g(Hg_i) = h'$ where $g_ig = h'g_k$. The choice of h is unique because we have fixed the representatives $g_1, \ldots, g_n$, consequently we can define the function $f_g : G/H \to H$ as required.

Now for $(h, Hg_i) \in H \times G/H$ and $g \in G$ we get

$$\phi(h, Hg_i)g = hg_ig$$

and

$$\phi((h, Hg_i)(f_g, s_g)) = \phi(hf_g(Hg_i), Hg_iH^Gg)$$
$$= \phi(hh', Hg_k)$$

where $g_ig = h'g_k$ and $Hg_ig = Hg_k$,

$$= hh'g_k = hg_ig.$$

Therefore ϕ is a covering. □

Notice that if $H = \{h_1, \ldots, h_m\}$ and $K = \{g_1, \ldots, g_n\}$ then the collection of subsets $\tau = \{h_1K, h_2K, \ldots, h_mK\}$ is a partition, for if $h_iK \cap h_jK \neq \varnothing$ then $x \in h_iK \cap h_jK \Rightarrow x = h_ig_l = h_jg_p$ for $g_l, g_p \in K$ so $Hg_l \cap Hg_p \neq \varnothing$ and thus $g_l = g_p$ and $h_i = h_j$.

This partition τ has the property that $\pi \cap \tau = 1_G$ for $h_iK \cap Hg_k = \{h_i \cdot g_k\}$ and there are $n \cdot m$ such distinct singletons. The order of

$G = |H| \cdot |G/H| = m \cdot n$ and so the partitions intersect in the identity. Consequently we could also use theorem 3.3.1 and corollary 3.3.2 which would eventually lead to the above result.

Corollary 3.5.3
If H is a normal subgroup of G then
$\mathcal{G} \leq \mathcal{H} \circ \mathcal{G}/\mathcal{H}$.

Proof This follows because G/H is now a group and $H^G = H$, therefore

$$(G/H, G/H^G) = (G/H, G/H) = \mathcal{G}/\mathcal{H}. \qquad \square$$

Let G be a finite group, then G possesses a composition series $G = G_n \supset G_{n-1} \supset \ldots \supset G_1 \supset G_0 = \{1\}$ where G_{i-1} is a normal subgroup of G_i and G_i/G_{i-1} is a simple group, for $i = 1, \ldots, n$. Using this fact and applying corollary 3.5.3 repeatedly we obtain the following important result:

Theorem 3.5.4
Let G be a finite group. There exist simple groups $K_1, \ldots, K_n$ such that

$$\mathcal{G} \leq \mathcal{K}_1 \circ \ldots \circ \mathcal{K}_n$$

and

$$\mathcal{K}_i \leq \mathcal{G} \quad \text{for } i = 1, \ldots, n.$$

Proof Clearly $\mathcal{G} \leq \mathcal{G}_{n-1} \circ \mathcal{G}_n/\mathcal{G}_{n-1}$ where $\mathcal{G}_n/\mathcal{G}_{n-1}$ is simple, so we put $\mathcal{K}_n = \mathcal{G}_n/\mathcal{G}_{n-1}$ and note that the canonical epimorphism of groups $\phi : G_n \to G_n/G_{n-1}$ is a covering $\mathcal{K}_n \leq \mathcal{G}_n$. Now $\mathcal{G}_{n-1} \leq \mathcal{G}_{n-2} \circ \mathcal{G}_{n-1}/\mathcal{G}_{n-2}$ and as before if we put $\mathcal{K}_{n-1} = \mathcal{G}_{n-1}/\mathcal{G}_{n-2}$ it is clear that $\mathcal{K}_{n-1} \leq \mathcal{G}_{n-1} \leq \mathcal{G}_n$ and $\mathcal{G} \leq \mathcal{G}_{n-2} \circ \mathcal{K}_{n-1} \circ \mathcal{K}_n$. Continuing in this way completes the process. $\square$

Recapping the results of these last two sections we have established:

Theorem 3.5.5
Let $\mathcal{M}$ be a permutation–reset machine, then

$$\mathbf{TS}(\mathcal{M}) \leq \mathcal{A} \circ \mathcal{A}_1 \circ \ldots \circ \mathcal{A}_m$$

where $\mathcal{A}$ is of the form $\prod^k \bar{\mathbf{2}}^{\cdot}$ and $\mathcal{A}_i = \mathcal{K}_i$ with $\mathcal{K}_i$ a simple group such that $\mathcal{K}_i \leq \mathcal{G}$, where $G = \mathbf{S}(\mathcal{M})$.

Proof First we use corollary 3.4.2 to obtain

$$\mathbf{TS}(\mathcal{M}) \leq \mathbf{TS}(\mathcal{N})^{\cdot} \circ \mathcal{G}$$

where $G = \mathbf{S}(\mathcal{M})$. Now $\mathbf{TS}(\mathcal{N})^{\cdot} \leq \overline{(Q, \varnothing)}^{\cdot} \leq \prod^{k} \bar{\mathbf{2}}^{\cdot}$ where $k = |Q| - 1$ and Q is the state set of $\mathcal{M}$ by example 3.3.

Now $\mathcal{G} \leq \mathcal{K}_1 \circ \ldots \circ \mathcal{K}_n$ where each K_i is a simple group such that $\mathcal{K}_i \leq \mathcal{G}$ for $i = 1, \ldots, n$. □

From chapter 2, the statement $\mathcal{K}_i \leq \mathcal{G}$ is the same as K_i divides G.

3.6 Connected transformation semigroups

A *connected* transformation semigroup $\mathcal{A} = (Q, S)$ is a transformation semigroup such that given $q, q_1 \in Q$ there exists $s \in S$ satisfying $q_1 = qs$. If $\mathcal{A} = (Q, G)$ is a transformation group then $\mathcal{A}$ is connected if and only if G is a transitive permutation group acting on Q.

A state machine $\mathcal{M}$ is *connected* if its transformation semigroup $\mathbf{TS}(\mathcal{M})$ is connected.

Connectedness is a useful property which we will now investigate.

A connected reset machine is closed and a connected permutation machine can be covered by the group machine defined by the group of the original machine rather than the group of all permutations of the states.

Theorem 3.6.1

Let $\mathcal{M} = (Q, \Sigma, F)$ be a connected machine.

(i) If $\mathcal{M}$ is a reset machine then $\mathbf{TS}(\mathcal{M}) = \overline{(Q, \varnothing)}$.

(ii) If $\mathcal{M}$ is a permutation machine then

$$\mathcal{M} \leq \mathbf{SM}(\mathcal{G})$$

where $G = \mathbf{S}(\mathcal{M})$ and

$$\mathbf{TS}(\mathcal{M}) \leq \mathcal{G}.$$

(iii) If $\mathcal{A} = (Q, G)$ is a connected transformation group then $\mathcal{A} \leq \mathcal{G}$.

Proof (i) is immediate.

(ii) Consider the function $\phi : G \to Q$ defined by

$$\phi(g) = q_0 F_\alpha$$

where q_0 is a fixed element of Q and $[\alpha] = g \in \mathbf{S}(\mathcal{M})$. (We used this function in the proof of 3.4.4 but with G equal to the group of all permutations of Q.) Then ϕ is surjective because of the connectivity of

$\mathcal{M}$. The rest of the proof of 3.4.4 can be adapted to this situation and we see that ϕ is thus a covering.

(iii) This is similar to (ii). □

We now examine connected transformation groups in more detail. These are transitive permutation groups. Suppose that $\mathcal{A} = (Q, G)$ is a connected transformation group and let $\pi = \{H_i\}_{i\in I}$ be a non-trivial admissible partition on $\mathcal{A}$. Consider a π-block H_i, then, given $g \in G$, we have

$$H_i \cdot g \subseteq H_j \quad \text{for some } j \in I.$$

Suppose that $q \in H_i \cap H_i g$ then $q \in H_i \cap H_j \Rightarrow H_i = H_j$. Therefore $\pi = \{H_i\}_{i\in I}$ is a primitive block system of the permutation group (Q, G). Consequently G is an imprimitive permutation group as G is transitive and π is non-trivial.

Let $H_1 = \{q_1, \ldots, q_r\}$ be a π-block and define $K = \{q \in G \mid H_1 g = H_1\}$. Then K is a subgroup of G. Suppose that $Kg_1, \ldots, Kg_s$ is a set of distinct right cosets of K in G such that

$$G = \bigcup_{j=1}^{s} Kg_j \quad \text{and} \quad Kg_j \cap Kg_l = \varnothing \quad \text{if } j \neq l.$$

Define

$$L_1 = \{q_1g_1, q_1g_2, \ldots, q_1g_s\}$$
$$L_2 = \{q_2g_1, q_2g_2, \ldots, q_2g_s\}$$
$$\vdots$$
$$L_r = \{q_rg_1, q_rg_2, \ldots, q_rg_s\}$$

and put $\tau = \{L_1, L_2, \ldots, L_r\}$. We note that τ is a partition of Q for if $L_i \cap L_j \neq \varnothing$ then we can find $q_ig_l = q_jg_m$ for suitable i, j, l, m. Then $q_i = q_j(g_m g_l^{-1}) \in H_1 \cap H_1(g_m g_l^{-1})$ and so $H_1 = H_1(g_m g_l^{-1})$ which implies that $g_m g_l^{-1} \in K$ and so $Kg_m = Kg_l$, $m = l$ and $i = j$. Furthermore if $q \in Q$ then there exists $g \in G$ such that $q = q_1 g$. Now $g = kg_j$ for some $k \in K$ and $j \in \{1, \ldots, s\}$, thus $q = q_1kg_j$. But $q_1k = q_i$ for some $i \in \{1, \ldots, r\}$ and so $q = q_ig_j \in L_i$. This establishes that τ is a partition of Q.

Now let $H_t \in \pi$ and suppose that $q \in H_t$. Then there exists $g \in G$ such that $q = q_1 g$ and as $g = kg_m$ for some $m \in \{1, \ldots, s\}$ we see that $q = g_1kg_m \in H_1g_m$. Therefore $H_1g_m \subseteq H_t$ and conversely $H_t \subseteq H_1g_m$. Hence $H_t = H_1g_m$ and we may write π as $\{H_1g_1, \ldots, H_1g_s\}$. Now $H_1g_m \cap L_n = \{q_ng_m\}$ and so $\pi \cap \tau = 1_Q$. We can now apply the procedure of 3.3.1 and 3.3.2 to this situation. First we calculate the transformation semigroup $\mathcal{A}/\langle\pi\rangle$. Each permutation $g \in G$ induces a permutation of the π-block which we will denote by $\bar{g}$. These mappings form a permutation group $\bar{G}$ on the set π and the mapping $\theta : G \to \bar{G}$ defined by $\theta(g) = \bar{g}$ for $g \in G$ is clearly a homomorphism. The kernel, N of θ, is the subgroup of all permutations in G that fix all the π-blocks, that is $g \in N$ if and only if $H_i g = H_i$ for all $i \in I$. It is fairly easy to see that

$$\mathcal{A}/\langle \pi \rangle \cong (\pi, G/N).$$

Instead of now calculating the transformation semigroup corresponding to the state machine $\mathcal{N}$ of 3.3.1 we will deduce the required result directly. First define

$$K_{H_1} = \{g \in K \mid q_i g = q_i \quad \text{for all } i \in \{1, \ldots, r\}\}.$$

Then K_{H_1} is a normal subgroup of K. Consider the transformation group $(\tau, K/K_{H_1})$ with the operation defined by $L_n \cdot K_{H_1} \cdot k = L_p$ where $q_n k = q_p$ and $n, p \in \{1, \ldots, r\}$ and $k \in K$. This operation is faithful for if $L_n K_{H_1} k = L_n K_{H_1} k'$ for all $n \in \{1, \ldots, r\}$ then

$$q_n k = q_n k' \quad \text{for all } n \in \{1, \ldots, r\}$$

and so

$$k(k')^{-1} \in K_{H_1} \quad \text{and} \quad K_{H_1} k = K_{H_1} k'.$$

We now establish the following result:

$$(Q, G) \leq (\tau, K/K_{H_1}) \circ (\pi, G/N).$$

Let $\phi : \tau \times \pi \to Q$ be defined by

$$\phi(L_n, H_1 g_m) = q_n g_m \quad \text{for } L_n \in \tau, H_1 g_m \in \pi.$$

Given $g \in G$ define the pair $(f_g, Ng) \in (K/K_{H_1})^{\pi} \times G/N$ by putting $f_g(H_1 g_m) = K_{H_1} k$ where $g_m g = k g_i$ for a unique $k \in K$ and $i \in \{1, \ldots, s\}$. Then

$$\begin{aligned} \phi(L_n f_g(H_1 g_m), H_1 g_m N g) &= \phi(L_n K_{H_1} k, H_1 g_m g) \\ &= \phi(L_p, H_1 g_i) \\ &\qquad \text{where } q_p = q_n k \text{ and } g_m g = k g_i \\ &= q_p g_i \\ &= q_n k g_i \\ &= q_n g_m g. \end{aligned}$$

Furthermore $\phi(L_n, H_1 g_m) \cdot g = q_n g_m g$ and so ϕ is a covering map. We restate this result in the following way.

Theorem 3.6.2

(Dilger [1976]) Let $\mathcal{A} = (Q, G)$ be a connected transformation group and π a non-trivial admissible partition on Q. Then

$$\mathcal{A} \leq (\tau, K/K_{H_1}) \circ (\pi, G/N).$$

Corollary 3.6.3

$\mathcal{A} \leq \mathcal{K}/\mathcal{K}_{H_1} \circ \mathcal{G}/\mathcal{N}$.

Proof Notice that both $(\tau, K/K_{H_1})$ and $(\pi, G/N)$ are connected transformation groups and so we may apply theorem 3.6.1(iii). □

Another useful fact is that both $\mathcal{G}/\mathcal{N} \leq \mathcal{G}$ and $\mathcal{K}/\mathcal{K}_{H_1} \leq \mathcal{G}$ by 2.4.2.

Example 3.7

Consider the following state machine

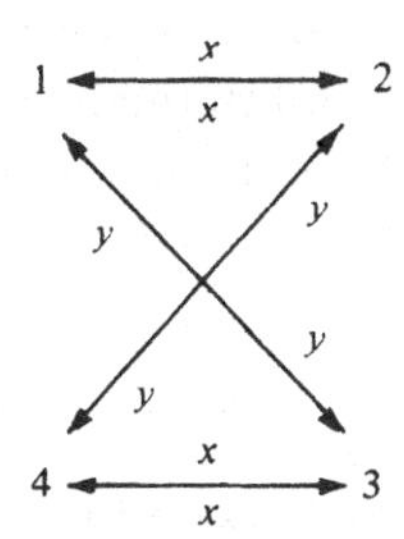

This yields a transformation group which is connected and possesses a non-trivial admissible partition, namely $\pi = \{\{1, 2\}, \{3, 4\}\}$. We will apply theorem 3.6.2. First we construct $\mathcal{A}/\langle\pi\rangle$, this is given by

$\{1, 2\}$ (loop x) $\longleftrightarrow$ $\{3, 4\}$ (loop x), arrows labelled y, y

which is isomorphic to $\mathbb{Z}_2$.

Now let $H_1 = \{1, 2\}$ and consider $K = \{g \in G \mid H_1 g = H_1\}$. The only inputs that can preserve H_1 in this way are the identity and powers of x. Since $x^2 = 1$ we have $K = \{1, x\}$. Put $G = Kg_1 \cup Kg_2 \cup \ldots \cup Kg_s$, with $g_1 = 1$. Clearly $g_2 = y$ is another coset representative and so

$$L_1 = \{1 \cdot 1, 1 \cdot y, 1 \cdot g_3, \ldots, 1 \cdot g_s\} = \{1, 3, \ldots.\}$$

$$L_2 = \{2 \cdot 1, 2 \cdot y, 2 \cdot g_3, \ldots, 2 \cdot g_s\} = \{2, 4, \ldots.\}$$

and this exhausts $Q = \{1, 2, 3, 4\}$ and so there are two cosets of K in G and $\tau = \{\{1, 3\}, \{2, 4\}\}$. Further $K_{H_1} = \{1\}$ and $(\tau, K/K_{H_1})$ is given by

$\{1, 3\}$ (loop 1) $\longleftrightarrow$ $\{2, 4\}$ (loop 1), arrows labelled x, x

again isomorphic to $\mathbb{Z}_2$. Thus $(Q, G) \leq \mathbb{Z}_2 \circ \mathbb{Z}_2$.

We have not actually calculated G but it is easily seen to be isomorphic to the Klein four-group V. Using a combination of theorems 3.6.1(ii) and 3.5.5 we can obtain the same decomposition, namely

$$(Q, G) = (\mathbf{4}, V) \leq \mathcal{V}$$

and

$$\mathcal{V} \leq \mathbb{Z}_2 \circ \mathbb{Z}_2$$

since $\{1\} \subset \mathbb{Z}_2 \subset V$ is a composition series with $V/\mathbb{Z}_2 \cong \mathbb{Z}_2$. However this approach does require a knowledge of the group G of the original machine. This may be large and cumbersome to evaluate and there are distinct advantages in using theorem 3.6.2. In Dilger [1976] an example illustrating this point can be found. If either of the transformation groups $(\tau, K/K_{H_1})$ or $(\pi, G/N)$ are imprimitive the theorem can be applied further.

3.7 Automorphism decompositions

Let $\mathcal{A} = (Q, S)$ be a transformation semigroup and let $\mathbf{Aut}_S Q$ denote the set of all automorphisms of $\mathcal{A}$. Thus $\gamma : Q \to Q$ belongs to $\mathbf{Aut}_S Q$ if and only if γ is bijective and $\gamma(qs) = \gamma(q) \cdot s$ for all $q \in Q$, $s \in S$. The identity function $1_Q : Q \to Q$ always belongs to $\mathbf{Aut}_S Q$, it may be the only element. The pair $\mathcal{A}^* = (Q, \mathbf{Aut}_S Q)$ is a transformation group under the action defined by

$$q\gamma = \gamma(q) \quad \text{for } q \in Q, \gamma \in \mathbf{Aut}_S Q,$$

where the set $\mathbf{Aut}_S Q$ is a group under the operation $*$ defined by $(\gamma * \gamma_1)(q) = \gamma_1(\gamma(q))$ for all $q \in Q$; $\gamma, \gamma_1 \in \Gamma$.

Theorem 3.7.1

Let $\mathcal{A} = (Q, S)$ be a transformation semigroup and suppose that $\Gamma = \mathbf{Aut}_S Q$. If $\Gamma \neq \{1\}$ and $\pi = \{H_i\}_{i \in I}$ is a set of distinct orbits of Q under Γ then π is an admissible partition.

Proof Let $H_i \in \pi$, and suppose that $q \in H_i$. If $s \in S$ then $qs \in H_j$ for some $j \in I$. Now let $q' \in H_i$ then $q' = \gamma(q)$ for some $\gamma \in \Gamma$ and $q's = \gamma(q)s = \gamma(qs)$ which shows that $q's \in H_j$. Hence $H_i s \subseteq H_j$ as required. □

Corollary 3.7.2

If $\mathcal{A} = (Q, S)$ is an irreducible transformation semigroup and $\mathbf{Aut}_S Q \neq \{1\}$ then $\mathcal{A}^*$ is a connected transformation group.

Proof The set of distinct orbits must consist of just one orbit, that is $\mathbf{Aut}_S Q$ is transitive on Q. □

From exercise 3.5, $\mathbf{Aut}_S Q$ is a primitive permutation group on Q.

Theorem 3.7.3

Let $\mathscr{A} = (A, S)$ be an irreducible transformation semigroup with $\mathbf{Aut}_S Q \neq \{1\}$ then $\mathscr{A} \leq \mathbb{Z}_p$ where p is a prime number and $|Q| = p$.

Proof We have established that $\mathscr{A}^* = (Q, \Gamma)$ is a transitive permutation group where $\Gamma = \mathbf{Aut}_S Q$. Let H_1 be a subgroup of Γ and suppose that $\{1\} \neq H_1 \subsetneq \Gamma$. Define the relation $\sim$ on Q by $q \sim q'$ if and only if $q' = h(q)$ for some $h \in H_1$. This equivalence relation defines an admissible partition π on $\mathscr{A}$ since $q \sim q'$ and $s \in S$ implies that $q's = h(q)s = h(qs)$ and so $qs \sim q's$. But $\mathscr{A}$ is irreducible and so $\pi = 1_Q$ or $\pi = \{Q\}$. The first conclusion leads to $H_1 = \{1\}$ and is excluded. Either Γ has no proper subgroups apart from $\{1\}$ and is thus cyclic of prime order or $\mathscr{A}_1^* = (Q, H_1)$ is a transitive permutation group. Now let H_2 be a proper non-trivial subgroup of H_2 and repeat the process. Again either $H_2 = \{1\}$ or $\mathscr{A}_2^* = (Q, H_2)$ is a transitive permutation group. Eventually we reach the position where there exists a transitive permutation group $\mathscr{A}_r^* = (Q, \mathbb{Z}_p)$ where p is a prime number. We construct the covering $\mathscr{A} \leq \mathbb{Z}_p$ as follows. Let $q_0 \in Q$ be fixed and define $\phi : \mathbb{Z}_p \to Q$ by

$$\phi(g) = q_0 g \quad \text{for } g \in \mathbb{Z}_p.$$

This is surjective since $\mathbb{Z}_p$ is transitive on Q. For $s \in S$ we define the element $h_s \in \mathbb{Z}_p$ by $q_0 s = q_0 h_s$ and then $\phi(g) \cdot s = (q_0 g)s = g(q_0 s) = g(q_0 h_s) = g(h_s(q_0)) = \phi(gh_s) = \phi(h_s g) = \phi(g * h_s)$ as $\mathbb{Z}_p$ is abelian.

Thus $\mathscr{A} \leq \mathbb{Z}_p$ with h_s covering s. Finally $|Q| \leq p$ and if $q \in Q$ then $\{g \in \mathbb{Z}_p \mid qg = q\} = \{1\}$ and so $Q = q\mathbb{Z}_p$ implies that $|Q| = p$. (Note that $|Q| \neq 1$ as $\mathbf{Aut}_S Q \neq \{1\}$.) □

This last result is a special case of the following theorem (Krohn, Langer & Rhodes [1967]).

Theorem 3.7.4

Let $\mathscr{A} = (Q, S)$ be a connected transformation semigroup and suppose that $\Gamma = \mathbf{Aut}_S Q$. If $\Gamma \neq \{1\}$ then

$$\mathscr{A} \leq \Gamma \circ \mathscr{A}/\langle \pi \rangle$$

where π is the set of distinct orbits of Q defined by Γ, and Γ is the transformation group defined by Γ.

Proof Let $Q = K_1 \cup \ldots \cup K_n$ be the decomposition of Q into the orbits defined by Γ, and let $k_i \in K_i$ be fixed orbit representatives. Then $Q = \bigcup_{i=1}^{n} k_i\Gamma$. The elements of Γ act as fixed point free permutations of Q, for if $\gamma \in \Gamma$, $q\gamma = q$ and we choose any $q' \in Q$ then $q' = qs$ for some $s \in S$ and $q'\gamma = qs\gamma = q\gamma s = qs = q'$ which makes γ equal to 1_Q. The partition $\pi = \{k_i\Gamma\}_{i=1,\ldots,n}$ is admissible. Putting $K = \{k_1, \ldots, k_n\}$, $\Gamma = \{\gamma_1, \ldots, \gamma_r\}$ and using the fact that the elements of Γ are fixed point free, we see that $\tau = \{K\gamma_1, \ldots, K\gamma_r\}$ is another partition satisfying the relationship $\pi \cap \tau = 1_Q$. We can then apply 3.3.2 to obtain a covering of A. However it is quicker to proceed directly.

Let $\phi : \Gamma \times \pi \to Q$ be defined by

$$\phi(\gamma, k_i\Gamma) = k_i\gamma^{-1} \quad \text{for } \gamma \in \Gamma, k_i\Gamma \in \pi.$$

This is a surjective function onto Q. Now choose any $s \in S$, then s defines an element $[s]$ of the semigroup of $\mathscr{A}/\langle\pi\rangle$. Let $f_s : \pi \to \Gamma$ be the function defined by

$$f_s(k_i\Gamma) = \bar{\gamma}^{-1}$$

where $k_i s \in K\bar{\gamma}$ for a unique $\bar{\gamma} \in \Gamma$.

Now choose $s \in S$, $\gamma \in \Gamma$, $k_i\Gamma \in \pi$ then

$$\begin{aligned} \phi((\gamma, k_i\Gamma)(f_s, [s])) &= \phi(\gamma * f_s(k_i\Gamma), k_i\Gamma[s]) \\ &= \phi(\gamma * (\bar{\gamma})^{-1}, k_t\Gamma) \quad \text{where } k_i s = k_t\bar{\gamma} \\ &= k_t(\gamma * (\bar{\gamma})^{-1})^{-1} \\ &= k_t\bar{\gamma}\gamma^{-1}. \end{aligned}$$

Also $\phi(\gamma, k_i\Gamma)s = k_i\gamma^{-1}s = k_i s\gamma^{-1} = k_t\bar{\gamma}\gamma^{-1}$ and this establishes the required covering. □

To extend this result to transformation semigroups which are not connected we introduce the following concept due to Shibata [1972].

Let $\mathscr{A} = (Q, S)$ be a transformation semigroup and suppose that $\varnothing \neq Q' \subseteq Q$ satisfies the condition that given $q', q_1' \in Q$ there exist $s, s' \in S$ such that $q_1' = q's$ and $q' = q_1's'$. We call Q' a *connected subset of* Q. A *stage* of Q is a maximal connected subset. Suppose that

$$Q = Q_1 \cup \ldots \cup Q_n$$

is a disjoint union of stages of Q, we call this a *stage description of* Q. Generally it is possible that $Q_i S \not\subseteq Q_i$ for some $i \in \{1, \ldots, n\}$, however in the situation where $Q_iS \subseteq Q_i$ for each $i \in \{1, \ldots, n\}$ we may establish the following result.

Theorem 3.7.5

Let $\mathscr{A} = (Q, S)$ be a transformation semigroup and $Q = Q_1 \cup \ldots \cup Q_n$ a stage description of Q. Suppose that $Q_iS \subseteq Q_i$ for each $i \in \{1, \ldots, n\}$. Let $\Gamma = \mathbf{Aut}_S Q$ and $N = \{\gamma \in \Gamma \mid Q_i\gamma = Q_i$ for each $i = 1, \ldots, n\}$. For each $i \in \{1, \ldots, n\}$ define $\bar{G}_i = \{\gamma \in \Gamma \mid Q_i\gamma = Q_i\}$ and $G_i = \{\gamma|_{Q_i} \mid \gamma \in \bar{G}_i\}$, that is the elements of G_i are the automorphisms of Q in $\bar{G}_i$ restricted to Q_i. Each stage Q_i generates a transformation semigroup $\mathscr{A}_i = (Q_i, S_i)$ where S_i is a quotient semigroup of S, $i = 1, \ldots, n$ and $G_i = \mathbf{Aut}_{S_i} Q_i$ for $i = 1, \ldots, n$. Let $Q_i = q_{i1}G_i \cup \ldots \cup q_{ir_i}G_i$ be an orbit decomposition for Q_i with respect to G_i $(i = 1, \ldots, n)$. Put

$$\pi = \bigcup_{i=1}^{n} \bigcup_{j=1}^{r_i} \{q_{ij}G_i\}.$$

This is an admissible partition on Q and

$$\mathscr{A} \leq \mathscr{N} \circ \mathscr{A}/\langle \pi \rangle.$$

Proof $G_i \subseteq \mathbf{Aut}_{S_i} Q_i$ is immediate. Let $h_i \in \mathbf{Aut}_{S_i} Q_i$, then $h_i : Q_i \to Q_i$. Define $l : Q \to Q$ by

$$l(q) = \begin{cases} h_i(q) & \text{if } q \in Q_i \\ q & \text{otherwise.} \end{cases}$$

It is clear that $l \in \Gamma$ and in fact $l \in \bar{G}_i$ and $h_i = l|_{Q_i}$. Thus $h_i \in G_i$. Now let $q_{ij}G_i \in \pi$ and $s \in S$, then $q_{ij}s \in Q_i$ and so $q_{ij}s = q_{ik}g_i$ for some $g_i \in G_i$ and $k \in \{1, \ldots, r_i\}$. Then $q_{ij}G_is = q_{ij}sG_i = q_{ik}g_iG_i = q_{ik}G_i$. Hence π is admissible. We now establish the covering $\mathscr{A} \leq \mathscr{N} \circ \mathscr{A}/\langle \pi \rangle$. Let $\phi : N \times \pi \to Q$ be defined by

$$\phi(n, q_{ij}G_i) = q_{ij}n^{-1}$$

where $n \in N$, $q_{ij}G_i \in \pi$. For $s \in S$ define $[s]$ to be the element of $\mathbf{S}(\mathscr{A}/\langle \pi \rangle)$ induced by s and define $f_s : \pi \to N$ by $f_s(q_{ij}G_i) = \bar{n}$ where $\bar{n} : Q \to Q$ is defined by

$$q\bar{n} = \begin{cases} qg_i^{-1} & \text{if } q \in Q_i \text{ and } q_{ij}s = q_{it}g_i \quad \text{for } g_i \in G_i,\ t \in \{1, \ldots, r_i\} \\ q & \text{otherwise.} \end{cases}$$

Now for $n \in N$, $q_{ij}G_i \in \pi$, $s \in S$ we have

$$\phi(n, q_{ij}G_i)s = q_{ij}n^{-1}s = q_{ij}sn^{-1}$$

and

$$\begin{aligned}
&\phi((n, q_{ij}G_i)(f_s, [s])) \\
&\quad = \phi(n * \bar{n}, q_{it}G_i) \quad \text{where } q_{ij}s = q_{it}g_i \text{ and } q\bar{n} = qg_i^{-1} \text{ for } q \in Q_i \\
&\quad = q_{it}(n * \bar{n})^{-1} \\
&\quad = q_{it}(\bar{n})^{-1}n^{-1} \\
&\quad = q_{ij}sg_i^{-1}(\bar{n})^{-1}n^{-1} \\
&\quad = q_{ij}sn^{-1} \quad \text{as required.}
\end{aligned}$$

□

Further results are possible in this direction but we will now turn our attention to a final method of decomposing transformation semigroups and state machines.

3.8 Admissible subset system decompositions

There is a natural generalization of the concept of an admissible partition of a state machine.

Let $\mathcal{M} = (Q, \Sigma, F)$ be a state machine and suppose that $\pi = \{H_i\}_{i \in I}$ is a collection of subsets of Q such that $Q = \bigcup_{i \in I} H_i$ and if $i \in I$ and $\sigma \in \Sigma$ there is $j \in I$ with $(H_i)F_\sigma \subseteq H_j$.

We call π an *admissible subset system for Q*. This definition is very similar to the definition of an admissible partition, but we no longer require the collection $\{H_i\}_{I \in I}$ to be mutually disjoint. (An *admissible subset system* for a transformation semigroup is defined analogously, so that if $\mathcal{A} = (Q, S)$ and $\pi = \{H_i\}_{i \in I}$ then $Q = \bigcup_{i \in I} H_i$ and given $i \in I$, $s \in S$ there exists $j \in I$ satisfying $H_is \subseteq H_j$.) Notice that if $i \in I$ and $\sigma \in \Sigma$ then there may be more than one element $j \in I$ such that $(H_i)F_\sigma \subseteq H_j$ so the uniqueness property associated with admissible partitions no longer applies in this situation. We can however construct a decomposition of $\mathcal{M}$ using an admissible subset system.

For each $i \in I$ and $\sigma \in \Sigma$ choose an element

$$j(i, \sigma) \in I$$

such that

$$(H_i)F_\sigma \subseteq H_{j(i,\sigma)}.$$

Now construct a machine $\mathcal{M}^* = (Q^*, \Sigma, F^*)$ where

$$Q^* = \{(q, H_i) \mid q \in H_i, i \in I\}$$

and

$$(q, H_i)F^*_\sigma = (qF_\sigma, H_{j(i,\sigma)}).$$

Define a partition π^* on $\mathcal{M}^*$ by

$$\pi^* = \{(H_i, H_i) \mid i \in I\}.$$

This is an admissible partition on $\mathcal{M}^*$ since

$$(H_i, H_i)F^*_\sigma \subseteq (H_{j(i,\sigma)}, H_{j(i,\sigma)})$$

for $\sigma \in \Sigma$, $i \in I$.

We are now in a position to prove our last decomposition result. As before we use $\mathbf{max}(\pi)$ to indicate the size of a maximal π-block.

Theorem 3.8.1

Let $\mathcal{M} = (Q, \Sigma, F)$ be a state machine and $\pi = \{H_i\}_{i \in I}$ an admissible subset system on Q. There is a state machine $\mathcal{N} = (Q', \Sigma', F')$ such that

$$\mathcal{M} \le \mathcal{N} \omega \mathcal{M}^*/\pi^* \quad \text{and} \quad |Q'| = \mathbf{max}(\pi).$$

Proof First we note that theorem 3.3.1 can be applied to the state machine $\mathcal{M}^*$ and so $\mathcal{M}^* \le \mathcal{N}\omega\mathcal{M}^*/\pi^*$ for some state machine $\mathcal{N}$. Next note that $\mathcal{M} \le \mathcal{M}^*$ under the covering $\phi : Q^* \to Q$ defined by $\phi(q, H_i) = q$ for each $(q, H_i) \in Q^*$. Then $(\phi(q, H_i))F_\sigma \subseteq qF_\sigma$ and $\phi((q, H_i)F^*_\sigma) = \phi((qF_\sigma, H_{j(i,\sigma)})) = qF_\sigma$ for $\sigma \in \Sigma$, $(q, H_i) \in Q^*$. Thus $\mathcal{M} \le \mathcal{M}^* \le \mathcal{N}\omega\mathcal{M}^*/\pi^*$. □

Corollary 3.8.2

$\mathbf{TS}(\mathcal{M}) \le \mathbf{TS}(\mathcal{N}) \circ \mathbf{TS}(\mathcal{M}^*/\pi^*)$.

Example 3.8

Consider the state machine $\mathcal{M}$ given by

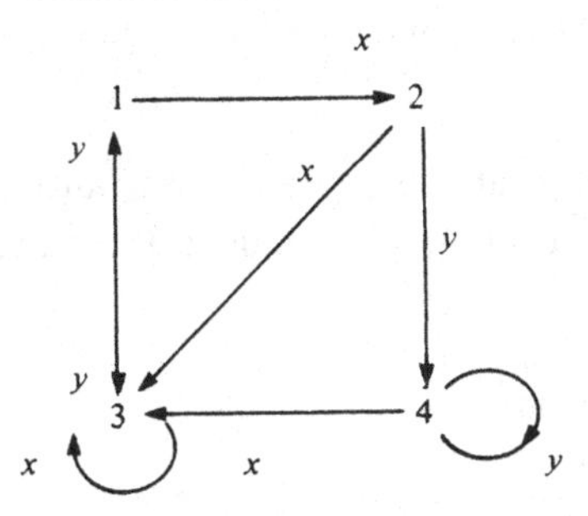

Let $H_1 = \{1, 2, 3\}$, $H_2 = \{2, 3, 4\}$, $H_3 = \{1, 3, 4\}$ and $\pi = \{H_1, H_2, H_3\}$. Then

$$(H_1)F_x = \{2, 3\}, (H_2)F_x = \{3\}, (H_3)F_x = \{2, 3\},$$

$$(H_1)F_y = \{1, 3, 4\}, (H_2)F_y = \{1, 4\}, (H_3)F_y = \{1, 3, 4\}.$$

Therefore π is an admissible subset system. Let us define

$$j(1, x) = 2, j(2, x) = 3, j(3, x) = 1,$$
$$j(1, y) = 3, j(2, y) = 3, j(3, y) = 3,$$

then an $\mathcal{M}^*$ may be defined.

Notice however that the semigroup of $\mathcal{M}^*$ is not equal to the semigroup of $\mathcal{M}$, for $F_{x^2} \equiv F_{yx^2}$ yet $(1, H_1)F^*_{x^2} = (2, H_2)F^*_x = (3, H_3)$ and $(1, H_1)F^*_{yx^2} = (3, H_3)F^*_{x^2} = (3, H_1)F^*_x = (3, H_2)$ and so $F^*_{yxx} \not\equiv F^*_{xx}$. However, in general $\mathbf{S}(\mathcal{M})$ is a quotient of $\mathbf{S}(\mathcal{M}^*)$ since $F^*_\alpha \equiv F^*_\beta \Rightarrow F_\alpha \equiv F_\beta$ for $\alpha, \beta \in \Sigma^*$.

The last result can now be applied to obtain a decomposition of an arbitrary state machine.

Let $\mathcal{M} = (Q, \Sigma, F)$ be a state machine with $|Q| = n$. Put $\pi = \{H \in \mathcal{P}(Q) \mid |H| = n-1\}$ where $\mathcal{P}(Q)$ denotes the set of all subsets of Q. This is a finite collection of proper subsets of Q and it is clearly admissible since $|(H)F_\sigma| \leq n-1$ for any $\sigma \in \Sigma$ and for any $H \in \pi$ and so $(H)F_\sigma \subseteq H'$ for some $H' \in \pi$. Then $\mathcal{M} \leq \mathcal{N}\omega\mathcal{M}^*/\pi^*$ where $\mathcal{N}$ has $n-1$ states by 3.8.1. Now either $|(Q)F_\sigma| = n$ or $|(Q)F_\sigma| < n$ for each $\sigma \in \Sigma$ and so either the input $\sigma \in \Sigma$ permutes the elements of Q and is thus a permutation in $\mathcal{M}^*/\pi^*$ or $(Q)F_\sigma \subseteq H$ for some $H \in \pi$ which implies that σ acts as a reset for $\mathcal{M}^*/\pi^*$. Thus $\mathcal{M}^*/\pi^*$ is a permutation–reset machine. This leads to:

Theorem 3.8.3

Let $\mathcal{M}$ be a state machine, then

$$\mathcal{M} \leq \mathcal{P}_1\omega_1\mathcal{P}_2\omega_2 \ldots \omega_{n-1}\mathcal{P}_n$$

where each $\mathcal{P}_i$ is a permutation–reset machine of smaller state size than $\mathcal{M}$.

Proof We simply apply the above process, first putting $\mathcal{P}_n = \mathcal{M}^*/\pi^*$ and then applying it again to $\mathcal{N}$ and continue in this way. □

Theorem 3.8.4

Let $\mathcal{M}$ be a state machine, then

$$\mathcal{M} \leq \mathcal{P}_1\omega_1\mathcal{P}_2\omega_2 \ldots \omega_{n-1}\mathcal{P}_n$$

where each $\mathcal{P}_i$ is either

(i) a simple grouplike machine, or
(ii) a two-state reset machine.

Proof We combine 3.8.3 with 3.5.6. □

Corollary 3.8.5

Let $\mathscr{A}=(Q,S)$ be a transformation semigroup, then $\mathscr{A}\leq \mathscr{B}_1\circ\mathscr{B}_2\circ\ldots\circ\mathscr{B}_n$ where each $\mathscr{B}_i$ is either

(i) of the form $\mathscr{G}_i$ for some simple group G_i, or
(ii) a finite direct product of $\bar{\mathbf{2}}^{\cdot}$.

Proof Apply 3.8.4 to $\mathbf{SM}(\mathscr{A})$ to get

$$\mathbf{SM}(\mathscr{A})\leq \mathscr{Z}_1\omega_1\mathscr{Z}_2\omega_2\ldots\omega_{n-1}\mathscr{Z}_n$$

$$\mathscr{A}=\mathbf{TS}(\mathbf{SM}(\mathscr{A}))\leq \mathbf{TS}(\mathscr{Z}_1)\circ\mathbf{TS}(\mathscr{Z}_2)\circ\ldots\circ\mathbf{TS}(\mathscr{Z}_n)$$

and if $\mathscr{Z}_i$ is a two-state reset machine then $\mathbf{TS}(\mathscr{Z}_i)\leq\bar{\mathbf{2}}^{\cdot}$ and if $\mathscr{Z}_i$ is a simple grouplike machine defined by the simple group G_i then $\mathbf{TS}(\mathscr{Z}_i)\leq\mathscr{G}_i$. □

This theorem is a special case of the Krohn–Rhodes theorem. We do not, as yet, know much about the simple groups that arise in the decomposition, in fact they divide the semigroup of the original machine, but the proof of this is a little involved. Furthermore the construction of the decomposition outlined in the above is very inefficient, much simpler and more practical decompositions are available. We will study these in chapter 4.

3.9 Complexity

If we start with an arbitrary transformation semigroup $\mathscr{A}=(Q,S)$ we can cover $\mathscr{A}$ with a wreath product involving transformation groups and transformation semigroups of reset machines. These latter transformation semigroups can be covered by direct products of the transformation semigroup $\bar{\mathbf{2}}^{\cdot}$ by 3.5.6. We introduce a class of transformation semigroups that is a generalization of the class of reset transformation semigroups. A transformation semigroup $\mathscr{A}=(Q,S)$ is *aperiodic* if

$$\mathscr{A}\leq\bar{\mathbf{2}}^{\cdot}\circ\bar{\mathbf{2}}^{\cdot}\circ\ldots\circ\bar{\mathbf{2}}^{\cdot}$$

that is if $\mathscr{A}$ can be covered by a *finite* wreath product of the transformation semigroup $\bar{\mathbf{2}}^{\cdot}$.

Now the proofs of 3.8.4 and 3.8.5 allow us to deduce that any transformation semigroup can be covered by a wreath product involving two types of transformation semigroups, namely

transformation groups of the form $\mathscr{G}$

and

aperiodic transformation semigroups.

Suppose that

$$\mathcal{A} = (Q, S) \text{ and } \mathcal{A} \le \mathcal{A}_1 \circ \mathcal{G}_1 \circ \mathcal{A}_2 \circ \mathcal{G}_2 \circ \ldots \circ \mathcal{A}_k \circ \mathcal{G}_k \circ \mathcal{A}_{k+1}$$

where $G_1, \ldots, G_k$ are groups and $\mathcal{A}_1, \ldots, \mathcal{A}_{k+1}$ are aperiodic transformation semigroups. Such a decomposition can be found using 3.8.5 but it may not be the only such decomposition involving groups and aperiodic transformation semigroups. The smallest value of k arising from such a decomposition is an indication of how complicated $\mathcal{A}$ is.

We define the *complexity of* $\mathcal{A}$, $\mathbf{C}(\mathcal{A})$ to be the smallest integer k such that

$$\mathcal{A} \le \mathcal{A}_1 \circ \mathcal{G}_1 \circ \ldots \circ \mathcal{A}_k \circ \mathcal{G}_k \circ \mathcal{A}_{k+1}$$

is a decomposition with the $\mathcal{A}_i$ aperiodic and the G_i groups.

A natural convention would be to define the complexity of $\mathcal{A}$ to be zero if $\mathcal{A}$ is aperiodic.

The groups $G_1, \ldots, G_k$ arising in the decomposition are free from any restrictions, they need not be simple nor divisors of S.

Notice that $\mathbf{1}^{\cdot}$ is aperiodic and for any transformation semigroup $\mathcal{A}$ we have $\mathcal{A} = \mathbf{1}^{\cdot} \circ \mathcal{A} = \mathcal{A} \circ \mathbf{1}^{\cdot}$. Thus the complexity of a transformation group $\mathcal{G}$ does not exceed 1 since $\mathcal{G} \le \mathbf{1}^{\cdot} \circ \mathcal{G} \circ \mathbf{1}^{\cdot}$.

Example 3.9

From examples 3.2 and 3.4 we see that the cyclic transformation semigroup $\mathcal{C}_{(3,3)} \le [(\bar{\mathbf{2}} \circ \mathcal{C}) \circ \mathcal{C}] \times \mathbb{Z}_3$. Since the transformation semigroup $\mathcal{C} \le \bar{\mathbf{2}}^{\cdot}$ we have $\mathcal{C}_{(3,3)} \le \mathcal{A}_1 \times \mathbb{Z}_3$ and so the complexity $\mathbf{C}(\mathcal{C}_{(3,3)}) \le 1$.

From example 3.5 this transformation semigroup is covered by $\bar{\mathbb{Z}}_2 \circ \mathbb{Z}_2$, and $\bar{\mathbb{Z}}_2 \le \bar{\mathbf{2}}^{\cdot} \circ \mathbb{Z}_2$ by 3.4.2 so we have complexity of at most 1.

Finally example 3.6 yields the covering by

$$(\bar{\mathbf{2}} \times \bar{\mathbf{2}})^{\cdot} \circ \mathbb{Z}_3 \le (\bar{\mathbf{2}}^{\cdot} \times \bar{\mathbf{2}}^{\cdot}) \circ \mathbb{Z}_3$$

again of complexity of at most 1.

One basic problem with calculating the complexity of a transformation semigroup is that, while it is usually easy enough to find upper bounds by describing a suitable decomposition, it is often far from easy to establish that no shorter such decompositions exist. Our first result in this direction will be to show that if G is any non-trivial group then $\mathbf{C}(\mathcal{G}) = 1$.

To achieve this we will assume that $\mathbf{C}(\mathcal{G}) = 0$, that is $\mathcal{G}$ is aperiodic. We first need the following result.

Theorem 3.9.1

Let $\mathcal{A} = (Q, S)$ be aperiodic and complete, there exists an integer $n \ge 0$ such that $s^{n+1} = s^n$ for all $s \in S$.

Proof We first establish that if $\mathscr{A} \leq \bar{\mathbf{2}}^{\cdot}$ then $s^2 = s$ for all $s \in S$. Suppose that $\bar{\mathbf{2}}^{\cdot} = (R, U)$ and $\phi : R \to Q$ is the covering partial function. Let $s \in S$, then there exists $u \in U$ such that

$$\phi(r)s \subseteq \phi(ru) \quad \text{for all } r \in R.$$

Let $q \in Q$, there exists $r \in R$ such that $q = \phi(r)$, then $qs = \phi(r)s = \phi(ru)$ since $\mathscr{A}$ is complete. Now $qs^2 = \phi(r) \cdot ss = \phi(ru)s = \phi(ru^2) = \phi(ru)$ since $u^2 = u$ for all $u \in U$.

Thus $qs^2 = qs$ and since q is arbitrary $s^2 = s$. Now let us assume that if $\mathscr{A} \leq (\bar{\mathbf{2}}^{\cdot})^k$, that is if $\mathscr{A} \leq \bar{\mathbf{2}}^{\cdot} \circ \ldots \circ \bar{\mathbf{2}}^{\cdot}$ where there are k elements in the product, then

$$s^{k+1} = s^k \quad \text{for all } s \in S.$$

Suppose that $\mathscr{A} \leq (\bar{\mathbf{2}}^{\cdot})^{k+1}$, we will show that $s^{k+2} = s^{k+1}$ for $s \in S$. Let $\mathscr{A} \leq \mathscr{B} \circ \bar{\mathbf{2}}^{\cdot}$ where $\mathscr{B} = (P, T) = (\bar{\mathbf{2}}^{\cdot})^k$. We know that for $t \in T$, $t^{k+1} = t^k$. If $\phi : P \times R \to Q$ is the covering partial function and for $s \in S$ we have (f_s, u_s) covering s with $f_s : R \to T$, then $\phi(p, r)s = \phi((p, r)(f_s, u_s))$. Choose any $q \in Q$, there exist $p \in P$, $r \in R$ with $q = \phi(p, r)$. Let $s \in S$, then

$$\begin{aligned}
qs^{k+2} &= \phi(p, r)s^{k+2} \\
&= \phi((p, r)(f_s, u_s)) \cdot s^{k+1} \\
&= \phi(pf_s(r), ru_s)s^{k+1} \\
&= \phi(pf_s(r) \cdot f_s(ru_s), ru_su_s)s^k \\
&= \phi(pf_s(r)f_s(ru_s), ru_s)s^k \\
&\vdots \\
&= \phi(pf_s(r)(f_s(ru_s))^{k+1}, ru_s) \\
&= \phi(pf_s(r)(f_s(ru_s))^k, ru_s) \\
&= \phi(pf_s(r), ru_s)s^k \\
&= \phi(p, r)s^{k+1} \\
&= qs^{k+1}.
\end{aligned}$$

Hence $s^{k+2} = s^{k+1}$.

The result now follows by induction. □

Corollary 3.9.2

Let G be a group. If $G \neq \{1\}$ then

$\mathbf{C}(\mathscr{G}) = 1$.

Proof We have $\mathbf{C}(\mathscr{G}) \leq 1$. Let $\mathbf{C}(\mathscr{G}) = 0$, then $\mathscr{G}$ is aperiodic and so there exists $n \geq 0$ such that $g^{n+1} = g^n$ for all $g \in G$. Then $g = 1$, a contradiction. □

Theorem 3.9.3

Let $\mathscr{A} = (Q, S)$, $\mathscr{B} = (P, T)$ be transformation semigroups.

(i) If $\mathscr{A} \leq \mathscr{B}$ then $\mathbf{C}(\mathscr{A}) \leq \mathbf{C}(\mathscr{B})$.

(ii) $\mathbf{C}(\mathscr{A} \circ \mathscr{B}) \leq \mathbf{C}(\mathscr{A}) + \mathbf{C}(\mathscr{B})$.

(iii) $\mathbf{C}(\mathscr{A} \times \mathscr{B}) \leq \max\{\mathbf{C}(\mathscr{A}), \mathbf{C}(\mathscr{B})\}$.

(iv) If π is an admissible partition on $\mathscr{A}$ then $\mathbf{C}(\mathscr{A}/\langle\pi\rangle) \leq \mathbf{C}(\mathscr{A})$.

(v) If π and τ are orthogonal partitions on $\mathscr{A}$ with $\pi \cap \tau = 1_Q$ then $\mathbf{C}(\mathscr{A}/\langle\pi\rangle) = \mathbf{C}(\mathscr{A})$ or $\mathbf{C}(\mathscr{A}/\langle\tau\rangle) = \mathbf{C}(\mathscr{A})$.

Proof (i) Let $\mathbf{C}(\mathscr{B}) = n$, then $\mathscr{B} \leq \mathscr{A}_1 \circ \mathscr{G}_1 \circ \ldots \circ \mathscr{G}_n \circ \mathscr{A}_{n+1}$ with the G_i groups and the $\mathscr{A}_i$ aperiodic transformation semigroups. Clearly $\mathscr{A} \leq \mathscr{B} \leq \mathscr{A}_1 \circ \mathscr{G}_1 \circ \ldots \circ \mathscr{G}_n \circ \mathscr{A}_{n+1}$ and so $\mathbf{C}(\mathscr{A}) \leq n$.

(ii) Let $\mathbf{C}(\mathscr{A}) = n$, $\mathbf{C}(\mathscr{B}) = m$, then

$$\mathscr{A} \leq \mathscr{A}_1 \circ \mathscr{G}_1 \circ \ldots \circ \mathscr{G}_n \circ \mathscr{A}_{n+1}$$

$$\mathscr{B} \leq \mathscr{A}'_1 \circ \mathscr{G}'_1 \circ \ldots \circ \mathscr{G}'_m \circ \mathscr{A}'_{m+1}$$

where $G_1, \ldots, G_n$, $G'_1, \ldots, G'_m$ are groups and $\mathscr{A}_1, \ldots, \mathscr{A}_{n+1}$, $\mathscr{A}'_1, \ldots, \mathscr{A}'_{m+1}$ are aperiodic.

Now $\mathscr{A} \circ \mathscr{B} \leq \mathscr{A}_1 \circ \mathscr{G}_1 \circ \ldots \circ \mathscr{G}_n \circ \mathscr{A}_{n+1} \circ \mathscr{A}'_1 \circ \mathscr{G}'_1 \circ \ldots \circ \mathscr{G}'_m \circ \mathscr{A}'_{m+1}$ and since $\mathscr{A}_{n+1} \circ \mathscr{A}'_1$ is aperiodic we have $\mathbf{C}(\mathscr{A} \circ \mathscr{B}) \leq n + m$.

(iii) Let $\max\{\mathbf{C}(\mathscr{A}), \mathbf{C}(\mathscr{B})\} = n$, and for the sake of argument let $\mathbf{C}(\mathscr{A}) = n$, $\mathbf{C}(\mathscr{B}) = m \leq n$.

Now

$$\mathscr{A} \leq \mathscr{A}_1 \circ \mathscr{G}_1 \circ \ldots \circ \mathscr{G}_n \circ \mathscr{A}_{n+1}$$

$$\mathscr{B} \leq \mathscr{A}'_1 \circ \mathscr{G}'_1 \circ \ldots \circ \mathscr{G}'_m \circ \mathscr{A}'_{m+1}$$

and

$$\begin{aligned}
\mathscr{A} \times \mathscr{B} &\leq (\mathscr{A}_1 \circ \mathscr{G}_1 \circ \ldots \circ \mathscr{G}_n \circ \mathscr{A}_{n+1}) \times (\mathscr{A}'_1 \circ \mathscr{G}'_1 \circ \ldots \circ \mathscr{G}'_m \circ \mathscr{A}'_{m+1}) \\
&\leq (\mathscr{A}_1 \times \mathscr{A}'_1) \circ [(\mathscr{G}_1 \circ \ldots \circ \mathscr{G}_n \circ \mathscr{A}_{n+1}) \\
&\qquad\qquad \times (\mathscr{G}'_1 \circ \ldots \circ \mathscr{G}'_m \circ \mathscr{A}'_{m+1})] \\
&\leq (\mathscr{A}_1 \times \mathscr{A}'_1) \circ (\mathscr{G}_1 \times \mathscr{G}'_1) \circ [(\mathscr{A}_2 \circ \ldots \circ \mathscr{G}_n \circ \mathscr{A}_{n+1}) \\
&\qquad\qquad \times (\mathscr{A}'_2 \circ \ldots \circ \mathscr{G}'_m \circ \mathscr{A}'_{m+1})] \\
&\;\;\vdots \\
&\leq (\mathscr{A}_1 \times \mathscr{A}'_1) \circ (\mathscr{G}_1 \times \mathscr{G}'_1) \circ (\mathscr{A}_2 \times \mathscr{A}'_2) \circ \ldots \circ (\mathscr{G}_m \times \mathscr{G}'_m) \\
&\qquad \circ (\mathscr{A}_{m+1} \times \mathscr{A}'_{m+1}) \circ \mathscr{G}_{m+1} \circ \mathscr{A}_{m+2} \circ \ldots \circ \mathscr{G}_n \circ \mathscr{A}_{n+1}
\end{aligned}$$

and so $\mathbf{C}(\mathscr{A} \times \mathscr{B}) \leq n$.

(iv) Since $\mathscr{A}/\langle\pi\rangle \leq \mathscr{A}$ we have $\mathbf{C}(\mathscr{A}/\langle\pi\rangle) \leq \mathbf{C}(\mathscr{A})$ by (i).

(v) $\mathscr{A} \leq \mathscr{A}/\langle\pi\rangle \times \mathscr{A}/\langle\tau\rangle$ and so $\mathbf{C}(\mathscr{A}) \leq \max(\mathbf{C}(\mathscr{A}/\langle\pi\rangle), \mathbf{C}(\mathscr{A}/\langle\tau\rangle))$. But $\mathbf{C}(\mathscr{A}/\langle\pi\rangle) \leq \mathbf{C}(\mathscr{A})$ and $\mathbf{C}(\mathscr{A}/\langle\tau\rangle) \leq \mathbf{C}(\mathscr{A})$ and so $\mathbf{C}(\mathscr{A}) = \mathbf{C}(\mathscr{A}/\langle\pi\rangle)$ or $\mathbf{C}(\mathscr{A}/\langle\tau\rangle)$. □

Example 3.10

We are now in a position to establish that all the examples in example 3.9 have complexity equal to 1. Thus $\mathbf{C}(\mathscr{C}_{3,3}) = \mathbf{C}(\mathbb{Z}_3) = 1$ by 3.9.3(v) and 3.9.2. From example 3.5 we see that $\alpha^2 = 1$ and so by 3.9.1 this transformation semigroup cannot be aperiodic. Finally in example 3.6, $\sigma^3 = 1$ and again this cannot be aperiodic.

Our next aim is to show that complexity is really a semigroup concept rather than a transformation semigroup or state machine concept. We will establish that if $\mathscr{A} = (Q, S)$ then $\mathbf{C}(\mathscr{A}) = \mathbf{C}(\mathscr{S})$.

First we need:

Theorem 3.9.4

Let $\mathscr{A} = (Q, S)$ be a transformation semigroup where S is not a monoid. Then

$$\mathscr{A} \le (Q, \varnothing)^{\cdot} \times (S^{\cdot}, S) \times \mathscr{C}.$$

Proof Let $\mathscr{C} = (\{a, b\}, \{\sigma\})$ where $a\sigma = a$, $b\sigma = a$.

Define $\phi : Q \times S^{\cdot} \times \{a, b\} \to Q$ by

$$\phi(q, s, a) = qs \quad \text{if } s \ne e, \text{ the adjoined identity in } S^{\cdot}$$

$$\phi(q, e, b) = q.$$

For $s' \in S$ let us consider the triple $(1_Q, s', \sigma)$. Then for $s' \in S$, $\phi(q, s, a) \cdot s' = qss'$ and

$$\phi((q, s, a) \cdot (1_Q, s', \sigma)) = \phi(q, ss', a\sigma) = qss'$$

furthermore

$$\phi(q, e, b) \cdot s' = qs'$$

and

$$\phi((q, e, b) \cdot (1_Q, s', \sigma)) = \phi(q, s', a) = qs'.$$

Thus $(1_Q, s', \sigma)$ covers s'. □

Theorem 3.9.5

Let $\mathscr{A} = (Q, S)$ be a transformation semigroup which is not a transformation monoid but such that S is a monoid. Then

$$\mathscr{A} \le (Q, \varnothing)^{\cdot} \times (S, S) \times \mathscr{C} \times \mathscr{C}.$$

Proof We first construct a semigroup S^* as follows.

Let $e \in S$ be the identity in S. Choose an element f which does not belong to S and form $S^* = S \cup \{f\}$. This is a semigroup when we extend the multiplication in S to S^* along with the identities $f \cdot s = s = s \cdot f$ for all $s \in S^*$. Furthermore (S^*, S) is a transformation semigroup. Now define $\phi : S \times \{a, b\} \to S^*$ by

$$\phi(s, a) = s \quad \text{for } s \in S$$
$$\phi(e, b) = f.$$

Then ϕ is a surjective partial function. We will show that, given $s' \in S$ the pair (s', σ) covers s'. Now

$$\phi(s, a) \cdot s' = ss' = \phi((s, a)(s', \sigma))$$

and

$$\phi(e, b) \cdot s' = fs' = s' = \phi(s', a) = \phi((e, b)(s', \sigma)).$$

Thus $(S^*, S) \leq (S, S) \times \mathscr{C}$.

We next show that $\mathscr{A} \leq (Q, \varnothing)^{\cdot} \times (S^*, S) \times \mathscr{C}$ by defining $\phi : Q \times S^* \times \{a, b\} \to Q$ by

$$\phi(q, s, a) = qs$$
$$\phi(q, f, b) = q.$$

This is surjective and if $s' \in S$ we will cover it with $(1_Q, s', \sigma)$ so that

$$\phi(q, s, a)s' = qss' = \phi((q, s, a)(1_Q, s', \sigma))$$

and

$$\phi(q, f, b)s' = qs' = \phi((q, s', a)) = \phi((q, f, b)(1_Q, s', \sigma)).$$

Finally $\mathscr{A} \leq (Q, \varnothing)^{\cdot} \times (S, S) \times \mathscr{C} \times \mathscr{C}$. □

Theorem 3.9.6
Let $\mathscr{A} = (Q, S)$ be a complete transformation semigroup, then $\mathbf{C}(\mathscr{A}) = \mathbf{C}(\mathscr{S})$.

Proof If $\mathscr{A}$ is a transformation monoid then by exercise 3.7 we have $\mathscr{A} \leq (Q, \varnothing)^{\cdot} \times \mathscr{S}$. Generally $\mathscr{A} \leq (Q, \varnothing)^{\cdot} \times \mathscr{S} \times \mathscr{C} \times \mathscr{C}$ from 3.9.4 and 3.9.5. Now suppose that $\mathbf{C}(\mathscr{S}) = n$ and $\mathscr{S} \leq \mathscr{A}_1 \circ \mathscr{G}_1 \circ \ldots \circ \mathscr{G}_n \circ \mathscr{A}_{n+1}$, then

$$\mathscr{A} \leq (Q, \varnothing)^{\cdot} \times (\mathscr{A}_1 \circ \mathscr{G}_1 \circ \ldots \circ \mathscr{G}_n \circ \mathscr{A}_{n+1}) \times \mathscr{C} \times \mathscr{C}$$
$$\leq ((Q, \varnothing)^{\cdot} \times \mathscr{A}_1) \circ \mathscr{G}_1 \circ \ldots \circ \mathscr{G}_n \circ (\mathscr{A}_{n+1} \times \mathscr{C} \times \mathscr{C})$$

and since $(Q, \varnothing)^{\cdot} \times \mathscr{A}_1$ and $\mathscr{A}_{n+1} \times \mathscr{C} \times \mathscr{C}$ are aperiodic we have $\mathbf{C}(\mathscr{A}) \leq n$. We now show that $\mathscr{S} \leq \prod^k \mathscr{A}$, the direct product of k copies of $\mathscr{A}$, where $k = |Q|$. Let $Q = \{q_1, \ldots, q_k\}$. Define $\phi : Q \times \ldots \times Q \to S^{\cdot}$ by

$$\phi((q_{i1}, \ldots, q_{ik})) = s \quad \text{if } q_j s = q_{ij}, j \in \{1, \ldots, k\}$$
$$\phi((q_1, \ldots, q_k)) = e$$

and

$$\phi((q_{i1}, \ldots, q_{ik})) = \varnothing \quad \text{otherwise.}$$

so that

$$\phi((q_1 s, \ldots, q_k s)) = s,$$
$$\phi((q_1, \ldots, q_k)) = e$$

and all other values of $\phi((q_{i1}, \ldots, q_{ik}))$ are undefined, where $q_{i1}, \ldots, q_{ik} \in Q$. Let $s' \in S$ and consider $(s', \ldots, s') \in S^k$. Then for $s' \in S$ we have $\phi((q_{i1}, \ldots, q_{ik})) \cdot s' = \varnothing$
or

$$\phi((q_{i1}, \ldots, q_{ik})) \cdot s' = ss' \subseteq \phi((q_{i1}s', \ldots, q_{ik}s')) = ss'$$

or

$$\phi((q_1, \ldots, q_k)) \cdot s' = es' = \phi((q_1 s_1', \ldots, q_k s')) = s'$$

and the covering is established.

Then by 3.9.3(iii) $\mathbf{C}(\mathcal{S}) \leq \mathbf{C}(\mathcal{A})$ and so

$$\mathbf{C}(\mathcal{S}) = \mathbf{C}(\mathcal{A}). \qquad \square$$

Finally we show that if $\mathcal{A}$ is incomplete then $\mathbf{C}(\mathcal{A}) = \mathbf{C}(\mathcal{A}^c)$. First we need the following result:

Theorem 3.9.7
If $\mathcal{A} \leq \mathcal{B}$ and $\mathcal{B}$ is complete then $\mathcal{A}^c \leq \mathbf{2}^{\cdot} \times \mathcal{B}$.

Proof Let $\mathcal{A} = (Q, S)$, $\mathcal{B} = (P, T)$, $\mathbf{2}^{\cdot} = (\{a, b\}, \{1\})$, $\mathcal{A}^c = (Q \cup \{r\}, S)$. Suppose that $\phi : P \to Q$ is the covering partial function giving $\mathcal{A} \leq \mathcal{B}$. Let $\psi : \{a, b\} \times P \to Q \cup \{r\}$ be defined by

$$\psi(a, p) = \phi(p)$$
$$\psi(b, p) = r, \quad p \in P.$$

If $s \in S$ let t be such that t covers s with respect to ϕ. Now for $s \in S$,

$$\psi(a, p)s = \phi(p)s \subseteq \phi(pt) = \psi(a, pt) = \psi((a, p)(1, t))$$
$$\psi(b, p)s = rs = r = \psi(b, pt) = \psi((b, p)(q, t)).$$

Hence ψ is the required covering. $\square$

Theorem 3.9.8
If $\mathcal{A} = (Q, S)$ is any transformation semigroup

$$\mathbf{C}(\mathcal{A}^c) = \mathbf{C}(\mathcal{A}).$$

Proof If $\mathscr{A}$ is complete, $\mathscr{A}^c = \mathscr{A}$. If $\mathscr{A}$ is not complete then $\mathscr{A} \leq \mathscr{A}^c$ and so $\mathbf{C}(\mathscr{A}) \leq \mathbf{C}(\mathscr{A}^c)$. Suppose that $\mathbf{C}(\mathscr{A}) = n$, then

$$\mathscr{A} \leq \mathscr{A}_1 \circ \mathscr{G}_1 \circ \ldots \circ \mathscr{G}_n \circ \mathscr{A}_{n+1}$$

and we may assume that all of the $\mathscr{A}_i$ and $\mathscr{G}_i$ are complete.

Hence by 3.9.7

$$\begin{aligned}\mathscr{A}^c &\leq \mathbf{2}^{\cdot} \times (\mathscr{A}_1 \circ \mathscr{G}_1 \circ \ldots \circ \mathscr{G}_n \circ \mathscr{A}_{n+1}) \\ &\leq (\mathbf{2}^{\cdot} \times \mathscr{A}_1) \circ \mathscr{G}_1 \circ \ldots \circ \mathscr{G}_n \circ \mathscr{A}_{n+1}\end{aligned}$$

and since $\mathbf{2}^{\cdot} \times \mathscr{A}_1$ is aperiodic we have

$$\mathbf{C}(\mathscr{A}^c) \leq n = \mathbf{C}(\mathscr{A}).$$

□

It can be shown that transformation semigroups of an arbitrary complexity exist. The proof of this fact is far from easy, it uses a considerable amount of the theory of semigroups and we do not have the space available for such a discussion. However, the interested reader may study the appropriate literature for the details (Eilenberg [1976]). We will close with the following unproved statement.

Let $n \geq 0$ be an integer, then $\mathbf{C}((\overline{2, S_2}) \circ (\overline{3, S_3}) \circ \ldots \circ (\overline{n, S_n})) = n - 1$, where S_k is the symmetric group on k symbols.

Now $(k, S_k) \leq \bar{\mathbf{k}}^{\cdot} \circ \mathscr{S}_k$ by 3.4.3, $\bar{\mathbf{k}}^{\cdot}$ is aperiodic and S_k is a group and so $\mathbf{C}((\overline{2, S_2}) \circ (\overline{3, S_3}) \circ \ldots \circ (\overline{n, S_n})) \leq n - 1$. It can be shown that $\mathbf{C}((\overline{2, S_2}) \circ (\overline{3, S_3}) \circ \ldots \circ (\overline{n, S_n})) = n - 1$.

Other important results in the theory of complexity also involve sophisticated techniques in semigroup theory and we shall have to leave the subject here.

The implications of complexity theory in the applications of state machines centre around the use of the complexity of a machine as a measure of how complicated the machine is. There are other possible measures, one of which involves finding the shortest chain of admissible partitions in the machine. We say that $\mathscr{M} = (Q, \Sigma, F)$ has an *admissible series* of length n if there exists a sequence

$$1_Q = \pi_0 < \pi_1 < \pi_2 < \ldots < \pi_n = \{Q\}$$

of admissible partitions on Q such that no admissible partition τ exists satisfying $\pi_i < \tau < \pi_{i+1}$ for $0 \leq i < n$.

The *dimension*, $\mathbf{D}(\mathscr{M})$, is then defined to be the smallest number that is the length of an admissible series of $\mathscr{M}$. Clearly $\mathbf{D}(\mathscr{M}) = 1$ if and only if $\mathscr{M}$ is irreducible. The dimension is a measure of the functional stability of the machine and is particularly valuable in some biological examples where functional stability is an evolutionary more important factor than minimal complexity. For example the machine of the Krebs cycle (example 2.10) is irreducible, it has complexity 2.

3.10 Exercises

3.1 Prove theorems 3.1.2 and 3.1.3.

3.2 Let Q be a finite set with $|Q| = n > 1$. Prove that

$$(Q, \varnothing) \leq \prod^{k} \bar{\mathbf{2}} = \bar{\mathbf{2}} \times \bar{\mathbf{2}} \times \ldots \times \bar{\mathbf{2}} \quad (k \text{ times})$$

where k, as a function of n, satisfies the formula

$$k(2) = 1$$

$$k(n) = 1 + k\{[(n+1)/2]\}, \quad n > 2.$$

Here $[(n+1)/2]$ means the largest integer less than or equal to $(n+1)/2$. Another way of expressing this is $k(n) = [\log_2(n-1)] + 1$.

3.3 Prove Theorem 3.3.3. Prove the state machine analogue of 3.5.2 and 3.5.3.

3.4 Let (Q, G) be a connected transformation group, let $q \in Q$ and put $H = \{g \in G \mid qg = q\}$. Prove that $(Q, G) \leq (G/H, G/H^G)$.

3.5 Prove that a connected transformation group is primitive if and only if it is irreducible.

3.6 Show that the set of all endomorphisms of an irreducible transformation semigroup $\mathcal{A} = (Q, S)$ equals $G \cup \{\bar{q}\}$ or G, where G is a group, according as $\mathcal{A}$ has a sink state q or not.

3.7 Let $\mathcal{A} = (Q, S)$ be a transformation monoid. Show $\mathcal{A} \leq (Q, \varnothing)^{\cdot} \times \mathcal{S}$.

3.8 If $\mathcal{A}$ is complete and $\mathcal{B} \leq \mathcal{A}$ then $\mathcal{B}^c \leq \mathbf{2}^{\cdot} \times \mathcal{A}$.

4
The holonomy decomposition

The aim of this chapter is the description of a method for decomposing an arbitrary transformation semigroup into a wreath product of 'simpler' transformation semigroups, namely aperiodic ones and transformation groups. The origin of this theory is the theorem due to Krohn and Rhodes which gave an algorithmic procedure for such a decomposition. There are now various proofs of this result extant, some are set in the theory of transformation semigroups and others are concerned with the theory of state machines. In the light of the close connections between the two theories forged in chapter 2 we can expect a similar correspondence between the two respective decomposition theorems. The proof of the decomposition theorem for state machines has the advantage that it can be motivated the more easily, but at the expense of some elegance. Recently Eilenberg has produced a new, and much more efficient, decomposition and it is this theory that we will now study. It is set in the world of transformation semigroups.

Before we embark on the details let us pause for a moment and consider how we could approach the problem of finding a suitable decomposition. Let $\mathcal{M} = (Q, \Sigma, F)$ be a state machine and let $|Q| = n$. Consider the collection π of all subsets of Q of order $n-1$. Then π is an admissible subset system, and we may construct a well-defined quotient machine $\mathcal{M}/\pi$. This state machine is a permutation–reset machine and $\mathcal{M}$ may be covered by a cascade product of $\mathcal{M}/\pi$ and a smaller state machine. We then repeat the procedure until we have $\mathcal{M}$ covered by a cascade product of permutation–reset machines. These can then be covered by cascade products of reset machines and group machines. This yields a Krohn–Rhodes type decomposition but the proof of the fact that the groups of the group machines are covered by the semigroup of $\mathcal{M}$ remains to be done, and is not easy.

There are other drawbacks with this approach. The admissible subset system π is very wasteful. One better choice for π would be the collection of all the maximal images of the machine $\mathcal{M}$. This is the direction that we take here, although we will develop the theory with reference to transformation semigroups rather than state machines. There are certain technical and notational advantages in this approach, but it would be possible to adapt much of the following theory to the state machine case.

We will prove the holonomy decomposition theorem which involves some quite difficult arguments. Once we have this result, however, we will be in a position to decompose a transformation semigroup much more efficiently than the traditional methods would allow.

The theory begins with a close study of admissible subset systems, their possible quotients and their coverings.

4.1 Relational coverings

If $\mathcal{A} = (Q, S)$ is a transformation semigroup and $\pi = \{H_i\}_{i\in I}$ is an admissible subset system then a quotient transformation semigroup $\mathcal{A}/\langle\pi\rangle$ may be defined in various ways. They are all based on the pair (π, S). We define an operation $\circledast: \pi \times S \to \pi$, which is an action of S on π by: $H_i \circledast s = H_j$ where H_j is chosen so that $H_is \subseteq H_j$, for $i, j \in I$, $s \in S$. (Since H_is may belong to more than one element of π we have to make a specific choice of one particular element of π. This means that the operation $\circledast$ is not always uniquely specified, there is a collection of such operations.) To make the triple $(\pi, S, \circledast)$ into a transformation semigroup we must now ensure that S acts faithfully on π under $\circledast$. This is done in the usual way by defining the relation $\sim$ on S by: $s \sim s_1 \Leftrightarrow H_i \circledast s = H_i \circledast s_1$ for all $H_i \in \pi$. Then $(\pi, S/\!\sim)$ becomes a transformation semigroup under the operation induced by $\circledast$. We will denote this transformation semigroup by $\mathcal{A}/\langle\pi\rangle$ when the operation $\circledast$ is understood. (We will also use the symbol $\circledast$ for the induced operation in $\mathcal{A}/\langle\pi\rangle$.)

Example 4.1

Let $\mathcal{A} = (Q, S)$ be the transformation monoid generated by the state machine:

	a	b	c
0	a	a	c
1	b	b	b

where $Q=\{a, b, c\}$, $S=\{\Lambda, 0, 1, 10\}$. The monoid S has the table:

S	Λ	0	1	10
Λ	Λ	0	1	10
0	0	0	1	10
1	1	10	1	10
10	10	10	1	10

If $\pi=\{\{a, b\}, \{b, c\}, \{a, c\}\}$ we may check that π is an admissible subset system for $\mathscr{A}$. Write $H_1=\{a, b\}$, $H_2=\{b, c\}$, $H_3=\{a, c\}$. There are several quotient transformation monoids that may be defined; we consider two possibilities:

$\circledast$	H_1	H_2	H_3
Λ	H_1	H_2	H_3
0	H_1	H_3	H_3
1	H_1	H_2	H_2
10	H_1	H_3	H_3

$\circledast'$	H_1	H_2	H_3
Λ	H_1	H_2	H_3
0	H_3	H_3	H_3
1	H_1	H_1	H_1
10	H_3	H_3	H_1

The monoids $S/\sim$ are $\{[\Lambda], [0], [1]\}$ and S respectively. Thus $\mathscr{A}/\langle\pi\rangle$ may be defined in several different ways. Similar things happen with transformation semigroups.

What is the connection between $\mathscr{A}$ and the various quotients $\mathscr{A}/\langle\pi\rangle$? Notice that we may define a function $\alpha: \pi \to \mathscr{P}(Q)$ by $\alpha(H_i)=H_i$ where $i \in I$. This defines a relation from the state set of any $\mathscr{A}/\langle\pi\rangle$ onto the state set of $\mathscr{A}$. We cannot expect, in general, that $\mathscr{A}/\langle\pi\rangle$ covers $\mathscr{A}$ in the traditional sense. But if we examine the properties of the relation α we see that $\alpha(H_i)s \subseteq \alpha(H_i \circledast [s])$ for all $s \in S$, $i \in I$. This is very similar to the requirement that $[s]$ covers s with respect to α, but now α is a relation and not necessarily a function. This leads us to the following concept.

Let $\mathscr{A}=(Q, S)$, $\mathscr{B}=(P, T)$ be transformation semigroups. A relation $\alpha: P \rightsquigarrow Q$ is called a *relational covering of $\mathscr{A}$ by $\mathscr{B}$* if

(i) α is surjective
(ii) given any $s \in S$ there exists a $t \in T$

such that

$$\alpha(p)\cdot s \subseteq \alpha(p \cdot t) \quad \text{for all } p \in P. \tag{*}$$

We then write $\mathscr{A} \lhd_\alpha \mathscr{B}$, or just $\mathscr{A} \lhd \mathscr{B}$ if a specific reference to the relation α is unnecessary. In the inequality (*) the element t is said to *cover s*.

If we recall that α is a completely additive relation the first condition yields $\bigcup_{p\in P}\alpha(p)=Q$. Each image $\alpha(p)$, $p\in P$, is a subset of Q and the collection of the distinct such images covers Q in the set-theoretical sense. Condition (ii) tells us, further, that the collection $\{\alpha(p)\mid p\in P\}$, of images under α, forms an admissible subset system. To see the connection between relational coverings, admissible subset systems and related quotient transformation semigroups we introduce a new concept.

Let $\mathscr{A}\lhd_\alpha\mathscr{B}$ be a relational covering, where $\mathscr{A}=(Q,S)$ and $\mathscr{B}=(P,T)$. Define T_α to be the set of elements of T that cover some element of S with respect to α, so $T_\alpha=\{t\in T\mid \exists s\in S$ such that $\alpha(p)s\subseteq\alpha(pt)$ for all $p\in P\}$. We say that the relation α is *close* if:

$$\alpha(p)=\alpha(p')\Rightarrow\alpha(pt)=\alpha(p't)\quad \text{for all } t\in T_\alpha,\text{ where } p,p'\in P.$$

It should be noted that although close coverings abound in the theory, by no means are all relational coverings close.

Theorem 4.1.1

Let $\mathscr{A}=(Q,S)$, $\mathscr{B}=(P,T)$ be transformation semigroups. If $\mathscr{A}\lhd_\alpha\mathscr{B}$ is a relational covering then $\{\alpha(p)\mid p\in P\}$ is an admissible subset system. If π is an admissible subset system and $\mathscr{A}/\langle\pi\rangle$ is a chosen quotient defined by π then $\mathscr{A}\lhd\mathscr{A}/\langle\pi\rangle$. If $\mathscr{A}\lhd_\alpha\mathscr{B}$ is a close relational covering and $\pi=\{\alpha(p),p\in P\}$ then there exists a quotient $\mathscr{A}/\langle\pi\rangle$ such that $\mathscr{A}/\langle\pi\rangle\le\mathscr{B}$. If π is a partition then $\mathscr{A}/\langle\pi\rangle\le\mathscr{B}$.

Proof The first two statements are immediate. Suppose now that $\mathscr{A}\lhd_\alpha\mathscr{B}$ is close. Let $\alpha(p)\in\pi$ and $s\in S$. There exists a $t\in T$ such that $\alpha(p)s\subseteq\alpha(pt)$ for all $p\in P$. We define an operation $\circledast$ of S on π by putting $\alpha(p)\circledast s=\alpha(pt)$ for each $\alpha(p)\in\pi$, $s\in S$. This operation is well-defined, for if $\alpha(p)=\alpha(p')$, $p,p'\in P$, then $\alpha(pt)=\alpha(p't)$ for all $t\in T_\alpha$. Hence $\alpha(p)\circledast s=\alpha(pt)=\alpha(p')\circledast s$. We now turn the pair (π,S) into a transformation semigroup $\mathscr{A}/\langle\pi\rangle$ by finding a quotient of S that acts faithfully on π. Then $\alpha:P\to\pi$ is a mapping such that $\mathscr{A}/\langle\pi\rangle\le\mathscr{B}$. If π is a partition then $\circledast$ is well-defined anyway.

Example 4.2

Let $\mathscr{A}$ and $\mathscr{B}$ be the transformation semigroups defined by the state machines

	a	b	c	d	e	f
0	a	c	b	e	c	c
1	c	b	a	b	f	f

and

	A	B	C	D	E
0	B	A	D	C	A
1	A	B	C	A	C

The relation $\alpha(A)=\alpha(B)=\{a, b, c\}$, $\alpha(C)=\alpha(E)=\{b, e, f\}$, $\alpha(D)=\{c, d\}$, defines a relational covering $\mathscr{A} \lhd_\alpha \mathscr{B}$ which is not close. If π is the admissible subset system defined by α then π is $\{\{a, b, c\}, \{b, e, f\}, \{c, d\}\}$. If we denote $\{a, b, c\}$ by X, $\{b, e, f\}$ by Y and $\{c, d\}$ by Z we can then find the possible quotient transformation semigroups $\mathscr{A}/\langle\pi\rangle$. There are two of these defined by

	X	Y	Z
0	X	X	Y
1	X	Y	X

and

	X	Y	Z
0	X	Z	Y
1	X	Y	X

In neither case does $\mathscr{A}/\langle\pi\rangle \leq \mathscr{B}$ hold, and so theorem 4.1.1 is not true for all relational coverings.

4.2 The skeleton and height functions

If $\mathscr{A}=(Q, S)$ is a transformation semigroup and $s \in S$ then $Qs=\{qs \mid q \in Q\}$ will be called the *image under s*. The collection of all these images constitutes a very important subset of the power set $\mathscr{P}(Q)$. In many ways the properties of this set of images reflect the structure of $\mathscr{A}$. A natural ordering exists on this set of images, but first we extend the set slightly. Define

$$\mathbf{I}(\mathscr{A})=\Big(\bigcup_{s\in S}\{Qs\}\Big)\cup\{Q\}\cup\Big(\bigcup_{q\in Q}\{\{q\}\}\Big)\cup\{\varnothing\}.$$

Thus $\mathbf{I}(\mathscr{A})$ consists of the set of all the images under the elements of S, together with the set Q and $\varnothing$ and the singleton subsets of Q. If $1_Q \in S$ then it is not necessary to adjoin the set Q in this way. Similarly each reset map in S will give rise to the appropriate singleton in Qs, so that if $1_Q \in S$ and $\mathscr{A}$ is closed then $\mathbf{I}(\mathscr{A})=\bigcup_{s\in S}\{Qs\}\cup\{\varnothing\}$.

Now let $A, B \in \mathbf{I}(\mathscr{A})$, we write $A \leq B$ if and only if either $A \subseteq B$ or $A \subseteq Bs$ for some $s \in S$. (The existence of the identity 1_Q in S will ensure that the first condition follows from the second.) This ordering on the set $\mathbf{I}(\mathscr{A})$ satisfies the following properties:

4.2.1(i) $A \leq A$
4.2.1(ii) $A \leq B$ and $B \leq C \Rightarrow A \leq C$.

We further define $A < B$ to mean $A \leq B$ but $B \not\leq A$. Then $(\mathbf{I}(\mathscr{A}), \leq)$ is seen to be a pre-ordered set. We call $(\mathbf{I}(\mathscr{A}), \leq)$ the *skeleton* of $\mathscr{A}$. Notice, however, that distinct transformation semigroups may have identical skeletons, so that although the skeleton gives us much information about $\mathscr{A}$ it cannot tell us everything about it.

In the usual way we construct an equivalence relation on $\mathbf{I}(\mathscr{A})$ by defining

$A \equiv B$ if and only if $A \leq B$ and $B \leq A$.

We immediately obtain the following information.

Proposition 4.2.1
If $A, B \in \mathbf{I}(\mathscr{A})$ then
(i) $A \leq B \Rightarrow |A| \leq |B|$
(ii) $A \equiv B \Rightarrow |A| = |B|$

Proof Both follow from the simple observation that $|Bs| \leq |B|$ since $|B|$ is finite. □

Example 4.3
Returning to the transformation semigroup $\mathscr{A}$ defined in example 4.1 we will calculate $\mathbf{I}(\mathscr{A})$. We get $\mathbf{I}(\mathscr{A}) = \{Q, \{a, c\}, \{a\}, \{b\}, \{c\}, \varnothing\}$. The ordering on $\mathbf{I}(\mathscr{A})$ will be displayed in diagrammatic form, with $A \leq B$ being replaced by an arrow $A \rightarrow B$. (We usually omit $\varnothing$ from the diagram.)

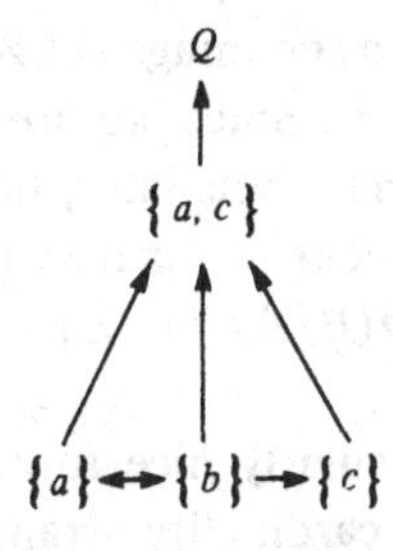

Then $\{a\} \equiv \{b\}$, but $\{b\} < \{c\}$ and $\{a\} < \{c\}$.

Example 4.4
Let $|Q| = n$ (n a positive integer) and let S be the set of all functions from Q to Q. Then $\mathscr{A} = (Q, S)$ is a transformation semigroup.

Now $\mathbf{I}(\mathscr{A}) = \mathscr{P}(Q)$ and if $A, B \subseteq Q$ then $A \leq B \Leftrightarrow |A| \leq |B|$, furthermore $A \equiv B \Leftrightarrow |A| = |B|$.

In this example we notice that $A \equiv B$ implies the existence of an element $s \in S$ such that $B = As$. An element $s' \in S$ also exists satisfying $A = Bs'$ and such that for any $a \in A$, $b \in B$, $ass' = a$, $bs's = b$. This last property is in fact a feature of any transformation semigroup.

Proposition 4.2.2

Let $\mathscr{A} = (Q, S)$ be a transformation semigroup. $A, B \in \mathbf{I}(\mathscr{A})$ with $A \equiv B$. There exist elements $s, \bar{s} \in S \cup \{1_Q\}$ such that

$$B = As, \quad A = B\bar{s},$$

and for all $a \in A$, $b \in B$,

$$as\bar{s} = a, \quad b\bar{s}s = b.$$

Proof Since $A \leq B$ and $B \leq A$ there exist elements $s, t \in S \cup \{1_Q\}$ such that $B \subseteq As$, $A \subseteq Bt$. As $|A| = |B|$ we have $B = As$ and $A = Bt$. Then $B = Bts$, $A = Ast$. Therefore ts and st are permutations on B and A respectively. If st is an 'identity' on A, that is, if $ast = a$ for all $a \in A$, then for each $b \in B$ there exists an $a_1 \in A$ such that $b = a_1 s$. Then $bts = a_1 sts = a_1 s = b$, and thus ts is an identity on B and we may choose $\bar{s} = t$. If st is not an identity on A then $(st)^n$ is an identity on A for some $n > 1$. Then we choose $\bar{s} = t(st)^{n-1}$ and note that $B\bar{s} = Bt(st)^{n-1} = A(st)^n = A$. Also, for $a \in A$, $as\bar{s} = a(st)^n = a$, and for $b \in B$, $b\bar{s}s = bt(st)^{n-1}s = a_1 st(st)^{n-1}s$ (where $b = a_1 s \in As$) $= a_1 s = b$. □

This result enables us to move easily between two equivalent images A and B. If any input $\bar{s}$ acts as a permutation on the image set A then $s'\bar{s}s$ will act as a permutation on the image set B. Since we are going to be interested in the permutations on these image sets later, this fact will prove useful. It will sometimes be convenient and suggestive to write s as (B/A) and $\bar{s}$ as (A/B). Then $as\bar{s} = a(B/A)\cdot(A/B) = a$ and $b\bar{s}s = b(A/B)\cdot(B/A) = b$ for $a \in A$, $b \in B$.

In the example 4.4 the skeleton has a particularly nice form, it is arranged naturally in 'layers' with sets of equal cardinality arranged in the same layer. It is tempting to expect that we can do the same with a more general transformation semigroup, with, perhaps the equivalent sets arranged in the same layer. This is in fact the case but we can do much more; we will define a general height function that maps the skeleton in an order-preserving fashion into the set of integers. The main requirements of the height function, apart from respecting the order on $\mathbf{I}(\mathscr{A})$, are that the function maps all singletons to zero and that there are 'no gaps in the image of it'. Formally we define a *height function* for $\mathscr{A}=(Q, S)$ to be any function $h:\mathbf{I}(\mathscr{A})\to\mathbb{Z}$ satisfying:

(i) $h(\{q\})=0$ for all $q\in Q$, $h(\varnothing)=-1$.

(ii) $A\equiv B\Rightarrow h(A)=h(B)$.

(iii) $A<B$ and $|B|>1\Rightarrow h(A)<h(B)$.

(iv) If $0\le i\le h(Q)$ then $\exists A\in\mathbf{I}(\mathscr{A})$ such that $h(A)=i$.

The last condition is just there to prevent the function from being 'too wasteful'. We cannot expect $A<B\Rightarrow h(A)<h(B)$ in all circumstances since we are also requiring the height of a singleton to be zero and this may conflict with the ordering on $\mathbf{I}(\mathscr{A})$. We have already seen example (4.3) where $\{a\}<\{c\}$ for $a, c\in Q$. This difficulty vanishes if the restriction $|B|>1$ is imposed.

There may be various functions that satisfy these conditions for a given transformation semigroup. One always exists and is constructed in the following way. Given the skeleton $(\mathbf{I}(\mathscr{A}), \le)$ of the transformation semigroup $\mathscr{A}$, let $A\in\mathbf{I}(\mathscr{A})$ with $|A|>1$. Suppose that $A_1<A_2<\ldots<A_n=A$ is the longest chain in $\mathbf{I}(\mathscr{A})$ satisfying $|A_1|>1$. We define $h(A)=n$. For $A\in I(\mathscr{A})$ satisfying $|A|=1$ we put $h(\mathscr{A})=0$ and $h(\varnothing)=-1$. In this way we may now associate an integer with every element of $\mathbf{I}(\mathscr{A})$, and so define a function $h:\mathbf{I}(\mathscr{A})\to\mathbb{Z}$.

To show that h satisfies the four conditions is fairly straightforward. For example let $A, B\in\mathbf{I}(\mathscr{A})$ with $A\equiv B$. If $|A|=|B|=1$ then $h(A)=h(B)=0$. Let $|A|>1, |B|>1$. Suppose that $h(A)=n, h(B)=m$ then there exist chains

$$A_1<A_2<\ldots<A_n=A,\quad B_1<B_2<\ldots<B_m=B$$

each of maximal length subject to the requirement $|A_1|\ne 1, |B_1|\ne 1$. Then $A\le B$ and $B\le A$, yielding $A_{n-1}<B$ and $B_{m-1}<A$. If $m>n$ we obtain the longer chain $B_1<\ldots<B_{m-1}<A$ of length m for A which is false. Similarly $n>m$ is incorrect and so $m=n$. The height function just defined is called the *minimal height function* for the transformation semigroup. The *height of* $\mathscr{A}$ is defined to be $h(Q)$ where h is this minimal height function (we sometimes write the height of $\mathscr{A}$ as $h(\mathscr{A})$ also).

Given the minimal height function various other height functions may then be defined. For example let $\{A_{ij}\}$ $(j=1,\ldots, m_i)$ be the distinct $\equiv$-equivalence classes of $\mathbf{I}(\mathscr{A})$ of minimal height $i\ge 1$. Put $l(A_{ij})=\sum_{k=1}^{i-1} m_k+j$. This yields a height function $l:I(\mathscr{A})\to\mathbb{Z}$ when we define $l(\{q\})=0$ for all $q\in Q$ and $l(\varnothing)=0$. It is a maximal height function in the sense that $l(Q)$ is a maximum, but is not unique with respect to this property in general, since it depends on the order in which the equivalence classes of height i are enumerated.

Now we consider an arbitrary transformation semigroup $\mathscr{A} = (Q, S)$ together with a height function $h: \mathbf{I}(\mathscr{A}) \to \mathbb{Z}$. Suppose that $\mathscr{B} = (P, T)$ is a transformation semigroup such that $\mathscr{A} \lhd_\alpha \mathscr{B}$. The subsets $\alpha(p)$, $p \in P$ may not be images in $\mathscr{A}$ but we will be particularly interested in the occasions when they are. We say that the relational covering $\mathscr{A} \lhd_\alpha \mathscr{B}$ is of *rank* i (with respect to h) if:

(i) $\alpha(p) \in \mathbf{I}(\mathscr{A})$ for all $p \in P$,

(ii) $h(\alpha(p)) \leq i$ for all $p \in P$ and $h(\alpha(p)) = i$ for at least one $p \in P$ where $0 \leq i \leq h(Q)$.

We will consider an example shortly, but before we do note that a relational covering of rank 0 is merely a covering since the image $\alpha(p)$ is a singleton for all $p \in P$, and thus α is a mapping. (The fact that the only elements of the skeleton that have height zero are singletons follows from condition (iii) of the height function.)

Example 4.5

Let $\mathscr{A} = (Q, S)$ be an arbitrary transformation semigroup with height function h, and suppose that $h(Q) = n$. Consider the 'proper maximal images' in $\mathbf{I}(\mathscr{A})$. Formally define

$$\mathbf{M}(Q) = \{A \in \mathbf{I}(\mathscr{A}) \mid A \neq Q \text{ and } A \subseteq C \subseteq Q \text{ with } C \in \mathbf{I}(\mathscr{A}) \Rightarrow \text{either } C = A \text{ or } C = Q\}.$$

We show that $\mathbf{M}(Q)$ forms an admissible subset system for $\mathscr{A}$. First note that if $q \in Q$ then either $\{q\} \in \mathbf{M}(Q)$ or $q \in A$ for some $A \in \mathbf{M}(Q)$ and so the subsets in $\mathbf{M}(Q)$ cover Q. Further let $A \in \mathbf{M}(Q)$ and $s \in S$. Since $A = Qs_1$ for some $s_1 \in S$ we have $As = Qs_1s \in \mathbf{I}(\mathscr{A})$ and either $As \in \mathbf{M}(Q)$ or $As \subseteq A_1$ for some $A_1 \in \mathbf{M}(Q)$. Therefore $\mathbf{M}(Q)$ is an admissible subset system for $\mathscr{A}$. Now consider the way in which the elements of S act on the subsets in $\mathbf{M}(Q)$. If s is a permutation of Q then s is also a permutation on the set $\mathbf{M}(Q)$. First let $A \in \mathbf{M}(Q)$ then $As \in \mathbf{I}(\mathscr{A})$ and let $C \in \mathbf{I}(\mathscr{A})$ with $As \subseteq C \subseteq Q$. If $s \neq 1_Q$ then $s^m = 1_Q$ for some $m > 1$ and we see that $A = Ass^{m-1} \subseteq Cs^{m-1} \subseteq Qs^{m-1} = Q$. Therefore $Cs^{m-1} = A$ or $Cs^{m-1} = Q$, that is $C = Cs^m = As$ or $C = Q$. Thus $As \in \mathbf{M}(Q)$. If s is not a permutation of Q then $Qs \subset Q$ and $Qs \subseteq A$ for some $A \in \mathbf{M}(Q)$. Now let $B \in \mathbf{M}(Q)$, then $Bs \subseteq Qs \subseteq A$ so s has the effect of sending each element of $\mathbf{M}(Q)$ to the element A. We now define a transformation semigroup using the elements of $\mathbf{M}(Q)$ as the states. Let $s \in S$, then either $Qs = Q$ or $Qs \subseteq B$

for some $B \in \mathbf{M}(Q)$. Suppose that for each s satisfying $Qs \neq Q$ we select a $B_s \in \mathbf{M}(Q)$ such that $Qs \subseteq B_s$, now define for $B' \in \mathbf{M}(Q)$ the operation:

$$B' \circledast s = \begin{cases} B's & \text{if } Qs = Q \\ B_s & \text{if } Qs \neq Q,\ Qs \subseteq B_s \text{ and } B_s \text{ is the chosen element} \\ & \text{of } \mathbf{M}(Q) \text{ associated with } s. \end{cases}$$

The pair $(\mathbf{M}(Q), S)$ gives rise to a transformation semigroup $(\mathbf{M}(Q), S/\sim)$ under the operation $\circledast$ in the same way as section 4.1. Therefore this transformation semigroup may be denoted by $\mathscr{A}/\langle\pi\rangle$ where $\pi = \mathbf{M}(Q)$.

If we define a relation $\alpha: \pi \rightsquigarrow Q$ by

$$\alpha(B) = B \quad \text{for } B \in \pi$$

we obtain a relational covering $\mathscr{A} \lhd_\alpha \mathscr{A}/\langle\pi\rangle$ of rank $n-1$. Because of the special choice of the quotient $\mathscr{A}/\langle\pi\rangle$ we may cover $\mathscr{A}/\langle\pi\rangle$ with a particularly useful transformation semigroup. Notice that the semigroup $S/\sim$ of $\mathscr{A}/\langle\pi\rangle$ is generated by a quotient of the maximal subgroup G of S and some reset elements. Then $S/\sim \subseteq Sg\langle(G/\sim) \cup (\bigcup_{B\in\pi} \bar{B})\rangle$ where each $\bar{B}$ is the reset map $\bar{B}: B' \to B$ for all $B' \in \pi$. The inclusion may be strict in some cases. The group $G/\sim$ is called the *holonomy group of Q* and it is to this that we now turn our attention.

4.3 The holonomy groups

Let $A \in \mathbf{I}(\mathscr{A})$ with $|A| > 1$. First we define the *maximal image space* (or *paving*) of A to be the set

$$\mathbf{M}(A) = \{B \in \mathbf{I}(\mathscr{A}) \mid B \subset A, B \neq A \text{ and } B \subseteq C \subseteq A \text{ with } C \in \mathbf{I}(\mathscr{A}) \Rightarrow \text{either } C = A \text{ or } C = B\}.$$

The elements B of $\mathbf{M}(A)$ are called the *maximal images* (or *bricks*) of A. The collection $\mathbf{M}(A)$ forms a covering of the set A, for if $a \in A$ then $\{a\} \in \mathbf{I}(\mathscr{A})$ and either $\{a\} \in \mathbf{M}(A)$ or $\{a\} \subseteq B$ for some $B \in \mathbf{M}(A)$.

Put $\mathbf{G}(A) = \{s \in S \mid As = A\}$, the set of all elements of S that act as permutations on the set A. Naturally $\mathbf{G}(A)$ may be empty. We have:

Proposition 4.3.1

For $A \in \mathbf{I}(\mathscr{A})$ with $|A| > 1$, each element of $\mathbf{G}(A)$ acts as a permutation on the set $\mathbf{M}(A)$.

Proof Let $s \in \mathbf{G}(A)$ and $B \in \mathbf{M}(A)$. Then $Bs \subset A$ and $Bs \in \mathbf{I}(\mathscr{A})$. Suppose that $Bs \subseteq C \subseteq A$. There exists some integer $n \geq 1$ such that s^n acts as the identity on A. Either $Bs = B$ or $B = Bss^{n-1} \subset Cs^{n-1} \subseteq As^{n-1} = A$. Therefore $B = Cs^{n-1}$ or $A = Cs^{n-1}$ and so $Bs = C$ or

$A = C$. Hence $Bs \in \mathbf{M}(A)$. Clearly $s^{n-1} \in \mathbf{G}(A)$ and the inverse of s on $\mathbf{M}(A)$ is s^{n-1}. □

By considering only the *distinct* permutations of $\mathbf{M}(A)$ given by the elements of $\mathbf{G}(A)$ we may define a transformation group $(\mathbf{M}(A), \mathbf{H}(A))$, providing of course that $\mathbf{G}(A) \neq \emptyset$. (Thus $\mathbf{H}(A)$ is a quotient of $\mathbf{G}(A)$.) If $\mathbf{G}(A) = \emptyset$ then we consider the *generalized* transformation group $(\mathbf{M}(A), \emptyset)$. This can only occur if S does not contain the identity 1_Q. In this case we will also write $\mathbf{H}(A) = \emptyset$.

The generalized transformation group $\mathscr{H}(A) = (\mathbf{M}(A), \mathbf{H}(A))$ is called the *holonomy transformation group of* A. The group $\mathbf{H}(A)$ is the *holonomy group of* A (if $\mathbf{H}(A) \neq \emptyset$).

We may, by referring back to example 4.5 notice that the transformation semigroup $\mathscr{A}/\langle\pi'\rangle$ chosen there may be covered by the closure of the *holonomy* transformation group of Q and thus $\mathscr{A} \lhd_\alpha \overline{\mathscr{H}(Q)}$ is of rank $n-1$ where n is the height of $\mathscr{A}$ with respect to the given height function. Notice also that the covering α is close.

Our next aim is to improve the relational covering constructed in example 4.5. Before we do this, however, we will dispose of three useful technical results concerned with maximal image spaces and holonomy groups.

Proposition 4.3.2

Let $\mathscr{A} = (Q, S)$ be a transformation semigroup, $h: \mathbf{I}(\mathscr{A}) \to \mathbb{Z}$ a height function and $h(Q) = n \geq 1$. Then we have:

(i) If $h(A) = n$ for some $A \in \mathbf{I}(\mathscr{A})$ then $A = Q$.
(ii) If $A, B \in \mathbf{I}(\mathscr{A})$ with $A \subsetneq B$ and $|B| > 1$ then $h(A) < h(B)$.
(iii) If $A, B \in \mathbf{I}(\mathscr{A})$ with $A \equiv B$ and $s, \bar{s} \in S$ satisfying the hypothesis of 4.2.2, then $\mathbf{M}(B) = \mathbf{M}(A)s$ and $\mathbf{M}(A) = \mathbf{M}(B)\bar{s}$.

Proof (i) $A \subseteq Q$ and so $h(A) \leq h(Q)$. Suppose that $A \subsetneq Q$, then $Q \leq A$ implies that $Q \subseteq As$ for some $s \in S$, so $|Q| \leq |As| \leq |A| \leq |Q|$ which yields a contradiction. Hence $A < Q$ and so $h(A) < h(Q)$. Therefore $h(A) = n \Rightarrow A = Q$.

(ii) This is proved in a similar way to (i).

(iii) Let $A = B\bar{s}$ and $B = As$ with $as\bar{s} = a$ for all $a \in A$, $b\bar{s}s = b$ for all $b \in B$. Consider $K \in \mathbf{M}(A)$, then $Ks \in \mathbf{I}(\mathscr{A})$. Let $Ks \subseteq C \subseteq B$ for some $C \in \mathbf{I}(\mathscr{A})$, then $K = Ks\bar{s} \subseteq C\bar{s} \subseteq B\bar{s} = A$ and $C\bar{s} \in \mathbf{I}(\mathscr{A})$ which implies $C\bar{s} = K$ or $C\bar{s} = A$ and thus $C = Ks$ or $C = B$. Therefore $Ks \in \mathbf{M}(B)$. Now let $L \in \mathbf{M}(B)$, then $L\bar{s}s = L$ and $L\bar{s} \subsetneq A$. Let us choose any $D \in \mathbf{I}(A)$ with $L\bar{s} \subseteq D \subseteq A$, then $L = L\bar{s}s \subseteq Ds \subseteq As = B$ and so $Ds = L$ or $Ds = B$, which gives $D = L\bar{s}$ or $D = A$. Therefore $\mathbf{M}(B) = \mathbf{M}(A) \cdot s$. The other result is proved similarly. □

Proposition 4.3.3

If $A \equiv B$ then $\mathcal{H}(A) \cong \mathcal{H}(B)$.

Proof $\mathcal{H}(A) = (\mathbf{M}(A), \mathbf{H}(A))$, $\mathcal{H}(B) = (\mathbf{M}(B), \mathbf{H}(B))$. There is a mapping $\phi: \mathbf{M}(A) \to \mathbf{M}(B)$ defined by $\phi(K) = Ks$ where $K \in \mathbf{M}(A)$ and $s \in S$ is such that $B = As$. Then $A = B\bar{s}$ for a suitable $\bar{s} \in S$ and so ϕ is invertible with inverse $\psi: \mathbf{M}(B) \to \mathbf{M}(A)$ defined by $\psi(L) = L\bar{s}$ for $L \in \mathbf{M}(B)$. Now we choose a $g \in \mathbf{G}(A)$ which satisfies $Ag = A$ and so $B(\bar{s}gs) = Ags = As = B$ which allows us to define a mapping $\xi: \mathbf{G}(A) \to \mathbf{G}(B)$ by $\xi(g) = \bar{s}gs$ for all $g \in G(A)$. The mapping $\eta: \mathbf{G}(B) \to \mathbf{G}(A)$ defined by $\eta(g') = sg'\bar{s}$ for all $g' \in \mathbf{G}(B)$ is the inverse of ξ. Finally we notice that for $g, g_1 \in \mathbf{G}(A)$,

$$B(\bar{s}gs)(\bar{s}g_1s) = B$$

and for $b \in B$,

$$b\xi(g)\xi(g_1) = b\bar{s}gs\bar{s}g_1s = b\bar{s}gg_1s = b\xi(gg_1)$$

and therefore ξ is an isomorphism of groups. The factor groups $\mathbf{H}(A)$ and $\mathbf{H}(B)$ must also be isomorphic since $A \equiv B$. □

Now we may return to the central problem of improving the relational covering of example 4.4. So we let $\mathcal{A} = (Q, S)$ be a transformation semigroup and $h: \mathbf{I}(\mathcal{A}) \to \mathbb{Z}$ a height function. Suppose that π is an admissible subset system with the property that each subset $A \in \pi$ also belongs to the skeleton $\mathbf{I}(\mathcal{A})$ and the maximum height of an element of π is i. Such an admissible subset system is said to have *rank* i. We may now produce an admissible subset system π' of rank $i-1$ with $\pi' \leq \pi$ (assuming that $i \geq 1$).

First let $\mathcal{X} = \{A \in \pi \mid h(A) < i\}$ and $\mathcal{Y} = \{A \in \pi \mid h(A) = i\}$. Then we define $\pi' = \mathcal{X} \cup (\bigcup_{Y \in \mathcal{Y}} \mathbf{M}(Y))$ where, as before, $\mathbf{M}(Y)$ is the maximal image space of Y. Clearly $\pi' \leq \pi$ and the rank of π' equals $i-1$; we have to establish the fact that π' is an admissible subset system, so we let $B \in \pi'$ and $s \in S$. If $B \in \mathcal{X}$ then $B \in \pi$ and so $Bs \subseteq A$ for some $A \in \pi$. This leads to two cases:

Case (i): $A \in \mathcal{X}$, in which case Bs is contained in a subset of π'.

Case (ii): $Bs \subseteq A$ with $A \in Y$. Since $Bs \leq B$ and $h(B) < i$ then $h(Bs) < i$ and so there exists $C \in \mathbf{M}(A)$ such that $Bs \subseteq C$. Therefore Bs is contained in an element of π'.

If $B \in \mathbf{M}(Y)$ for some $Y \in \mathscr{Y}$ then $Bs \subseteq Ys$. We have three further cases to consider:

Case (iii): $Ys \subseteq A$ for some $A \in \mathscr{X}$ and then $Bs \subseteq A$.

Case (iv): $Ys \subsetneq A$ for some $A \in \mathscr{Y}$, then $Ys \subseteq C$ for some $C \in \mathbf{M}(A)$ and so $Bs \subseteq C$ with $C \in \pi'$.

Case (v): $Ys \in \mathscr{Y}$. Now $Ys \leq Y$ and $h(Ys) = h(Y) = i$, hence $Ys \equiv Y$. By 4.3.2(iii) $\mathbf{M}(Ys) = \mathbf{M}(Y)s$ and so $Bs \in \mathbf{M}(Y')$ and thus $Bs \in \pi'$. This result is stated as:

Theorem 4.3.4

Let $\mathscr{A} = (Q, S)$ be a transformation semigroup and $h: \mathbf{I}(\mathscr{A}) \to \mathbb{Z}$ a height function. Let π be an admissible subset system of rank i. There exists an admissible subset system π' of rank $i-1$ with $\pi' \leq \pi$.

In this way we may start with an admissible subset system of $\mathscr{A}$ of rank equal to the height of $\mathscr{A}$ and successively reduce the rank of the covering. First let π^n be the trivial admissible subset system $\{Q\}$ where $n = h(Q)$ then $(\pi^n)'$ is the admissible subset system of example 4.4. Putting $\pi^{n-1} = (\pi^n)'$, $\pi^{n-2} = (\pi^{n-1})', \ldots$ we obtain a sequence $\pi^n > \pi^{n-1} > \pi^{n-2} > \ldots$ of admissible subset systems. We call this the *derived sequence* of $\mathscr{A}$. The rank of π^n equals the height n of $\mathscr{A}$, the rank of π^{n-1} equals $n-1$, and in general π^j has rank j. We will use the derived sequence as a means of defining a relational covering of $\mathscr{A}$ using transformation groups. Notice that if we can associate a suitable relational covering of $\mathscr{A}$ with the admissible subset system π^j then there may be a natural candidate for a relational covering associated with π^{j-1}.

Suppose that $\pi^j = \mathscr{X} \cup \mathscr{Y}$ where the elements of $\mathscr{X}$ are of height less than j and the height of the elements of $\mathscr{Y}$ is j. The definition of π^{j-1} is $\mathscr{X} \cup \bigcup_{Y \in \mathscr{Y}} \mathbf{M}(Y)$. The sets $\mathbf{M}(Y)$ are the underlying sets of the holonomy transformation groups $\mathscr{H}(Y)$. We have seen that when $j = n$ the closure of the holonomy transformation group $\mathscr{H}(Q)$ yields a relational covering of $\mathscr{A}$ of rank $n-1$. (Here n is the height of Q.) Can we build on this to produce an inductive method of generating relational coverings of smaller rank?

Let $1 \leq j \leq n$ where $n = h(Q)$. The set of elements of the skeleton of height j may be partitioned by the equivalence relation $\equiv$. Let $A_1^j, \ldots, A_{r_j}^j$ be a set of representatives of the distinct equivalence classes. We form the holonomy transformation groups $\mathscr{H}(A_1^j), \ldots, \mathscr{H}(A_{r_j}^j)$ and then take their join $\mathscr{H}(A_1^j) \vee \ldots \vee \mathscr{H}(A_{r_j}^j)$. This is also denoted by $\mathscr{H}_j^{\vee}(\mathscr{A})$. The state set of this transformation semigroup must be defined with care, for we need the individual state sets of $\mathscr{H}(A_k^j)$ to be disjoint if we are going to form the join. To ensure this we will consider the state set of $\mathscr{H}(A_k^j)$ to be $\{k\}\times\mathbf{M}(A_k^j)$. So a typical element of the state set of $\mathscr{H}_j^\vee(\mathscr{A})$ is (k, B_k^j) where $1\le k\le r_j$, $B_k^j\in\mathbf{M}(A_k^j)$.

We now state our main inductive result.

Theorem 4.3.5

Let $\mathscr{A}\lhd_{\alpha_j}\mathscr{B}$ be a relational covering of rank j such that the image of α_j is π^j. There exists a relational covering $\mathscr{A}\lhd_{\alpha_{j-1}}\overline{\mathscr{H}_j^\vee(\mathscr{A})\circ\mathscr{B}}$ such that

(i) the rank of α_{j-1} is $j-1$,
(ii) the image of α_{j-1} is π^{j-1}.

Proof Let $\mathscr{A}=(Q, S)$ and $\mathscr{B}=(P, T)$. If $s\in S$ there exists $t_s\in T$ such that

$$\alpha_j(p)\cdot s\subseteq\alpha_j(p\cdot t_s)\quad\text{for all } p\in P.$$

To define $\alpha_{j-1}:(\bigcup_{k=1}^{r_j}(\{k\}\times\mathbf{M}(A_k^j)))\times P\rightsquigarrow Q$ consider an element $((k, B_k^j), p)\in(\bigcup_{k=1}^{r_j}(\{k\}\times\mathbf{M}(A_k^j)))\times P$ and put

$$\alpha_{j-1}((k, B_k^j), P)=\begin{cases}\alpha_j(p) & \text{if } h(\alpha_j(p))<j\\ B_k^j\cdot(\alpha_j(p)/A_k^j) & \text{if } \alpha_j(p)\equiv A_k^j\\ \varnothing & \text{otherwise.}\end{cases}$$

From this definition we note that $\alpha_{j-1}((k, B_k^j), p)$ is an element of the skeleton of $\mathscr{A}$ of height less than j. Furthermore $B_k^j\cdot(\alpha_j(p)/A_k)\in\mathbf{M}(\alpha_j(p))$. Since the image of α_j is π^j, suppose that $Z\in\pi^{j-1}$. If $Z\in\pi^j$ we have $h(Z)<j$ and $Z=\alpha_j(p)$ for some $p\in P$ and so $\alpha_{j-1}((k, B_k^j), p)=\alpha_j(p)=Z$ for any $(k, B_k^j)\in\bigcup_{k=1}^{r_j}(\{k\}\times\mathbf{M}(A_k^j))$. Writing π^j as $\mathscr{X}\cup\mathscr{Y}$ as in 4.3.4 we have $\pi^{j-1}=\mathscr{X}\cup(\bigcup_{Y\in\mathscr{Y}}\mathbf{M}(Y))$. If $Z\in\mathbf{M}(Y)$ for some $Y\in\mathscr{Y}$ then $Y\in\pi^j$ and $Y=\alpha_j(p)$ for some $p\in P$. Now $Y\equiv A_k$ for some $1\le k\le r_j$ and $Y=A_k\cdot(Y/A_k)$. Then $Z=B\cdot(Y/A_k)$ for some $B\in\mathbf{M}(A_k)$ and $Z=\alpha_{j-1}((k, B), p)$. Hence the image of α_{j-1} equals π^{j-1}.

The proof of the fact that α_{j-1} is a relational covering now occupies our attention for a few paragraphs. The crucial part is the definition of the element of the action semigroup of $\overline{\mathscr{H}_j^\vee(\mathscr{A})\circ\mathscr{B}}$ that will cover a given element of S.

Let $s\in S$ and suppose that $t_s\in T$ covers s with respect to the relational covering α_j. Thus $\alpha_j(p)\cdot s\subseteq\alpha_j(p\cdot t_s)$ for all $p\in P$. Now the action semigroup of $\overline{\mathscr{H}_j^\vee(\mathscr{A})\circ\mathscr{B}}$ consists of the set of all ordered pairs (f, t) where $t\in T$ and $f:P\to(\bigvee_{k=1}^{r_j}\mathbf{H}(A_k^j))$. Having chosen our element $s\in S$ we define a function $f_s : P \to \overline{(\bigvee_{k=1}^{r_j} \mathbf{H}(A_k^j))}$ in the following way. Let $p \in P$, three possibilities arise:

Case (i): $\alpha_j(p \cdot t_s) \in \mathscr{X}$, whence $f_s(p)$ is chosen arbitrarily.

Case (ii): $\alpha_j(p \cdot t_s) \in \mathscr{Y}$ and $\alpha_j(p) \cdot s \subsetneq \alpha_j(p \cdot t_s)$ then $\alpha_j(p \cdot t_s) \equiv A_k^j$ for some $1 \le k \le r_j$. Now

$$\alpha_j(p) \cdot s \cdot (A_k^j/\alpha_j(p \cdot t_s)) \subsetneq \alpha_j(p \cdot t_s) \cdot (A_k^j/\alpha_j(p \cdot t_s)) = A_k^j$$

and so $\alpha_j(p) \cdot s \cdot (A_k^j/\alpha_j(p \cdot t_s)) \subseteq B'$ for some $B' \in \mathbf{M}(A_k^j)$. We put $f_s(p) = \overline{(k, B')}$, the constant map.

Case (iii): $\alpha_j(p \cdot t_s) \in \mathscr{Y}$ and $\alpha_j(p) \cdot s = \alpha_j(p \cdot t_s)$, then $\alpha_j(p) \equiv \alpha_j(p \cdot t_s)$ since $\alpha_j(p) \cdot s \le \alpha_j(p)$ and yet $\alpha_j(p)$ is of height j at most.

Now let $\alpha_j(p) \equiv A_k^j$ for some $1 \le k \le r_j$, then

$$A_k^j \cdot (\alpha_j(p)/A_k^j) \cdot s \cdot (A_k^j/\alpha_j(p \cdot t_s)) = \alpha_j(p) \cdot s \cdot (A_k^j/\alpha_j(p \cdot t_s))$$
$$= \alpha_j(p \cdot t_s)) \cdot (A_k^j/\alpha_j(p \cdot t_s)) = A_k^j.$$

The element $(\alpha_j(p)/A_k^j) \cdot s \cdot (A_k^j/\alpha_j(p \cdot t_s))$ defines an element h of the holonomy group $\mathbf{H}(A_k^j)$ and so we put

$$f_s(p) = h.$$

This defines the function $f : P \to \overline{\bigvee_{k=1}^{r_j} \mathbf{H}(A_k^j)}$. What remains is the task of showing that (f_s, t_s) covers s with respect to α_{j-1}. Let $((l, B_l^j), p) \in (\bigcup_{k=1}^{r_j} (\{k\} \times \mathbf{M}(A_k^j))) \times P$, we will prove that

$$\alpha_{j-1}((l, B_l^j), p) \cdot s \subseteq \alpha_{j-1}(((l, B_l^j), p) \cdot (f_s, t_s)). \qquad (*)$$

In case (i), where $\alpha_j(p \cdot t_s) \in \mathscr{X}$, we have

$$\alpha_{j-1}((l, B_l^j), p) \cdot s = \alpha_j(p) \cdot s \subseteq \alpha_j(p \cdot t_s) \quad \text{if } \alpha_j(p) \in \mathscr{X},$$

and

$$\alpha_{j-1}((l, B_l^j), p) \cdot s \subseteq \alpha_j(p) \cdot s \subseteq \alpha_j(p \cdot t_s)$$

in all other cases.

Since $\alpha_j(p \cdot t_s) \in \mathscr{X}$, $f_s(p)$ is arbitrary and

$$\alpha_{j-1}(((l, B_l^j), p) \cdot (f_s, t_s)) = \alpha_{j-1}((l, B_l^j) \cdot f_s, p \cdot t_s))$$
$$= \alpha_j(p \cdot t_s).$$

Therefore the inequality $(*)$ holds in this case.

In case (ii) $\alpha_j(p \cdot t_s) \in \mathscr{Y}$ and $\alpha_j(p) \cdot s \subsetneq \alpha_j(p \cdot t_s)$. As before $\alpha_{j-1}((l, B_l^j), p) \cdot s \subsetneq \alpha_j(p) \cdot s$. Now $f_s(p) = \overline{(k, B')}$ where $B' \in \mathbf{M}(A_k^j)$ and $\alpha_j(p) \cdot s \cdot (A_k^j/\alpha_j(p \cdot t_s)) \subseteq B'$, and so

$$\alpha_{j-1} \cdot (((l, B_l^j), p) \cdot (f_s, t_s)) = \alpha_{j-1}((k, B'), p \cdot t_s)$$
$$= B' \cdot (\alpha_j(p \cdot t_s)/A_k^j).$$

Now

$$\alpha_j(p)\cdot s = \alpha_j(p)\cdot s\cdot(A^j_k/\alpha_j(p\cdot t_s))\cdot(\alpha_j(p\cdot t_s)/A^j_k)$$
$$\subseteq B'\cdot(\alpha_j(p\cdot t_s)/A^j_k)$$

and so (*) holds again.

In case (iii), $\alpha_j(p\cdot t_s)\in\mathscr{Y}$ and $\alpha_j(p)\cdot s = \alpha_j(p\cdot t_s)$. If

$$\alpha_j(p)\equiv A^j_l$$

then

$$\alpha_{j-1}(((l, B^j_l), p)\cdot(f_s, t_s))$$
$$= \alpha_{j-1}((l, B^j_l\cdot(\alpha_j(p)/A^j_l)\cdot s\cdot(A^j_l/\alpha_j(p\cdot t_s))), p\cdot t_s)$$
$$= B^j_l\cdot(\alpha_j(p)/A^j_l)\cdot s\cdot(A^j_l/\alpha_j(p\cdot t_s))\cdot(\alpha_j(p\cdot t_s)/A^j_l)$$
$$= B^j_l\cdot(\alpha_j(p)/A^j_l)\cdot s$$
$$= \alpha_{j-1}((l, B^j_l), p)\cdot s$$

and so (*) holds. Finally if $\alpha_j(p)\not\equiv A^j$ then

$$\alpha_{j-1}((l, B^j_l), p) = \varnothing$$

and so

$$\alpha_{j-1}((l, B^j_l), p)\cdot s \subseteq \alpha_{j-1}(((l, B^j_l), p)\cdot(f_s, t_s))$$

as required. □

Theorem 4.3.6

Let $\mathscr{A}$ be a transformation semigroup and $h:\mathbf{I}(\mathscr{A})\to\mathbb{Z}$ a height function then

$$\mathscr{A}\leq\overline{\mathscr{H}_1^\vee(\mathscr{A})}\circ\overline{\mathscr{H}_2^\vee(\mathscr{A})}\circ\ldots\circ\overline{\mathscr{H}_n^\vee(\mathscr{A})}$$

where $n = h(\mathscr{A})$.

Proof We have already established in example 4.5 the relational covering $\mathscr{A}\lhd_{\alpha_{n-1}}\overline{\mathscr{H}(Q)}$ and recalling that $\mathscr{H}(Q) = \mathscr{H}_n^\vee(\mathscr{A})$ we have $\mathscr{A}\lhd_{\alpha_{n-1}}\mathscr{H}_n^\vee(\mathscr{A})$. The above theorem leads to the relational coverings,

$$\mathscr{A}\lhd_{\alpha_{n-2}}\overline{\mathscr{H}_{n-1}^\vee(\mathscr{A})}\circ\overline{\mathscr{H}_n^\vee(\mathscr{A})}$$
$$\mathscr{A}\lhd_{\alpha_{n-3}}\overline{\mathscr{H}_{n-2}^\vee(\mathscr{A})}\circ\overline{\mathscr{H}_{n-1}^\vee(\mathscr{A})}\circ\overline{\mathscr{H}_n^\vee(\mathscr{A})}$$
$$\vdots$$
$$\mathscr{A}\lhd_{\alpha_0}\overline{\mathscr{H}_1^\vee(\mathscr{A})}\circ\ldots\circ\overline{\mathscr{H}_n^\vee(\mathscr{A})},$$

where α_0 is of rank 0 and is thus a covering (i.e. a partial function). □

It is sometimes useful to have alternative holonomy decomposition theorems, especially those involving products rather than joins of the holonomy transformation groups.

As before let $A_1^j, \ldots, A_{r_j}^j$ be a set of representatives of the distinct equivalence classes of the elements of the skeleton of height j. Define

$$\mathcal{H}_j^\times(\mathcal{A}) = \prod_{k=1}^{r_j} \mathcal{H}(A_k^j),$$

the direct product of the individual holonomy transformation groups. Notice that $\mathcal{H}_j^\times(\mathcal{A})$ is a transformation group. (We need only adjoin an identity to those generalized holonomy transformation groups $\mathcal{H}(A_k^j) = (\mathbf{M}(A_k^j), \mathbf{H}(A_k^j))$ where $\mathbf{H}(A_k^j) = \varnothing$, so that the direct product may be defined in a suitable way.)

Theorem 4.3.7

Let $\mathcal{A} \lhd_{\beta_j} \mathcal{B}$ be a relational covering of rank j such that the image of β_j is π^j. There exists a relational covering $\mathcal{A} \lhd_{\beta_{j-1}} \mathcal{H}_j^\times(\mathcal{A}) \circ \mathcal{B}$ such that

(i) the rank of β_{j-1} is $j-1$,
(ii) the image of β_{j-1} is π^{j-1}.

Proof As before let $\mathcal{A} = (Q, S)$, $\mathcal{B} = (P, T)$ we define $\beta_{j-1} : (\prod_{k=1}^{r_j} \mathbf{M}(A_k^j)) \times P \rightsquigarrow Q$ by

$$\beta_{j-1}((B_1^j, \ldots, B_{r_j}^j), p) = \begin{cases} \beta_j(p) & \text{if } h(\beta_j(p)) < j \\ B_k^j \cdot (\beta_j(p)/A_k^j) & \text{if } \beta_j(p) \equiv A_k^j \end{cases}$$

where $(B_1^j, \ldots, B_{r_j}^j) \in \prod_{k=1}^{r_j} \mathbf{M}(A_k^j)$, $p \in P$.

With respect to this relation we will establish the covering condition

$$\beta_{j-1}((B_1^j, \ldots, B_{r_j}^j), p) \cdot s \subseteq \beta_{j-1}(((B_1^j, \ldots, B_{r_j}^j), p) \cdot (g_s, t_s))$$

where $s \in S$, $t_s \in T$ is such that $\beta_j(p) \cdot s \subseteq \beta_j(p \cdot t_s)$ and $g_s : P \to \prod_{k=1}^{r_j} \mathbf{H}(A_k^j)$ is chosen suitably. The definition of g_s is taken in three cases; let $p \in P$.

Case (i): $\beta_j(p \cdot t_s)$ is of height less than j, in which case $g_s(p)$ may be defined arbitrarily.

Case (ii): $\beta_j(p \cdot t_s)$ is of height j and $\beta_j(p) \cdot s \subsetneq \beta_j(p \cdot t_s)$. Now $\beta_j(p \cdot t_s) \equiv A_k^j$ for some $k \in \{1, \ldots, r_j\}$ and so $\beta_j(p) \cdot s \cdot (A_k^j/\beta_j(p \cdot t_s)) \subsetneq \beta_j(p \cdot t_s) \cdot (A_k^j/\beta_j(p \cdot t_s)) = A_k^j$ and thus $\beta_j(p) \cdot s \cdot (A_k^j/\beta_j(p \cdot t_s)) \subseteq B'$ for some $B' \in \mathbf{M}(A_k^j)$. Define $g_s(p)$ to be the transformation that sends the k-th coordinate of an element $(B_1^j, \ldots, B_{r_j}^j)$ of $\prod_{l=1}^{r_j} \mathbf{M}(A_l^j)$ to B' and is arbitrarily defined on the other coordinates, thus

$$(B_1^j, \ldots, B_k^j, \ldots, B_{r_j}^j) \cdot g_s(p) = (*, \ldots, B', \ldots, *).$$

Case (iii): $\beta_j(p \cdot t_s) = \beta_j(p) \cdot s \equiv A_k^j$ for some $k \in \{1, \ldots, r_j\}$. As before (proof of 4.3.5) the element

$$(\beta_j(p)/A_k^j) \cdot s \cdot (A_k^j/\beta_j(p \cdot t_s))$$

defines an element h of the holonomy group $\mathbf{H}(A^j_k)$. The definition of $g_s(p)$ is then taken to be any element of the semigroup of $\overline{\mathscr{H}^{\times}_j(\mathscr{A})}$ that acts like h on the k-th coordinate, thus

$$(B^j_1, \ldots, B^j_k, \ldots, B^j_{r_j}) \cdot (g_s(p)) = (*, \ldots, B^j_k \cdot h, \ldots, *).$$

We now show that (g_s, t_s) covers s, and to do this let

$$((B^j_1, \ldots, B^j_k, \ldots, B^j_{r_j}), p) \in \left(\prod_{k=1}^{r_j} \mathbf{M}(A^j_k) \right) \times P.$$

In case (i) $\beta_j(p \cdot t_s)$ is of height j and $\beta_{j-1}((B^j_1, \ldots, B^j_{r_j}), p) \cdot s \subseteq \beta_j(p) \cdot s \subseteq \beta_j(p \cdot t_s)$ since β_j is a relational covering and t_s covers s. Now

$$\begin{aligned} \beta_{j-1}(((B^j_1, \ldots, B^j_{r_j}), p) \cdot (g_s, t_s)) &= \beta_{j-1}(((B^j_1, \ldots, B^j_{r_j}) \cdot g_s, pt_s)) \\ &= \beta_j(p \cdot t_s) \end{aligned}$$

and so

$$\beta_{j-1}((B^j_1, \ldots, B^j_{r_j}), p) \cdot s \subseteq \beta_{j-1}(((B^j_1, \ldots, B^j_{r_j}), p) \cdot (g_s, t_s))$$

in this case.

In case (ii) $\beta_j(p \cdot t_s) \equiv A^j_k$ for some $k \in \{1, \ldots, r_j\}$ and $\beta_j(p) \cdot s \subsetneq \beta_j(p \cdot t_s)$. Then

$$\beta_{j-1}(((B^j_1, \ldots, B^j_{r_j}), p) \cdot (g_s, t_s)) = \beta_{j-1}(((*, \ldots, B', \ldots, *), p \cdot t_s))$$

where

$$B' \in \mathbf{M}(A^j_k) \quad \text{and} \quad \beta_j(p) \cdot s \cdot (A^j_k / \beta_j(p \cdot t_s)) \subseteq B'.$$

Thus

$$\beta_{j-1}(((B^j_1, \ldots, B^j_{r_j}), p) \cdot (g_s, t_s)) = B'(\beta_j(p \cdot t_s)/A^j_k).$$

Now

$$\begin{aligned} \beta_{j-1}((B^j_1, \ldots, B^j_{r_j}), p) \cdot s &\subseteq \beta_j(p) \cdot s \\ &= \beta_j(p) \cdot s \cdot (A^j_k / \alpha_j(p \cdot t_s)) \cdot (\beta_j(p \cdot t_s)/A^j_k) \\ &\subseteq B'(\beta_j(p \cdot t_s)/A^j_k) \\ &= \beta_{j-1}(((B^j_1, \ldots, B^j_{r_j}), p) \cdot (g_s, t_s)). \end{aligned}$$

Finally for case (iii), $\beta_j(p \cdot t_s) \equiv \beta_j(p) \cdot s \equiv A^j_k$ and

$$\begin{aligned} &\beta_{j-1}(((B^j_1, \ldots, B^j_{r_j}), p) \cdot (g_s, t_s)) \\ &\quad = \beta_{j-1}(((*, \ldots, B^j_k \cdot h, \ldots, *), p \cdot t_s)) \\ &\quad = B^j_k \cdot h \cdot (\beta_j(p \cdot t_s)/A^j_k) \\ &\quad = B^j_k(\beta_j(p)/A^j_k) \cdot s \cdot (A^j_k/\beta_j(p \cdot t_s)) \cdot (\beta_j(p \cdot t_s)/A^j_k) \\ &\quad = B^j_k(\beta_j(p)/A^j_k) \cdot s \\ &\quad = \beta_{j-1}((B^j_1, \ldots, B^j_{r_j}), p)) \cdot s. \end{aligned}$$

Therefore, in all cases we have (g_s, t_s) covering s. Hence β_{j-1} is a relational covering and the rest of the theorem follows in a similar way to theorem 4.3.5 □

Theorem 4.3.8

Let $\mathscr{A}$ be a transformation semigroup and $h:\mathbf{I}(\mathscr{A})\to\mathbb{Z}$ a height function, then

$$\mathscr{A}\leq\overline{\mathscr{H}_1^\times(\mathscr{A})}\circ\overline{\mathscr{H}_2^\times(\mathscr{A})}\circ\ldots\circ\overline{\mathscr{H}_n^\times(\mathscr{A})}\quad\text{where } n=h(\mathscr{A}).$$

Theorem 4.3.9

Let $\mathscr{A}$ be a transformation semigroup and $h:\mathbf{I}(\mathscr{A})\to\mathbb{Z}$ a height function. Suppose that $h(\mathscr{A})=n$ and π^{n-1} is the first non-trivial element of the derived sequence of $\mathscr{A}$. Then

(i) $\mathscr{A}\leq\overline{\mathscr{H}_1^\vee(\mathscr{A})}\circ\ldots\circ\overline{\mathscr{H}_{n-1}^\vee(\mathscr{A})}\circ\mathscr{A}/\langle\pi^{n-1}\rangle$

(ii) $\mathscr{A}\leq\overline{\mathscr{H}_1^\times(\mathscr{A})}\circ\ldots\circ\overline{\mathscr{H}_{n-1}^\times(\mathscr{A})}\circ\mathscr{A}/\langle\pi^{n-1}\rangle$

where $\mathscr{A}/\langle\pi^{n-1}\rangle$ is a suitably chosen quotient transformation semigroup.

Proof In the discussion of example 4.5 we defined a quotient transformation semigroup $\mathscr{A}/\langle\pi\rangle$. In the context of this section π is denoted by π^{n-1} and so the result follows from the fact that $\mathscr{A}/\langle\pi^{n-1}\rangle$ yields a relational covering of $\mathscr{A}$ of rank $n-1$. □

Theorems 4.3.5 and 4.3.7 are known as the *holonomy reduction theorems* and theorems 4.3.6 and 4.3.8 as the *holonomy decomposition theorems.*

It is possible to use theorem 4.3.6 to deduce another decomposition theorem. First let $\mathscr{H}_j^+(\mathscr{A})=\mathscr{H}(A_1^j)+\ldots+\mathscr{H}(A_{r_j}^j)$, where $A_1^j,\ldots,A_{r_j}^j$ are representatives of the skeletal elements of height j with respect to some height function on the transformation semigroup $\mathscr{A}$. We have that $\mathscr{H}_j^\vee(\mathscr{A})\leq\mathscr{H}_j^+(\mathscr{A})$ and so we may establish:

Theorem 4.3.10

Let $\mathscr{A}$ be a transformation semigroup and $h:\mathbf{I}(\mathscr{A})\to\mathbb{Z}$ a height function. Then

$$\mathscr{A}\leq\overline{\mathscr{H}_1^+(\mathscr{A})}\circ\overline{\mathscr{H}_2^+(\mathscr{A})}\circ\ldots\circ\overline{\mathscr{H}_n^+(\mathscr{A})}$$

where n is the height of $\mathscr{A}$.

We will examine some examples later, but for the moment there are some remarks that are worth making. First of all this decomposition is a considerable improvement on the original decomposition theorem for transformation semigroups obtained by Krohn and Rhodes [1965]. The increase in efficiency is easily demonstrated by examining some of their examples. Notice that the groups involved in the holonomy transformation semigroups are immediately seen to be divisors of the semigroup S. However, Eilenberg noted that it was still possible to obtain even better decompositions than those given by the holonomy decomposition theorem and it is this aspect of the theory that we turn to next. Although the proof of the holonomy decomposition theorem given here is the same as Eilenberg's we have stressed the admissible subset systems and the derived sequence. The reason for this will become apparent in the next section where we introduce an 'improved' holonomy decomposition theorem, which in many cases gives a more efficient decomposition than the standard holonomy decomposition theorem.

4.4 An 'improved' holonomy decomposition and examples

We begin with an example to motivate the discussion.

Example 4.6

If we recall example 4.1 we note that the transformation monoid $\mathscr{A}$ has the following skeleton:

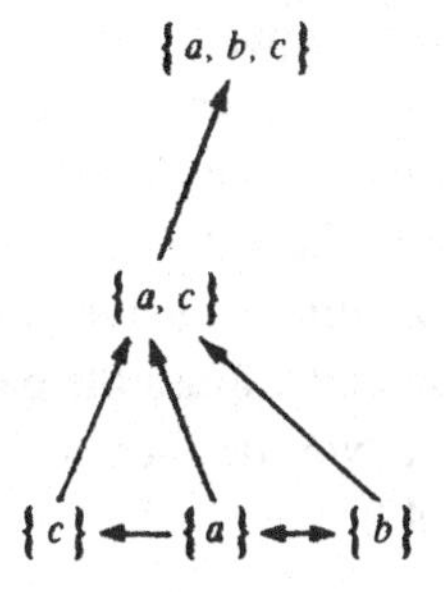

There is a unique height function, $h(Q)=2$, $h(\{a, c\})=1$, $h(\{a\})=h(\{b\})=h(\{c\})=0$, $h(\varnothing)=-1$. The first derived admissible subset system π' consists of the two sets $\{a, c\}$, $\{b\}$. Then $\mathscr{A}/\langle\pi'\rangle$ is given by

0, Λ
1, Λ
{a, c}
{b}
1
0

and so $\mathscr{A}/\langle\pi'\rangle \cong \bar{\mathbf{2}}$. Also $\mathscr{H}_2^\vee(\mathscr{A}) = \mathscr{H}_2^\times(\mathscr{A}) \cong (\pi', \{\Lambda\})$ and $\overline{\mathscr{H}_2^\vee(\mathscr{A})} \cong \bar{\mathbf{2}}^\cdot$.

The derived system $\pi^0 = \{\{a\}, \{b\}, \{c\}\}$.

$$\mathcal{H}_1^{\vee}(\mathcal{A}) = \mathcal{H}_1^{\times}(\mathcal{A}) \cong (\mathbf{M}(\{a, c\}), \{\Lambda\})$$
$$\cong \mathbf{2}^{\cdot}.$$

Then the holonomy decomposition theorem yields

$$\mathcal{A} \leq \bar{\mathbf{2}}^{\cdot} \circ \bar{\mathbf{2}}^{\cdot}.$$

Notice, however, that π' is an *orthogonal* admissible partition, since $\pi \cap \tau = 1$ where $\tau = \{\{a, b\}, \{c\}\}$ and τ is an admissible partition. Then by theorem 3.2.1

$$\mathcal{A} \leq \mathcal{A}/\langle\tau\rangle \times \mathcal{A}/\langle\pi'\rangle = \mathcal{A}/\langle\tau\rangle \times \bar{\mathbf{2}}^{\cdot}.$$

We may check that $\mathcal{A}/\langle\tau\rangle = \mathcal{C}^{\cdot}$ and thus $\mathcal{A} \leq \mathcal{C}^{\cdot} \times \bar{\mathbf{2}}^{\cdot}$, which is better than $\mathcal{A} \leq \bar{\mathbf{2}}^{\cdot} \circ \bar{\mathbf{2}}^{\cdot}$ for two reasons. Firstly $\mathcal{C}^{\cdot} \leq \bar{\mathbf{2}}^{\cdot}$ and secondly direct products are much more efficient than wreath products. (Actually if we examine the proof of theorem 4.3.5 it becomes clear that the wreath product in this example may be replaced by the direct product since the definition of f given by the theorem yields an identity in this example.)

The behaviour of this example gives us a method of approaching the problem of improving the holonomy decomposition theorem. If $\mathcal{A} = (Q, S)$ is an arbitrary transformation semigroup and $h: \mathbf{I}(\mathcal{A}) \to \mathbb{Z}$ is a given height function we may define the derived sequence $\pi^n > \pi^{n-1} > \ldots > \{1\}$ of admissible subset systems. Suppose that π^p is the largest orthogonal admissible partition in the sequence that is non-trivial. Naturally such a π^p may not exist and if this is the case the following theory will not lead to an improved decomposition. However in many cases there is such a π^p. Now let τ be an orthogonal admissible partition such that $\tau \cap \pi^p = \{1\}$. From theorem 3.2.1 we deduce that $\mathcal{A} \leq \mathcal{A}/\langle\tau\rangle \times \mathcal{A}/\langle\pi^p\rangle$ and from theorems 4.1.1 and 4.3.5

$$\mathcal{A}/\langle\pi^p\rangle \leq \overline{\mathcal{H}_{p+1}^{\vee}(\mathcal{A})} \circ \overline{\mathcal{H}_{p+2}^{\vee}(\mathcal{A})} \circ \ldots \circ \overline{\mathcal{H}_n^{\vee}(\mathcal{A})}$$

where p is the rank of π^p and $n = h(\mathcal{A})$. Therefore

$$\mathcal{A} \leq \mathcal{A}/\langle\tau\rangle \times (\overline{\mathcal{H}_{p+1}^{\vee}(\mathcal{A})} \circ (\overline{\mathcal{H}_{p+2}^{\vee}(\mathcal{A})} \circ \ldots \circ \overline{\mathcal{H}_n^{\vee}(\mathcal{A})}).$$

We can now apply the holonomy decomposition theorem to the transformation semigroup $\mathcal{A}/\langle\tau\rangle$ and continue the process. So we choose a height function for $\mathcal{A}/\langle\tau\rangle$, look at its derived sequence and see if a largest non-trivial orthogonal admissible partition lies in this sequence. If one exists we repeat the above process, using theorems 3.2.1 and 4.3.5 to obtain an 'improved' decomposition of $\mathscr{A}/\langle\tau\rangle$. This procedure is repeated as many times as is necessary to obtain a complete decomposition of $\mathscr{A}$. That this decomposition of $\mathscr{A}$ is better than the holonomy decomposition may be established, but the details are rather complex.

Suppose that $\mathscr{A}=(Q, S)$ is a transformation semigroup with $h:\mathbf{I}(\mathscr{A})\to \mathbb{Z}$ a height function. Let $\pi^n>\pi^{n-1}>\ldots>\{1\}$ be the derived sequence and suppose that π^p $(0<p<n)$ is an orthogonal admissible partition. Let $\pi^p\cap\tau=\{1\}$ and write $\mathscr{A}/\langle\tau\rangle=\mathscr{B}$. There is a natural epimorphism (f, g): $\mathscr{A}\to\mathscr{B}$ defined as follows. Let $T=S/\sim$ be the semigroup of $\mathscr{B}$, then each $t\in T$ corresponds to a set of elements from S. The function g will send an element $s\in S$ to the equivalence class it belongs to, and this is an element of T. Similarly f will send an element of Q to the τ-class it belongs to and the result is a homomorphism of $\mathscr{A}$ onto $\mathscr{B}$. In many cases the homomorphism (f, g) allows us to transfer information about the transformation semigroup $\mathscr{A}$ to the transformation semigroup $\mathscr{B}$ although it is not always a straightforward procedure. For example the skeleton $\mathbf{I}(\mathscr{B})$ can be constructed using $\mathbf{I}(\mathscr{A})$ and (f, g). Similarly a derived sequence may be induced on $\mathscr{B}$ from $\mathscr{A}$, but it is no longer so well behaved and this complicates matters. Similarly a height function on $\mathscr{A}$ may be used in some cases to define a height function on $\mathscr{B}$ but many difficulties arise in this theory. Some of these questions are examined in the exercises at the end of this chapter. We can now look at some examples.

Example 4.7

Let $\mathscr{A}=(Q, S)$ be the transformation semigroup defined by the following graph:

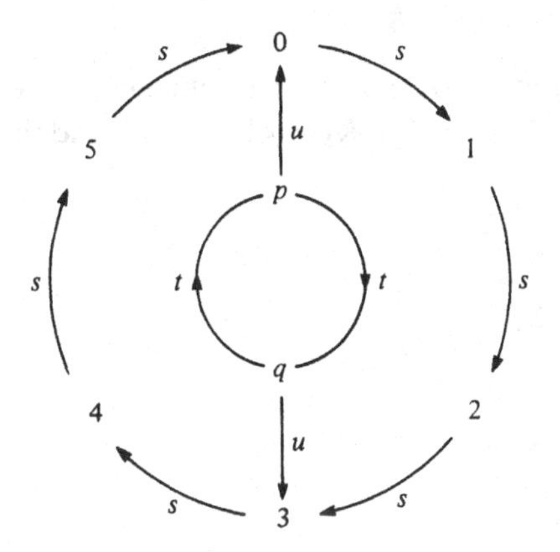

The skeleton is

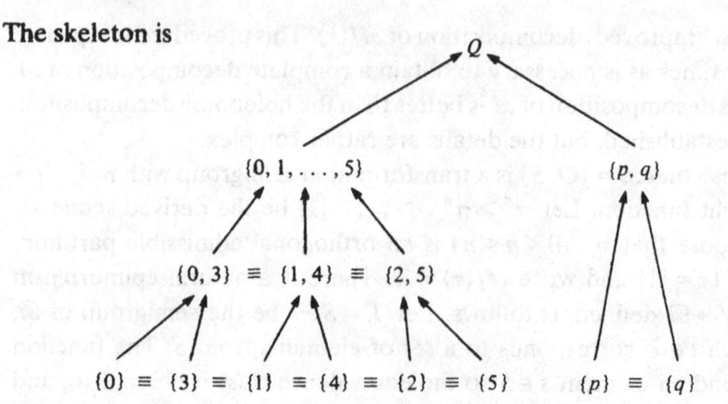

(with $\varnothing$ omitted).
If we choose the minimal height function then $h(Q)=3$ and

$$\pi^2=\{\{0,1,\dots,5\},\{p,q\}\},\ \pi^1=\{\{0,3\},\{1,4\},\{2,5\},\{p\},\{q\}\}.$$

The holonomy decomposition is

$$\mathcal{A}\leq\overline{\mathcal{H}(\{0,3\})}\circ\overline{\mathcal{H}(\{0,1,\dots,5\})\vee\mathcal{H}(\{p,q\})}\circ\overline{\mathcal{H}(Q)}.$$

Now

$$\mathcal{H}(Q)=(\{\{0,1,\dots,5\},\{p,q\}\},\varnothing)\cong\mathbf{2}$$
$$\mathcal{H}(\{0,1,\dots,5\})=(\{\{0,3\},\{1,4\},\{2,5\}\},\{s,s^2,s^3\})\cong\mathbb{Z}_3$$
$$\mathcal{H}(\{p,q\})=(\{\{p\},\{q\}\},\{t,t^2\})\cong\mathbb{Z}_2$$
$$\mathcal{H}(\{0,3\})=(\{\{0\},\{3\}\},\{s^3,s^6\})\cong\mathbb{Z}_2$$

and so

$$\mathcal{A}\leq\bar{\mathbb{Z}}_2\circ(\overline{\mathbb{Z}_2\vee\mathbb{Z}_3})\circ\bar{\mathbf{2}}.$$

Now notice, however, that π^2 is a partition and if $\tau=\{\{0,p\},\{3,q\},\{1\},\{2\},\{4\},\{5\}\}$ then τ is an admissible partition and $\tau\cap\pi^2=\{1\}$. Hence we may deduce that

$$\mathcal{A}\leq\mathcal{A}/\langle\tau\rangle\times\mathcal{A}/\langle\pi^2\rangle\leq\mathcal{A}/\langle\tau\rangle\times\bar{\mathbf{2}}.$$

Now we consider the skeleton of $\mathcal{A}/\langle\tau\rangle$. For convenience we will put $K_1=\{0,p\}$, $K_2=\{3,q\}$, $K_3=\{1\}$, $K_4=\{2\}$, $K_5=\{4\}$, $K_6=\{5\}$ and the state table for $\mathcal{A}/\langle\tau\rangle$ is:

	s	t	u
K_1	K_3	K_2	K_1
K_2	K_5	K_1	K_2
K_3	K_4	$\varnothing$	$\varnothing$
K_4	K_2	$\varnothing$	$\varnothing$
K_5	K_6	$\varnothing$	$\varnothing$
K_6	K_1	$\varnothing$	$\varnothing$

The skeleton is:

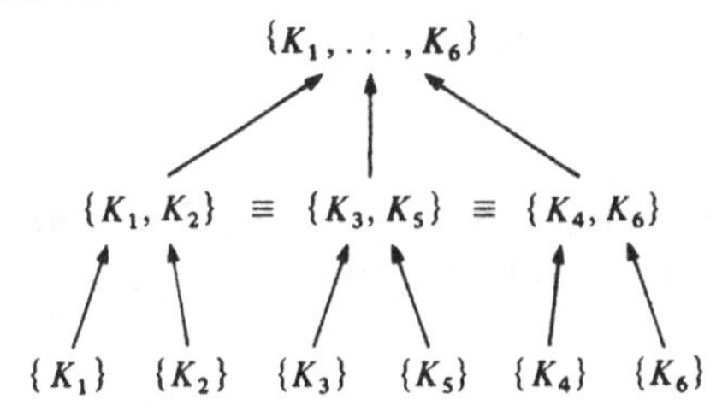

The first derived admissible subset system is:

$$\rho^1 = \{\{K_1, K_2\}, \{K_3, K_5\}, \{K_4, K_6\}\},$$

since there is only one height function available and that yields $h(\mathscr{A}/\langle\tau\rangle) = 2$. Now ρ^1 is an orthogonal partition since the partition $\xi = \{\{K_1, K_4, K_5\}, \{K_2, K_3, K_6\}\}$ is admissible and $\xi \cap \rho' = \{1\}$. Therefore

$$\mathscr{A}/\langle\tau\rangle \leq (\mathscr{A}/\langle\tau\rangle)/\langle\xi\rangle \times (\mathscr{A}/\langle\tau\rangle)/\langle\rho'\rangle$$

and

$$(\mathscr{A}/\langle\tau\rangle)/\langle\rho'\rangle \leq \overline{\mathscr{H}_2^{\vee}(\mathscr{A}/\langle\tau\rangle)} \cong \overline{(\rho', \{s, s^2, s^3\})}$$
$$\cong \bar{\mathbb{Z}}_3.$$

Also, $(\mathscr{A}/\langle\tau\rangle)/\langle\xi\rangle$ is

u $\{K_1, K_4, K_5\}$ $\xrightarrow{s,\,t}$ $\xleftarrow{s,\,t}$ $\{K_2, K_3, K_6\}$ u

and so $(\mathscr{A}/\langle\tau\rangle)/\langle\xi\rangle \leq \bar{\mathbb{Z}}_2$.

Hence $\mathscr{A}/\langle\tau\rangle \leq \bar{\mathbb{Z}}_2 \times \bar{\mathbb{Z}}_3$ and $\mathscr{A} \leq \bar{\mathbb{Z}}_2 \times \bar{\mathbb{Z}}_3 \times \bar{\mathbf{2}}$.

Decomposing $\mathscr{A}$ with respect to another height function does not really alter things much although the basic holonomy decomposition will not be the same. For example, let $h_1 : \mathbf{I}(\mathscr{A}) \to \mathbb{Z}$ be defined by $h_1(Q) = 4$, $h_1(\{0, 1, \ldots, 5\}) = 3$, $h_1(\{p, q\}) = 2$, $h_1(\{0, 3\}) = 1$ etc. giving us $\mathscr{A} \leq \bar{\mathbb{Z}}_2 \circ \bar{\mathbb{Z}}_2 \circ \bar{\mathbb{Z}}_3 \circ \bar{\mathbf{2}}$ for the basic decomposition, but our improved decomposition is $\mathscr{A} \leq \bar{\mathbb{Z}}_2 \times \bar{\mathbb{Z}}_3 \times \bar{\mathbf{2}}$ since we have not changed τ when changing the height function. Theorem 3.4.3 enables us to replace each transformation group of the form $(\mathbf{M}(A), \mathbf{H}(A))$ by the wreath product $(\mathbf{M}(A), \varnothing)^{\cdot} \circ (\mathbf{H}(A), \mathbf{H}(A))$ and so we have $\mathscr{A} \leq (\bar{\mathbf{2}}^{\cdot} \circ \mathbb{Z}_2) \times (\bar{\mathbf{3}}^{\cdot} \circ \mathbb{Z}_3) \times \bar{\mathbf{2}}$. Now apply the identity $(\mathscr{A} \circ \mathscr{B}) \times (\mathscr{C} \circ \mathscr{D}) \leq (\mathscr{A} \times \mathscr{C}) \circ (\mathscr{B} \times \mathscr{D})$ to obtain

$$\begin{aligned}
\mathscr{A} &\leq [(\bar{\mathbf{2}}^{\cdot} \times \bar{\mathbf{3}}^{\cdot}) \circ (\mathbb{Z}_2 \times \mathbb{Z}_3)] \times \bar{\mathbf{2}} \\
&= [(\bar{\mathbf{2}}^{\cdot} \times \bar{\mathbf{3}}^{\cdot}) \circ (\mathbb{Z}_2 \times \mathbb{Z}_3)] \times (\bar{\mathbf{2}} \circ \mathbf{1}^{\cdot}) \\
&\leq (\bar{\mathbf{2}}^{\cdot} \times \bar{\mathbf{3}}^{\cdot} \times \bar{\mathbf{2}}) \circ (\mathbb{Z}_2 \times \mathbb{Z}_3 \times \mathbf{1}^{\cdot}) \\
&= (\bar{\mathbf{2}}^{\cdot} \times \bar{\mathbf{3}}^{\cdot} \times \bar{\mathbf{2}}) \circ \mathbb{Z}_6.
\end{aligned}$$

This decomposition is in the form of a wreath product $\mathcal{A} \circ \mathcal{G}$ where $\mathcal{A}$ is aperiodic and $\mathcal{G}$ is a transformation group. Thus $\mathbf{C}(\mathcal{A}) \leq 1$.

Example 4.8

Consider the transformation semigroup $\mathcal{A} = (Q, S)$ where $Q = \{1, 2, 3\}$, $S = \{\alpha, \beta, \alpha^2, \beta^2, \beta\alpha, \beta^2\alpha\}$ and with generating graph:

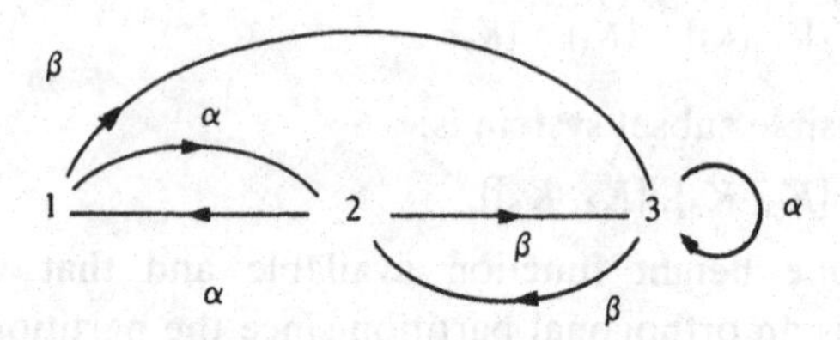

The skeleton is:

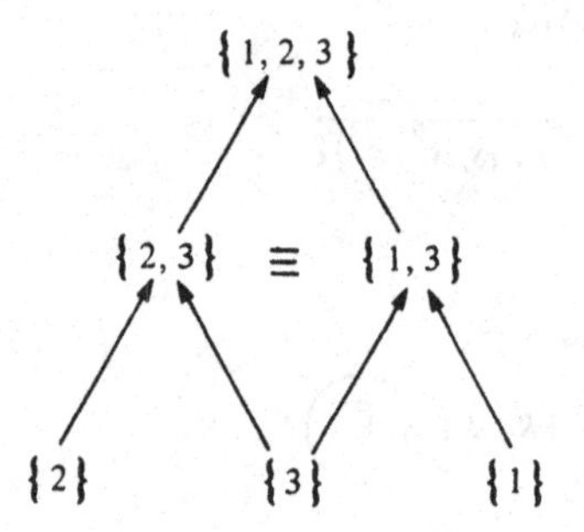

and so the height is 2. Then $\pi' = \{\{2, 3\}, \{1, 3\}\}$ which is not a partition and so the basic holonomy decomposition theorem must be used.

$$\mathcal{H}_n^\vee(\mathcal{A}) = \mathcal{H}(Q) = (\{\{2, 3\}, \{1, 3\}\}, \{\alpha, \alpha^2\}) \cong \mathbb{Z}_2,$$

$$\mathcal{H}(\{2, 3\}) = (\{\{2\}, \{3\}\}, \{\beta, \beta^2\}) = \mathbb{Z}_2 = \mathcal{H}_1^\vee(\mathcal{A}).$$

Thus $\mathcal{A} \leq \bar{\mathbb{Z}}_2 \circ \bar{\mathbb{Z}}_2$.

Example 4.9

Let $\mathcal{A} = (Q, S)$ where $Q = \{0, 1, 2, 3\}$, $S = \{a, b, c, d\}$ given by the table:

	0	1	2	3
a	0	2	0	2
b	2	0	2	0
c	1	3	1	3
d	3	1	3	1

The skeleton is:

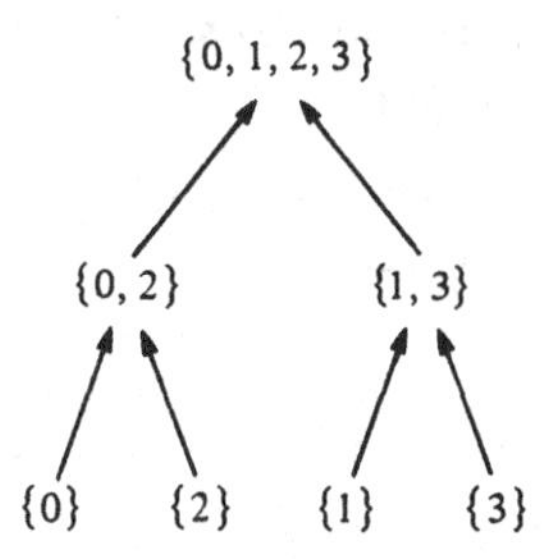

and so $h(Q)=2$. Then $\pi'=\{\{0,2\},\{1,3\}\}$ which is a partition but is not orthogonal. Therefore we apply the basic holonomy decomposition theorem.

$$\mathcal{H}_2(\mathcal{A})=\mathcal{H}(Q)=(\{\{0,2\},\{1,3\}\},\varnothing)\cong\mathbf{2}$$

$\mathcal{H}(\{0,2\})=(\{\{0\},\{2\}\},\varnothing)=\mathbf{2}=\mathcal{H}(\{1,3\})$ and so $\mathcal{H}_1^{\vee}(\mathcal{A})=\mathbf{2}\vee\mathbf{2}$ and

$$\mathcal{A}\le(\overline{\mathbf{2}\vee\mathbf{2}})\circ\bar{\mathbf{2}}.$$

Example 4.10

We now consider the cyclic transformation semigroup $\mathscr{C}_{(p,r)}$. This may be defined by the diagram:

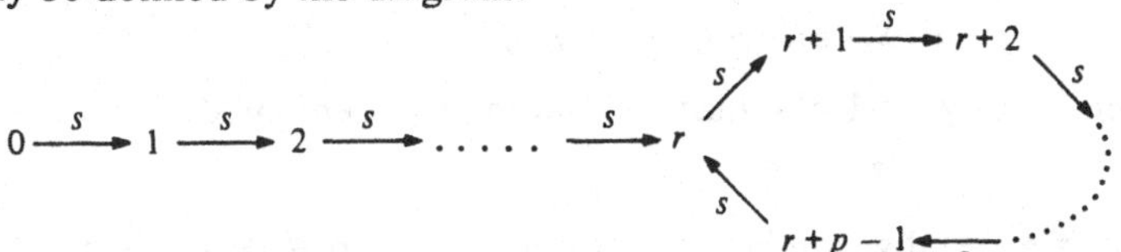

The skeleton is:

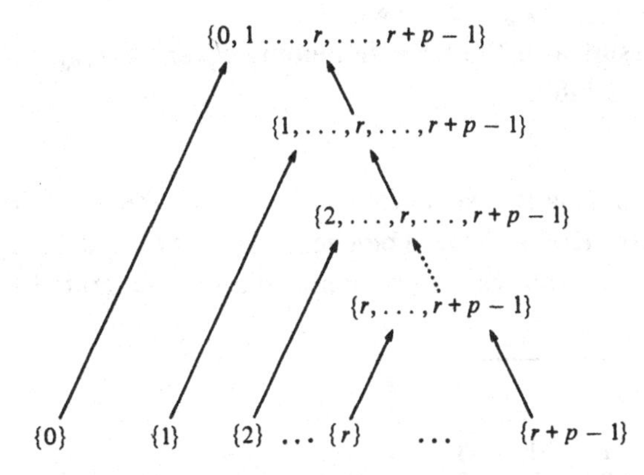

The height of the transformation semigroup is $r+1$ if $p>1$, and is r if $p=1$. The derived sequence when $p>1$ is easily calculated from the skeleton and the one of particular interest is $\pi^1 = \{\{0\}, \{1\}, \ldots, \{r-1\}, \{r, \ldots, r+p-1\}\}$. This is an orthogonal partition. To see this suppose that $r > p$ and $r = qp + t$ where $0 \leq t < p$, and put

$$\begin{aligned}\tau = \{&\{0, p, \ldots, (q+1)p\}, \{1, p+1, \ldots, (q+1)p+1\}, \ldots,\\ &\{t-1, p+t-1, \ldots, r+p-1\}, \{t, p+t, \ldots, r\},\\ &\{t+1, p+t+1, \ldots, qp+t+1\}, \ldots,\\ &\{p-1, 2p-1, \ldots, (q+1)p-1\}\}.\end{aligned}$$

Then $\pi^1 \cap \tau = \{1\}$ and τ is admissible. Then

$$\mathscr{C}_{(p,r)} \leq \mathscr{A}/\langle\tau\rangle \times \mathscr{A}/\langle\pi^1\rangle.$$

Clearly $\mathscr{A}/\langle\tau\rangle \cong \mathbf{Z}_p$ and $\mathscr{A}/\langle\pi'\rangle = \mathscr{C}_{(1,r)}$ and so $\mathscr{C}_{(p,r)} \leq \mathbf{Z}_p \times \mathscr{C}_{(1,r)}$.

If $r = p$ put $\tau = \{\{0, r\}, \{1, r+1\}, \ldots, \{r-1, 2r-1\}\}$ then $\pi^1 \cap \tau = \{1\}$ and τ is admissible. In this case

$$\mathscr{C}_{(p,r)} \leq \mathbf{Z}_r \times \mathscr{C}_{(1,r)}$$

as before.

Finally let $r < p$ and choose $\tau = \{\{r\}, \{r+1\}, \ldots, \{p-1\}, \{0, p\}, \{1, p+1\} \ldots \{r-1, r+p-1\}\}$.

As before $\pi^1 \cap \tau = \{1\}$ and τ is admissible, hence $\mathscr{A} \leq \mathbf{Z}_p \times \mathscr{C}_{(1,r)}$.

In all cases the partition τ is defined by the relation

$$q \sim q' \Leftrightarrow q' \equiv q (\mathrm{mod}\, p), \quad q, q' \in Q.$$

Then for any $r \geq 1$, the basic holonomy theorem yields

$$\mathscr{C}_{(1,r)} \leq \bar{\mathbf{2}}^{r-1} \circ \mathscr{C}.$$

(The $\mathscr{C}$ arises if we replace the holonomy group $\mathscr{H}_r(\mathscr{C}_{(1,r)})$ by the quotient of the first element of the derived sequence – we use theorem 4.3.9(i).) Finally we obtain $\mathscr{C}_{(p,r)} \leq \mathbf{Z}_p \times (\bar{\mathbf{2}}^{r-1} \circ \mathscr{C})$.

Contrast this result with the basic holonomy covering $\mathscr{C}_{(p,r)} \leq \bar{\mathbf{Z}}_p \circ \bar{\mathbf{2}}^r$, which is clearly very inferior.

Example 4.11

This transformation semigroup arises from the study of the tricarboxylic acid (or Krebs) cycle in biochemistry. Let $Q = \{0, 1, 2, 3, 4\}$, and let $\mathscr{A} = (Q, S)$ be the transformation semigroup generated by the table:

	0	1	2	3	4
a	1	1	2	3	4
b	0	2	3	3	0
c	0	1	2	4	4

The transformation semigroup is irreducible and so we must apply the basic holonomy decomposition. The skeleton is:

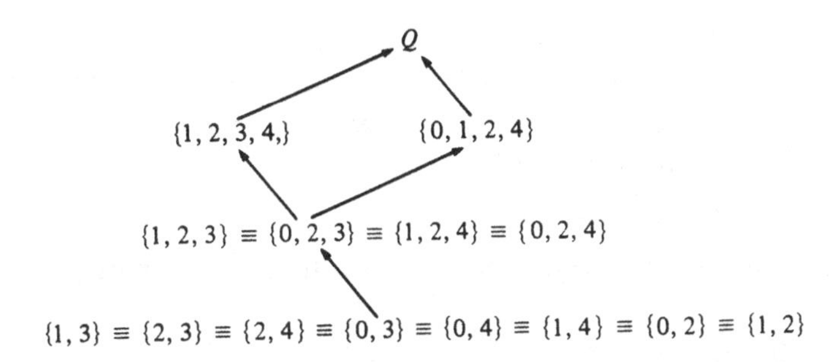

where the singletons and $\varnothing$ are omitted and also some of the lines indicating the $\leqslant$ relation are left out for clarity. The minimal height function h yields $h(Q)=4$. The derived sequence is

$$\pi^3=\{\{1,2,3,4\},\{0,1,2,4\},\{0,2,3\}\}$$

$$\pi^2=\{\{1,2,3\},\{1,2,4\},\{0,2,3\},\{1,2,4\},\{0,2,4\}\}$$

$$\pi^1=\{\{1,3\},\{2,3\},\{1,2\},\{0,3\},\{2,4\},\{1,4\},\{0,4\},\{0,2\}\}.$$

Now $\mathscr{A}/\langle\pi^3\rangle\cong\bar{\mathbf{3}}$, and the holonomy groups are given by:

$$\mathscr{H}_3^\vee(\mathscr{A})=\mathscr{H}(\{1,2,3,4\})\vee\mathscr{H}(\{0,1,2,4\})\cong\mathbf{2}^\cdot\vee\mathbf{2}^\cdot,$$

$$\mathscr{H}_2^\vee(\mathscr{A})\cong\mathscr{H}(\{0,2,3\})\cong\mathbb{Z}_3,\ \mathscr{H}_1^\vee(\mathscr{A})\cong\mathscr{H}(\{0,3\})\cong\mathbb{Z}_2$$

and so

$$\mathscr{A}\leq\bar{\mathbb{Z}}_2\circ\bar{\mathbb{Z}}_3\circ(\bar{\mathbf{2}}^\cdot\vee\bar{\mathbf{2}}^\cdot)\circ\bar{\mathbf{3}}.$$

Does this decomposition have a biochemical interpretation?

4.5 The Krohn–Rhodes decomposition

The first major result in the algebraic decomposition theory of transformation semigroups was due to Krohn and Rhodes and appeared in 1965. Since then many versions of this result have appeared. Some are set in the context of state machines, some in the context of Mealy machines and of course some occur in treatments of the theory of transformation semigroups. We will state the most relevant form of the theorem here.

Theorem 4.5.1 (Krohn–Rhodes decomposition theorem)

Let $\mathscr{A}=(Q,S)$ be a transformation semigroup. Then $\mathscr{A}$ may be covered by a finite wreath product of transformation semigroups of the following two types:

(i) aperiodic transformation semigroups,
(ii) transformation groups (G, G) where G is a finite simple group and $G|S$.

Proof Choose a maximal height function $h: \mathbf{I}(\mathscr{A}) \to \mathbb{Z}$ for $\mathscr{A}$. We first suppose that $S \neq \varnothing$. Now $\mathscr{H}_i^\vee(\mathscr{A}) = \mathscr{H}(A)$ for some $A \in \mathbf{I}(\mathscr{A})$. By the basic holonomy theorem (4.3.7)

$$\mathscr{A} \leq \overline{\mathscr{H}_1^\vee(\mathscr{A})} \circ \overline{\mathscr{H}_2^\vee(\mathscr{A})} \circ \ldots \circ \overline{\mathscr{H}_n^\vee(\mathscr{A})}$$

where $n = h(Q)$. Applying theorem 3.4.3 to each component $\overline{\mathscr{H}_i^\vee(\mathscr{A})}$ yields

$$\overline{\mathscr{H}_i^\vee(\mathscr{A})} \leq \overline{(P_i, \varnothing)}^{\cdot} \circ \mathscr{H}_i$$

where $\mathscr{H}_i^\vee(\mathscr{A}) = (P_i, H_i)$, P_i is the maximal image space of the element of height i in the skeleton and H_i is the holonomy group of this element. We now apply example 3.3 to get $\overline{(P_i, \varnothing)}^{\cdot} \leq \prod^k \bar{\mathbf{2}}^{\cdot}$ where $|P_i| \leq 2^k$ and theorem 3.5.4 to get $\mathscr{H}_i \leq \mathscr{G}_{i1} \circ \ldots \circ \mathscr{G}_{im}$ where each G_{ij} is a finite simple group, $j = 1, \ldots, m$ and $G_{ij} | H_i$. Finally note that $H_i | S$ by the construction of the holonomy groups.

In the case where $S = \varnothing$ then $\mathscr{A} = (Q, \varnothing) \leq \prod^k \bar{\mathbf{2}}^{\cdot}$ by example 3.3. □

The corresponding theorem for state machines says that any state machine may be covered by direct and cascade products involving two-state reset machines and simple grouplike state machines whose groups are covered by the semigroup of the original machine. Proofs of theorems of this type were developed by Zeiger and then Ginzburg (see Ginzburg [1968].) Their approach was briefly sketched in the opening paragraphs of this chapter, but it has the serious disadvantage of leading to a very inefficient and lengthy decomposition process. For example, their original method would require the use of a computer to obtain a decomposition of the example 4.11.

The development of the holonomy decomposition theorem was achieved by Eilenberg [1976]. His original definition of a height function contained a small inadequacy (condition (iii)). This failing was indicated to me by T. Keville [1978]. The approach we take here, using the derived sequence, is not very far removed from Eilenberg's own treatment. The advantage of our method occurs in the 'improvement' of the holonomy decomposition in section 4.4.

The applications of this theory are likely to prove important in all those areas where automata theory can be used to model discrete systems. We have already noted how an important metabolic pathway can be compared with some simple cyclic groups and aperiodic transformation semigroups. There are many other examples, in the biological and psychological sciences, and also in computer science and engineering.

All the results of this chapter may be applied to 'pure' semigroup theory also. Recall that if S is any semigroup then $(S^{\cdot}, S)$ is a transformation semigroup, and so the decomposition theory may be applied to $(S^{\cdot}, S)$. However, it is often possible to obtain better results, in this situation, by using the internal structure and techniques available in semigroup theory. In particular, the skeleton is essentially just the collection of principal left ideals of S with the appropriate relation. Since there are many other algebraic structures in a semigroup, we may as well make use of them. Of particular note is the depth decomposition theorem for semigroups due to Tilson. (See Eilenberg [1976].) This often leads to a shorter decomposition of a semigroup compared to the basic holonomy decomposition applied to the semigroup.

4.6 Exercises

4.1 Prove that the minimal height function defined in section 4.2 satisfies all the requirements of the definition of a height function.

4.2 Calculate the skeleton and the height function for the transformation semigroup (Q, S) represented by:

	1	2	3
a	1	1	1
b	1	2	3

where $Q=\{1, 2, 3\}$, $S=\{a, b\}$.

4.3 Construct the skeleton and the derived sequence for the transformation semigroup $\bar{\mathbf{2}}^{\cdot} \vee \bar{\mathbf{2}}^{\cdot}$. Show that π^1 is orthogonal and hence establish that $\bar{\mathbf{2}}^{\cdot} \vee \bar{\mathbf{2}}^{\cdot} \leq \bar{\mathbf{2}}^{\cdot} \times \mathbf{2}^{\cdot}$.

4.4 Decompose example 4.4 and hence show that this transformation semigroup may be covered by a wreath product of aperiodic transformation semigroups and symmetric permutation groups.

4.5 Let $\mathscr{A}=(Q, S)$ be a state machine. Show that $\mathbf{I}(\mathscr{A})=\mathbf{I}(\mathscr{A}^{\cdot})=\mathbf{I}(\bar{\mathscr{A}})$ and $\mathbf{I}(\mathscr{A}^c)=\{B \cup \{z\} \mid B \in \mathbf{I}(\mathscr{A})\}$. If $A \in \mathbf{I}(\mathscr{A})$ show that the holonomy transformation group of A regarded as an image of $\mathscr{A}^{\cdot}$ equals $\mathscr{H}(\mathscr{A})^{\cdot}$ and the holonomy transformation group of $\mathscr{A}$ regarded as an image of $\bar{\mathscr{A}}$ equals $\mathscr{H}(\mathscr{A})$. If $A \in \mathbf{I}(\mathscr{A}^c)$ and $h(A) > 1$ then $\mathbf{H}(\mathscr{A}) = \mathbf{H}(B)$ where $A = B \cup \{z\}$ and $B \in \mathbf{I}(\mathscr{A})$. If $h(A) = 1$ then $\mathbf{H}(A) = \mathbf{H}(B) + \mathbf{1}$.

4.6 Let τ be an admissible partition on the transformation semigroup $\mathscr{A} = (Q, S)$ and $(f, g): \mathscr{A} \to \mathscr{A}/\langle\tau\rangle$ the natural epimorphism. Show that $\mathbf{I}(\mathscr{A}/\langle\tau\rangle) = f(\mathbf{I}(\mathscr{A}))$ (as sets).

4.7 Let $h: I(\mathscr{A}) \to \mathbb{Z}$ be a height function with $h(Q) = n$, and $\pi^n > \pi^{n-1} > \ldots > 1$ the derived sequence. Suppose that π^i is an orthogonal partition for some $1 < i < n$ and let $\pi^i \cap \tau = 1$ with τ an admissible partition. Let $(f, g): \mathscr{A} \to \mathscr{A}/\langle\tau\rangle$ be the natural epimorphism. Show that $f(\pi^i) \geq f(\pi^{i-1}) \geq \ldots \geq f(1)$ is a sequence of admissible subset systems in $\mathscr{A}/\langle\tau\rangle$.

4.8 With the notation of 4.7 define a function $k: \mathbf{I}(\mathscr{A}/\langle\tau\rangle) \to \mathbb{Z}$ by $k(B) = \inf\{h(A) \mid A \in \mathbf{I}(\mathscr{A}),\ f(A) = B,\ B \in \mathbf{I}(\mathscr{A}/\langle\tau\rangle)\}$. Prove that $k(B) = \inf\{j \mid B \in f(\pi^j)\}$.

5

Recognizers

We have seen how Mealy machines can be used to model the connections between inputs and outputs of complex systems, and how to decompose the underlying state machines that are central to this procedure. There is another area in which state machines play a major role. In the development of computer systems it is important to distinguish between certain *sequences* of inputs. The computer must be able to recognize those instructions that are compatible with its system and these instructions will take the form of input words from an input alphabet.

This chapter is concerned with the mathematical theory of recognizers; these are state machines that are able to *discriminate* between two disjoint sets of input words. The foundations for this theory, initially developed by S. Ç. Kleene in 1956, had an important influence on the construction of compilers for computers. It is also of independent mathematical interest and is closely related to the study of languages and psycholinguistics.

5.1 Automata or recognizers

Let $\mathcal{M} = (Q, \Sigma, F)$ be a state machine (as usual Q and Σ are finite and F is a partial function, $F: Q \times \Sigma \to Q$). Let $i \in Q$ be a fixed state called the *initial state* and suppose that $T \subseteq Q$ is a set of states called the *set of terminal states.*

The collection $\mathfrak{M} = (\mathcal{M}, i, T)$ is called an *automaton* or a *recognizer.* We will use the second term since automaton is often used as a generic noun to describe all types of machine. The recognizer is able to distinguish between certain types of word from the monoid Σ^*. For example, let $\alpha \in \Sigma^*$, then $iF_\alpha \in Q$ or $iF_\alpha = \varnothing$ and we say that $\mathfrak{M}$ *recognizes* α if and only if $iF_\alpha \in T$. The set Σ^* is partitioned into two disjoint subsets, the set of words recognized by $\mathfrak{M}$ and the set of words not recognized by $\mathfrak{M}$. The set of words of Σ^* recognized by $\mathfrak{M}$ is called the *behaviour* of $\mathfrak{M}$ and is denoted by $|\mathfrak{M}|$. Thus $|\mathfrak{M}| = \{\alpha \in \Sigma^* \mid iF_\alpha \in T\}$.

One major aim is the characterization of the subsets of Σ^* that can arise as the behaviour of a recognizer. We shall see that some subsets of Σ^* can never be the behaviour of a recognizer. Another fact that will soon become apparent is that different recognizers can have the same behaviour.

We need some straightforward notation for describing subsets of Σ^*. Let $A \subseteq \Sigma^*$, $B \subseteq \Sigma^*$, with $A \neq \emptyset$, $B \neq \emptyset$, we define

$$A \cdot B = \{\alpha \in \Sigma^* \mid \alpha = ab;\ a \in A,\ b \in B\},$$

$$A^+ = \{\alpha \in \Sigma^* \mid \alpha = a_1 \cdot \ldots \cdot a_n;\ a_i \in A,\ 1 \leq i \leq n,\ n > 0\},$$

$$A^* = A^+ \cup \{\Lambda\}.$$

Examples 5.1

These examples will be described by using directed graphs to describe the recognizer with the initial state indicated by a bold arrow, unlabelled, and pointing towards the state. The terminal states are shown with a bold arrow, unlabelled, pointing away from the state. Let $\Sigma = \{0, 1\}$ in all of these examples.

0, 1

(i) q

Here $Q = T = \{q\}$. The initial state is also q. Any word from Σ^* will be recognized by this machine and so $|\mathfrak{M}| = \Sigma^*$.

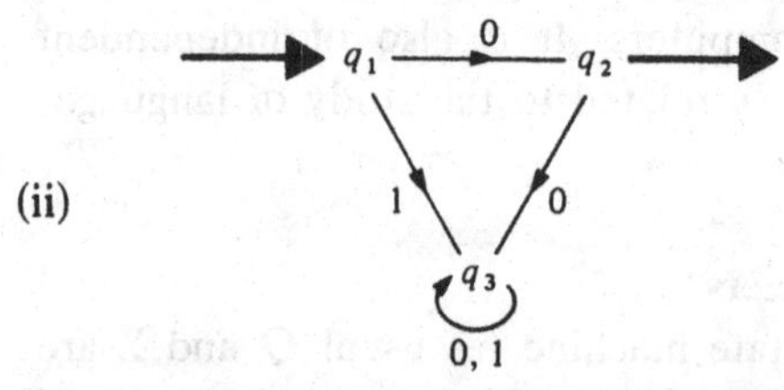

Then the behaviour, $|\mathfrak{M}|$, is $\{0\}$. (Notice that 01 is not recognized since $q_1F_{01} = \emptyset$.)

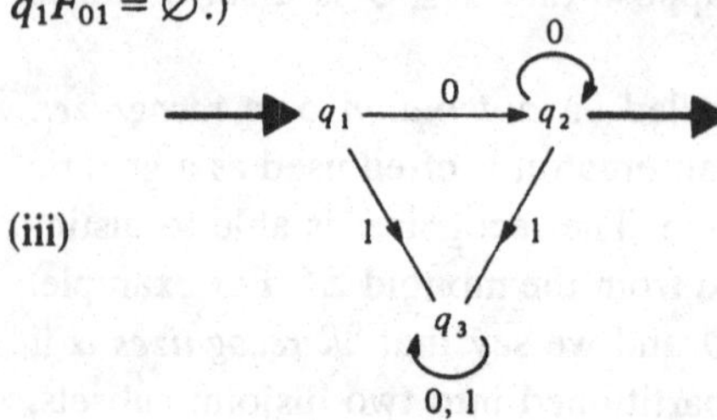

This recognizer has behaviour $\{0\}^+$ (which can be written as $\{0\}^* \cdot \{0\}$ or $\{0\} \cdot \{0\}^*$ or $\{0\}^* \backslash \{\Lambda\}$).

(iv)

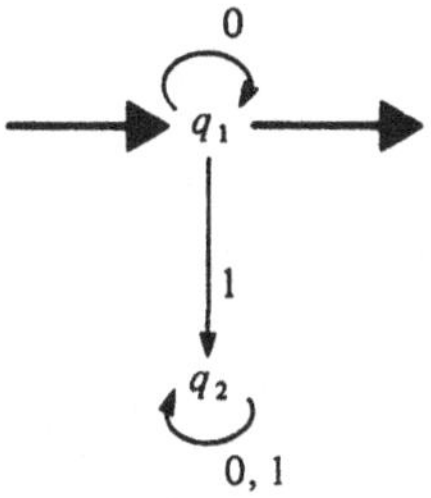

This has behaviour $\{0\}^*$.

(v)

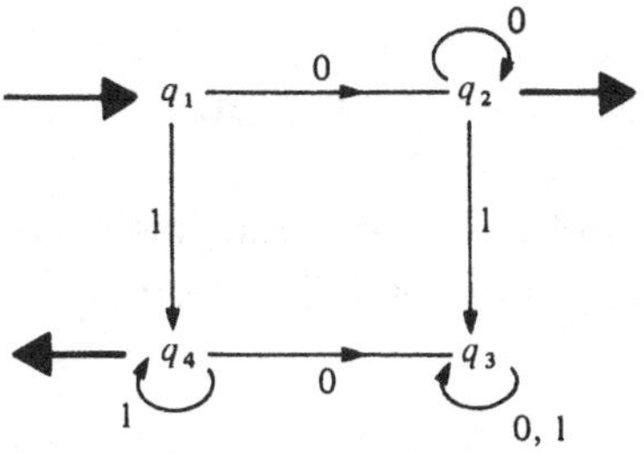

The behaviour of this machine is $\{0^+\} \cup \{1\}^+$.

(vi)

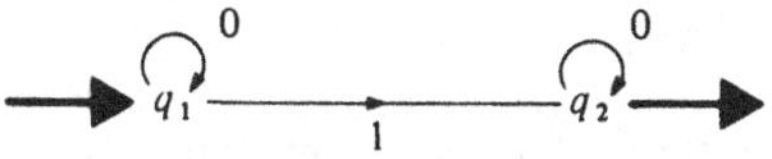

The behaviour is $\{0\}^* \cdot \{1\} \cdot \{0\}^*$, that is all words in Σ^* containing precisely one occurrence of 1.

(vii)

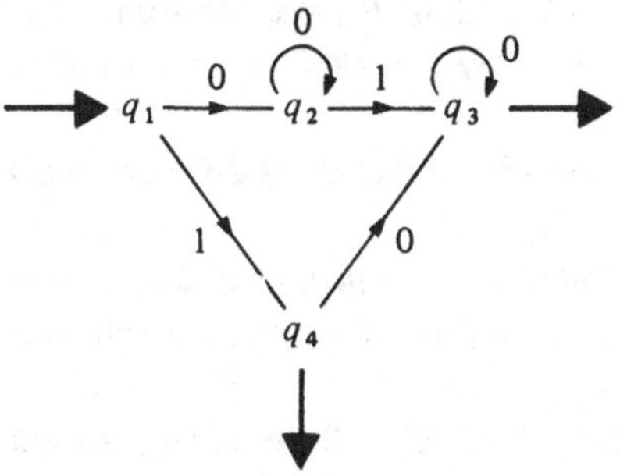

This has behaviour $\{0\}^+ \cdot \{1\} \cdot \{0\}^* \cup \{1\} \cup \{1\} \cdot \{0\}^+$ which is the set of all words of Σ^* containing precisely one occurrence of 1, that is the behaviour is equal to $\{0\}^* \cdot \{1\} \cdot \{0\}^*$ as in (vi).

(viii)

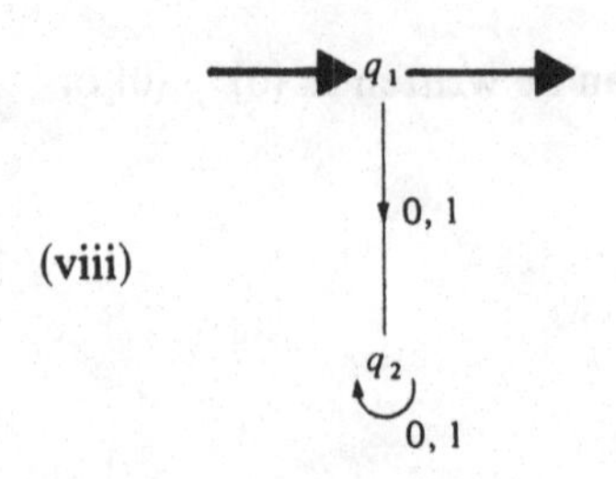

This recognizer has behaviour $\{\Lambda\}$.

(ix) $\longrightarrow q_1 \circlearrowleft$ 0, 1 $\quad q_2 \longrightarrow$

The behaviour of this recognizer is $\varnothing$.

We will meet other examples as we proceed further.

The concept of the completion of a state machine has been studied in chapter 2. We can immediately deduce the following result:

Theorem 5.1.1

Let $\mathcal{M} = (Q, \Sigma, F)$ be a state machine, $i \in Q$ and $T \subseteq Q$. Consider the completion $\mathcal{M}^c$ of $\mathcal{M}$, and let $\mathfrak{M}^c = (\mathcal{M}^c, i, T)$, then

$|\mathfrak{M}^c| = |\mathfrak{M}|$.

Proof We assume first that $\mathcal{M}$ is not complete and let $\mathcal{M}^c = (Q \cup \{x\}, \Sigma, F')$ where $x \notin Q$, $xF'_\sigma = x$, $qF'_\sigma = x$ if $qF_\sigma = \varnothing$ and $qF'_\sigma = qF_\sigma$ if $qF_\sigma \neq \varnothing$. We now assume that $|\mathfrak{M}| \neq \varnothing$. Let $\alpha \in |\mathfrak{M}|$, then $iF_\alpha \in T$. Since $iF_\alpha \neq \varnothing$ we may deduce that $iF'_\alpha = iF_\alpha \in T$ and so $\alpha \in |\mathfrak{M}^c|$. Now let $\alpha \in |\mathfrak{M}^c|$, then $iF'_\alpha \in T$. If $iF'_\alpha \neq iF_\alpha$ then $iF'_\beta = x$ for some $\beta \in \Sigma^*$ such that $\alpha = \beta\gamma$, $\gamma \in \Sigma^*$. But then $iF'_\alpha = xF'_\gamma = x \notin T$ and so $\alpha \notin |\mathfrak{M}^c|$. Hence $|\mathfrak{M}^c| = |\mathfrak{M}|$.

If $|\mathfrak{M}| = \varnothing$ then $iF_\alpha \notin T$ for any $\alpha \in \Sigma^*$ and so $iF'_\alpha \notin T$ for any $\alpha \in \Sigma^*$. $\square$

The recognizer $\mathfrak{M}^c$ will be called the *completion* of $\mathfrak{M}$. It is clear from theorem 5.1.1 that we will lose very little if we concentrate our studies on complete recognizers.

Let Σ be a finite set, a subset A of Σ^* will be called *recognizable* if there exists a recognizer $\mathfrak{M}$ such that A is the behaviour of $\mathfrak{M}$, that is if $A = |\mathfrak{M}|$ for some recognizer $\mathfrak{M}$. We say that $\mathfrak{M}$ *recognizes A*.

Given a recognizable subset A it is usually possible to find many distinct recognizers that recognize A and one of our tasks in the next section is to construct a standard complete recognizer that recognizes A and is also the 'most efficient' recognizer with this property. We will now explain the term 'most efficient'.

Intuitively a recognizer $\mathfrak{M}$ recognizing the set $A \subseteq \Sigma^*$ would be considered efficient if there were no 'wasted states'. For example, suppose that $\mathfrak{M} = (\mathcal{M}, i, T)$ where $\mathcal{M} = (Q, \Sigma, F)$ and consider the set of states

$$R = \{iF_\alpha \mid \alpha \in \Sigma^*\}.$$

R then consists of all those states of Q that can be reached from the initial state i. These are the only states that can influence the behaviour $|\mathfrak{M}| = A$ and consequently if $R \subsetneq Q$ there will be some states in Q that will never feature in our discussions about A. If $\mathfrak{M}$ has the property that $R = Q$ we will call $\mathfrak{M}$ *accessible.* Given a recognizer $\mathfrak{M} = (\mathcal{M}, i, T)$ we can remove the states in the set $Q \backslash R$, and obtain an accessible recognizer which clearly has the same behaviour as $\mathfrak{M}$. This is called the *accessible part,* $\mathfrak{M}^a$, of $\mathfrak{M}$. Thus

$$\mathfrak{M}^a = (\mathcal{M}^a, i, T)$$

where $\mathcal{M}^a = (R, \Sigma, F^a)$, $R = \{iF_\alpha \mid \alpha \in \Sigma^*\}$, and $qF^a_\alpha = qF_\alpha$ for $q \in R$, $\alpha \in \Sigma^*$. Note that $|\mathfrak{M}^a| = |\mathfrak{M}|$. It is clear that if $\mathfrak{M}$ is complete then $\mathfrak{M}^a$ is also complete.

Another way in which states may be redundant is if there are states in the recognizer that never lead to a terminal state. Thus if $q \in Q$ and $qF_\alpha \notin T$ for all $\alpha \in \Sigma^*$ then q can never lie on a successful 'route' from the initial state i to a final state in T. Consider the set S, of all states that can lead to a terminal state, so that

$$S = \{q \mid qF_\alpha \in T \quad \text{for some } \alpha \in \Sigma^*\}.$$

If $S = Q$ we call $\mathfrak{M}$ *coaccessible.* The *coaccessible* part of a recognizer $\mathfrak{M} = (\mathcal{M}, i, T)$ is defined to be $\mathfrak{M}^b = (\mathcal{M}^b, i, T)$ where

$$\mathcal{M}^b = (S, \Sigma, F^b),$$
$$S = \{q \mid qF_\alpha \in T \text{ for some } \alpha \in \Sigma^*\}$$

and

$$qF^b_\alpha = qF_\alpha \text{ for } q \in S, \alpha \in \Sigma^*.$$

Clearly $|\mathfrak{M}^b| = |\mathfrak{M}|$. A recognizer $\mathfrak{M} = (\mathcal{M}, i, T)$ is called *trim* if it is both accessible and coaccessible.

Example 5.2
Let $\Sigma = \{0, 1\}$, $Q = \{q_1, q_2, q_3, q_4, q_5\}$.

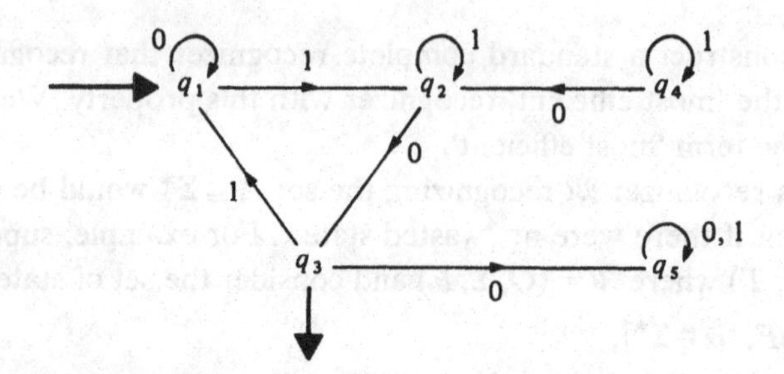

This defines a complete recognizer $\mathfrak{M} = (\mathcal{M}, q_1, \{q_3\})$. Then $\mathfrak{M}^a$ is given by

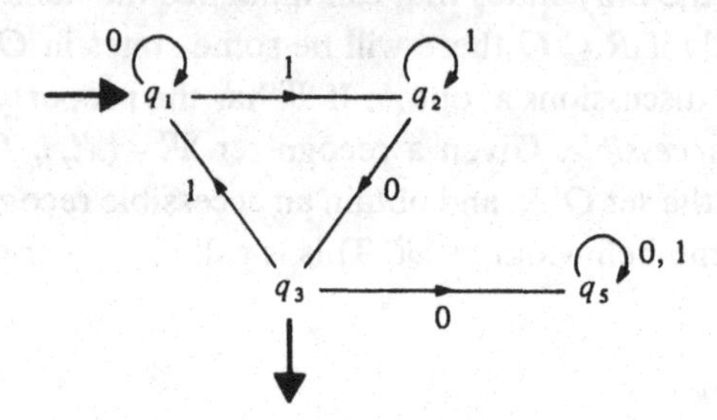

$(\mathfrak{M}^a)^b$ is given by

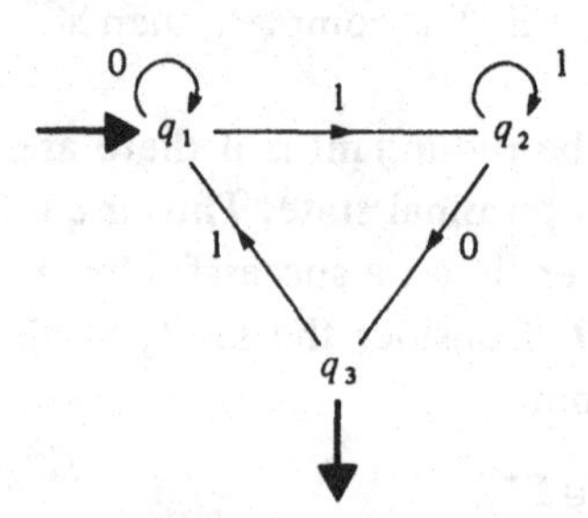

and this is a trim recognizer which satisfies $|(\mathfrak{M}^a)^b| = |\mathfrak{M}|$, but is no longer complete.

We could equally well have constructed $(\mathfrak{M}^b)^a$ and this would have produced the same machine.

Our final task for this section is the introduction of some useful notation.

Let $\mathcal{M} = (Q, \Sigma, F)$ be a complete state machine and suppose that $q \in Q$, $\alpha \in \Sigma^*$; define $q * \alpha = qF_\alpha$, and then for $A \subseteq \Sigma^*$, $S \subseteq Q$ we have:

$$q * A = \{q * \alpha \mid \alpha \in A\}$$

$$S * \alpha = \{q * \alpha \mid q \in S\}$$

$$S * A = \{q * \alpha \mid q \in S, \alpha \in A\}.$$

$$q * \alpha^{-1} = \{p \in Q \mid q = p * \alpha\}$$
$$q * A^{-1} = \{p \in Q \mid q = p * \alpha \text{ for some } \alpha \in A\}$$
$$S * \alpha^{-1} = \{p \in Q \mid p * \alpha \in S\}$$
$$S * A^{-1} = \{p \in Q \mid p * \alpha \in S \text{ for some } \alpha \in A\}.$$

If $A \subseteq \Sigma^*$ and $B \subseteq \Sigma^*$, $a \in A$ and $b \in B$ then define

$$a \cdot b^{-1} = \{\alpha \in \Sigma^* \mid \alpha b = a\}$$
$$a^{-1} \cdot b = \{\alpha \in \Sigma^* \mid a\alpha = b\}$$
$$a^{-1} \cdot B = \{\alpha \in \Sigma^* \mid a\alpha \in B\}$$
$$a \cdot B^{-1} = \{\alpha \in \Sigma^* \mid \alpha b = a \text{ for some } b \in B\}$$
$$A \cdot b^{-1} = \{\alpha \in \Sigma^* \mid \alpha b \in A\}$$
$$A^{-1} \cdot b = \{\alpha \in \Sigma^* \mid a\alpha = b \text{ for some } a \in A\}$$
$$A \cdot B^{-1} = \{\alpha \in \Sigma^* \mid \alpha b \in A \text{ for some } b \in B\}$$
$$A^{-1} \cdot B = \{\alpha \in \Sigma^* \mid a\alpha \in B \text{ for some } a \in A\}.$$

With $\mathcal{M} = (Q, \Sigma, F)$ and $p, q \in Q$ we put $q^{-1} \circ p = \{\alpha \in \Sigma^* \mid p = q * \alpha\}$, that is the set of words that 'send q to p'.

If $R, S \subseteq Q$ then we let

$$q^{-1} \circ R = \{\alpha \in \Sigma^* \mid q * \alpha \in R\}$$
$$S^{-1} \circ R = \{\alpha \in \Sigma^* \mid q * \alpha \in R \text{ for some } q \in S\}.$$

Some elementary results can now be stated, their proof will be left as exercises. Some useful identities are to be found in exercise 5.8.

Proposition 5.1.2

Let $\mathcal{M} = (Q, \Sigma, F)$ be a state machine, $A, B, C \subseteq \Sigma^*$ and $S \subseteq Q$.

(i) $(S * A) * B = S * (A \cdot B)$

(ii) $(S * A^{-1}) * B^{-1} = S * (B \cdot A)^{-1}$

Proposition 5.1.3

Let $\mathcal{M} = (Q, \Sigma, F)$ and $\mathfrak{M} = (\mathcal{M}, i, T)$ and $A = |\mathfrak{M}|$ then

$$A = i^{-1} \circ T.$$

If $q = i * \alpha$, $\alpha \in \Sigma^*$, then

$$q^{-1} \circ T = \alpha^{-1} A.$$

Proof Recall that
$$i^{-1} \circ T = \{\alpha \in \Sigma^* \mid i * \alpha \in T\}$$
$$= \{\alpha \in \Sigma^* \mid iF_\alpha \in T\}$$
$$= A.$$

Now let $q = i * \alpha$, then

$$\begin{aligned} q^{-1} \circ T &= \{\beta \in \Sigma^+ \mid q * \beta \in T\} \\ &= \{\beta \in \Sigma^* \mid (i * \alpha) * \beta \in T\} \\ &= \{\beta \in \Sigma^* \mid iF_{\alpha\beta} \in T\} \\ &= \{\beta \in \Sigma^* \mid \alpha\beta \in A\} \\ &= \alpha^{-1}A. \end{aligned}$$

□

Proposition 5.1.4

Let $\mathcal{M} = (Q, \Sigma, F)$ and $\mathfrak{M} = (\mathcal{M}, i, T)$, then $\mathfrak{M}$ is accessible if and only if $Q = i * (\Sigma^*)$ and $\mathfrak{M}$ is coaccessible if and only if $Q = T * (\Sigma^*)^{-1}$.

Proof This follows from the definitions since

$$R = \{iF_\alpha \mid \alpha \in \Sigma^*\} = \{i * \alpha \mid \alpha \in \Sigma^*\} = i * (\Sigma^*)$$

and

$$\begin{aligned} S &= \{q \mid qF_\alpha \in T \text{ for some } \alpha \in \Sigma^*\} \\ &= \{q \mid q * \alpha \in T \text{ for some } \alpha \in \Sigma^*\} \\ &= T * (\Sigma^*)^{-1}. \end{aligned}$$

□

5.2 Minimal recognizers

Let Σ be a finite set and $A \subseteq \Sigma^*$. If A is recognizable then there exists a recognizer

$$\mathfrak{M} = (\mathcal{M}, i, T) \quad \text{where } \mathcal{M} = (Q, \Sigma, F) \quad \text{and} \quad A = |\mathfrak{M}|.$$

We shall now construct a recognizer with behaviour equal to A directly.

Let us consider all subsets of Σ^* of the form

$$\alpha^{-1} \cdot A = \{\beta \in \Sigma^* \mid \alpha\beta \in A\},$$

where $\alpha \in \Sigma^*$. Put Q_A to be the *set* of all such subsets, noting that this may include the empty set, $\varnothing$.

Thus $Q_A = \{\alpha^{-1} \cdot A \mid \alpha \in \Sigma^*\}$ and clearly $A \in Q_A$ since $A = \Lambda^{-1} \cdot A$. The state function $F^A : Q_A \times \Sigma \to Q_A$ is defined by

$$\left.\begin{aligned} (\alpha^{-1} \cdot A)F^A_\sigma &= (\alpha\sigma)^{-1} \cdot A \\ \varnothing F^A &= \varnothing \end{aligned}\right\} \text{ for } \sigma \in \Sigma.$$

Put $i_A = A$ and define $T_A = \{\alpha^{-1} \cdot A \in Q_A \mid \alpha \in A\}$. (Note that $\alpha^{-1} \cdot A \in T_A \Leftrightarrow \Lambda \in \alpha^{-1} \cdot A$.) This defines a state machine

$$\mathcal{M}_A = (Q_A, \Sigma, F^A)$$

and a recognizer

$$\mathfrak{M}_A = (\mathcal{M}_A, i_A, T_A)$$

once we have established that $F^A: Q_A \times \Sigma \to Q_A$ is a well-defined mapping and Q_A is a finite set.

Note that if $A = \varnothing$ then $Q_A = \{\varnothing\}$, $i_A = \varnothing$ and $T_A = \varnothing$ (that is there are no final states).

Theorem 5.2.1

If $A \subseteq \Sigma^*$ is recognizable then $\mathfrak{M}_A$ is a recognizer with the property that $|\mathfrak{M}_A| = A$.

Proof Let $\alpha^{-1} \cdot A$, $\gamma^{-1} \cdot A \in Q_A$ with $\alpha^{-1} \cdot A = \gamma^{-1} \cdot A \neq \varnothing$. If $\sigma \in \Sigma$ then

$$\begin{aligned}(\alpha^{-1} \cdot A)F^A_\sigma &= (\alpha\sigma)^{-1} \cdot A \\ &= \sigma^{-1} \cdot (\alpha^{-1}A) \text{ by exercise 5.8.} \\ &= \sigma^{-1} \cdot (\gamma^{-1} \cdot A) \\ &= (\gamma\sigma)^{-1} \cdot A \\ &= (\gamma^{-1} \cdot A)F^A_\sigma\end{aligned}$$

and so F^A is a well-defined function.

Next we show that Q_A is finite. Let $\alpha^{-1} \cdot A \in Q_A$ and put $q = i\alpha$ where $\mathfrak{M} = (\mathcal{M}, i, T)$ is a recognizer that recognizes A. Now

$$\begin{aligned}\alpha^{-1} \cdot A &= \{\beta \in \Sigma^* \mid \alpha\beta \in A\} \\ &= \{\beta \in \Sigma^* \mid i\alpha\beta \in T\} \\ &= \{\beta \in \Sigma^* \mid q\beta \in T\} \\ &= q^{-1} \circ T.\end{aligned}$$

Since Q is finite there can only be a finite number of sets of this form and so Q_A is finite.

If $a \in A$ then $a^{-1}A \in T_A$ and so $AF^A_a \in T_A$ which means that $a \in |\mathfrak{M}_A|$. Now let $x \in |\mathfrak{M}_A|$, then $AF^A_x \in T_A$ which gives $x^{-1} \cdot A \in T_A$. Suppose that $x^{-1} \cdot A = a^{-1} \cdot A$ where $a \in A$, then $a\Lambda = a \in A$ and so $\Lambda \in a^{-1} \cdot A = x^{-1} \cdot A$.

Therefore $x\Lambda \in A$ and so $x \in A$, proving that $|\mathfrak{M}_A| \subseteq A$. Consequently

$$|\mathfrak{M}_A| = A. \qquad \square$$

Note that $\mathfrak{M}_A$ is complete and accessible, but will not be coaccessible if $\varnothing \in Q_A$.

Examples 5.3

Let $\Sigma = \{0, 1\}$

(i) $A = \{0\} \cdot \{1\}^* \cup \{1\}^*$, $0^{-1}A = \{1\}^* = 1^{-1}A$, $(0^2)^{-1}A = \emptyset$, etc. and $Q_A = \{A, 0^{-1}A, \emptyset\}$ with state function:

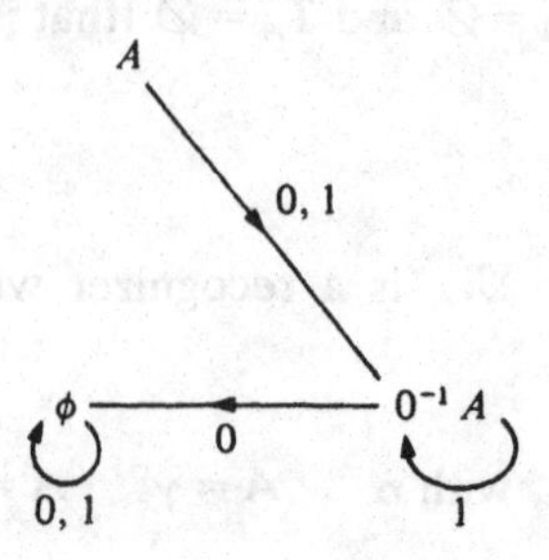

$T_A = \{A, 0^{-1}A\}$ and so the complete recognizer $\mathfrak{M}_A$ is:

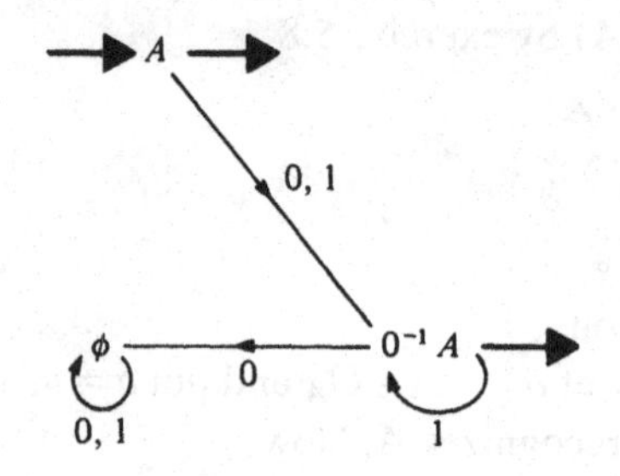

(ii) $A = \{0\}^* \cdot \{1\} \cdot \{0\}^*$, $1^{-1}A = \{0\}^*$, $Q_A = \{A, \{0\}^*, \emptyset\}$, $T_A = \{1^{-1}A\}$.

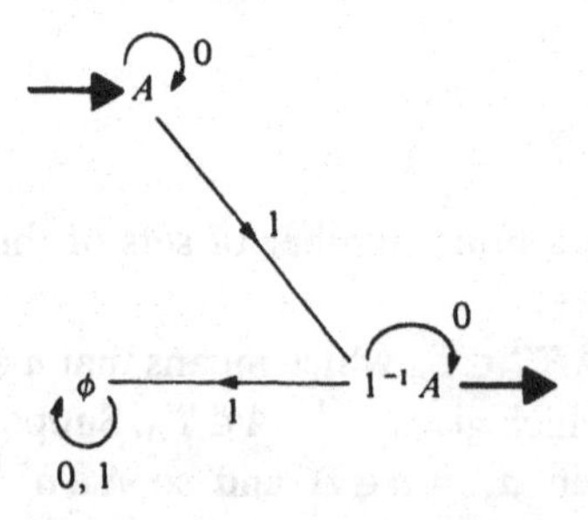

(the completion of example 5.1 (vi))

The recognizer $\mathfrak{M}_A$ has the following minimality property.

Theorem 5.2.2

Let $A \subseteq \Sigma^*$ be recognizable and suppose that $\mathfrak{M} = (\mathcal{M}, i, T)$, where $\mathcal{M} = (Q, \Sigma, F)$, is a complete accessible recognizer with behaviour A. There exists a function $f: Q \to Q_A$ such that:

(i) $f(i) = i_A$,

(ii) $f^{-1}(T_A) = T$,

(iii) $(f(q))F_\sigma^A = f(qF_\sigma)$ for all $q \in Q$, $\sigma \in \Sigma$,

(iv) f is surjective.

Proof Define $f: Q \to Q_A$ by

$$f(q) = q^{-1} \circ T = \{\alpha \in \Sigma^* \mid q * \alpha \in T\}.$$

We must first show that $f(q) \in Q_A$. Since Q is accessible there exists $\beta \in \Sigma^*$ such that $q = iF_\beta = i * \beta$. Then

$$\begin{aligned} \beta^{-1} \cdot A &= \{\gamma \in \Sigma^* \mid \beta\gamma \in A\} \\ &= \{\gamma \in \Sigma^* \mid i * (\beta\gamma) \in T\} \\ &= \{\gamma \in \Sigma^* \mid (i * \beta) * \gamma \in T\} \\ &= \{\gamma \in \Sigma^* \mid q * \gamma \in T\} \\ &= q^{-1} \circ T. \end{aligned}$$

Thus f is a function. Then

(i) $f(i) = i^{-1} \circ T = \{\alpha \in \Sigma^* \mid i * \alpha \in T\} = A = i_A$.

(ii) Let $f(q) \in T_A$, then $q^{-1} \circ T = a^{-1}A$ for some $a \in A$ and so $\Lambda \in q^{-1} \circ T$, that is $q * \Lambda \in T$ and so $q \in T$. Hence $f^{-1}(T_A) \subseteq T$. Now for $t \in T$ we have $f(t) = t^{-1} \circ T$ and since $\Lambda \in t^{-1} \circ T$ we see that $t^{-1} \circ T \in T_A$. Thus $f(T) \subseteq T_A$ and so $f^{-1}(T_A) = T$.

(iii)
$$\begin{aligned} (f(q))F_\sigma^A &= (q^{-1}T)F_\sigma^A \\ &= (\beta^{-1}A)F_\sigma^A \quad \text{if } q = i * \beta \\ &= (\beta\sigma)^{-1}A \\ &= \{\gamma \in \Sigma^* \mid \beta\sigma\gamma \in A\} \\ &= \{\gamma \in \Sigma^* \mid i\beta\sigma\gamma \in T\} \\ &= \{\gamma \in \Sigma^* \mid q\sigma\gamma \in T\} \\ &= (q\sigma)^{-1} \circ T = f(qF_\sigma) \quad \text{for } q \in Q, \sigma \in \Sigma. \end{aligned}$$

(iv) Let $s^{-1}A \in Q_A$ where $s \in \Sigma^*$, then put $p = is \in Q$ and note that $p^{-1} \circ T = s^{-1}A$ and so $s^{-1}A = f(p)$.

Therefore f is surjective. □

We can now regard the recognizer $\mathfrak{M}_A$ as being the *minimal complete recognizer* of the recognizable subset A, where the term 'minimal' refers to the properties described in theorem 5.2.2, in particular (iv) implies that $|Q_A| \leq |Q|$. If we try to construct the recognizer $\mathfrak{M}_A$ in the case where A is not recognizable we will find that the set of states Q_A is no longer finite and so $\mathfrak{M}_A$ will not then be a recognizer according to our definition. This is examined in the next theorem.

Theorem 5.2.3

Let $A \subseteq \Sigma^*$, then A is recognizable if and only if the collection $\{\beta^{-1}A \mid \beta \in \Sigma^*\}$ is finite.

Proof If A is recognizable then the proof of 5.2.1 establishes that Q_A is finite and so $\{\beta^{-1}A \mid \beta \in \Sigma^*\}$ is finite. Clearly if $\{\beta^{-1}A \mid \beta \in \Sigma^*\}$ is finite then we may construct the recognizer $\mathfrak{M}_A$ which will then establish the fact that A is recognizable. □

5.3 Recognizable sets

The examples of recognizable sets that we have already seen will now be augmented by developing general techniques for constructing more recognizable sets from given recognizable sets.

Notice first that the following are examples of recognizable sets where Σ is a given finite set and $\sigma \in \Sigma$.

$$\{\sigma\}, \{\Lambda\}, \emptyset, \Sigma^*.$$

Now suppose that A, B are recognizable subsets of Σ^*. We will show that $A \cup B, A \cdot B, A^*, \Sigma^* \backslash A, A \cap B, A^+$ are also recognizable. The basic technique is the same in all cases, namely that we construct a recognizer with the desired property using recognizers of A and B. The machines so formed may not be minimal in the sense of 5.2.2. but that is irrelevant here.

Theorem 5.3.1

Let $A, B \subseteq \Sigma^*$. If A and B are recognizable then $A \cup B$ is also recognizable.

Proof Let $\mathfrak{M} = (\mathcal{M}, i, T)$, $\mathfrak{M}' = (\mathcal{M}', i', T')$ be recognizers with $\mathcal{M} = (Q, \Sigma, F)$, $\mathcal{M}' = (Q', \Sigma, F')$ and such that $|\mathfrak{M}| = A$, $|\mathfrak{M}'| = B$. Consider $\mathcal{M} \vee \mathcal{M}' = (Q \times Q', \Sigma, \bar{F})$ where $(q, q')\bar{F}_\sigma = (qF_\sigma, q'F'_\sigma)$ for $\sigma \in \Sigma$, $q \in Q$, $q' \in Q'$.

Let $\mathfrak{M} \vee \mathfrak{M}' = (\mathcal{M} \vee \mathcal{M}', (i, i'), (T \times Q') \cup (Q \times T'))$. We show that $|\mathfrak{M} \vee \mathfrak{M}'| = A \cup B$. Let $\gamma \in |\mathfrak{M} \vee \mathfrak{M}'|$, then $(i, i')\bar{F}_\gamma \in (T \times Q') \cup (Q \times T')$ so $(iF_\gamma, i'F'_\gamma) \in (T \times Q') \cup (Q \times T')$ and either $iF_\gamma \in T$ or $i'F'_\gamma \in T'$, that is either $\gamma \in A$ or $\gamma \in B$, and so $\gamma \in A \cup B$.

Now let $\gamma \in A \cup B$, then either $qF_\gamma \in T$ or $q'F'_\gamma \in T'$. If $qF_\gamma \in T$ then $(q, q')\bar{F}_\gamma = (qF_\gamma, q'F'_\gamma) \in T \times Q'$ and if $q'F'_\gamma \in T'$ then $(q, q')\bar{F}_\gamma = (qF_\gamma, q'F'_\gamma) \in Q \times T'$ and in either case $\gamma \in |\mathfrak{M} \vee \mathfrak{M}'|$. □

Theorem 5.3.2

Let $A, B \subseteq \Sigma^*$. If A and B are recognizable sets then $A \cdot B$ is also recognizable.

Proof Let $\mathfrak{M} = (\mathcal{M}, i, T)$, $\mathfrak{M}' = (\mathcal{M}', i', T')$ be recognizers with $\mathcal{M} = (Q, \Sigma, F)$, $\mathcal{M}' = (Q', \Sigma, F')$ and such that $|\mathfrak{M}| = A$, $|\mathfrak{M}'| = B$.

Consider $\mathcal{M} \triangle \mathcal{M}' = (Q \times \mathcal{P}(Q'), \Sigma, F^\triangle)$ where

$$(q, P)F_\sigma = \begin{cases} (qf_\sigma, PF_\sigma) & \text{if } qF_\sigma \notin T \\ (qF_\sigma, PF_\sigma \cup \{i'\}) & \text{if } qF_\sigma \in T \end{cases}$$

for $q \in Q$, $P \in \mathcal{P}(Q')$, $\sigma \in \Sigma$ (here $PF_\sigma = \{pF_\sigma \mid p \in P\}$).

Now put $T^\triangle = \{(q, P) \mid q \in Q, P \in \mathcal{P}(Q'), P \cap T' \neq \varnothing\}$ and examine the recognizer.

$$\mathfrak{M} \triangle \mathfrak{M}' = (\mathcal{M} \triangle \mathcal{M}', (i, \varnothing), T^\triangle).$$

Let $\alpha \in A \cdot B$, then $\alpha = a \cdot b$ for some $a \in A$, $b \in B$ and $iF_\alpha \in T$. Now

$$\begin{aligned} (i, \varnothing)F_\alpha^\triangle &= (i, \varnothing)F_{ab}^\triangle \\ &= (iF_a, i')F_b^\triangle \\ &= (iF_{ab}, P) \quad \text{where } i'F_b \in P, P \in \mathcal{P}(Q') \\ &\in T^\triangle. \end{aligned}$$

Hence $A \cdot B \subseteq |\mathfrak{M} \triangle \mathfrak{M}'|$.

Now let $\beta \in |\mathfrak{M} \triangle \mathfrak{M}'|$, then $(i, \varnothing)F_\beta^\triangle \in T^\triangle$, so that $(iF_\beta, P) \in T^\triangle$ for some $P \in \mathcal{P}(Q)$. Clearly $P \neq \varnothing$ and so there exists $\gamma, \delta \in \Sigma^*$ such that $\beta = \gamma \cdot \delta$ and $iF_\gamma \in T$. We call γ an initial segment of β and note that $\gamma \in A$. Let $C = \{\gamma \in \Sigma^* \mid \beta = \gamma \cdot \delta \text{ for some } \delta \in \Sigma^* \text{ and } \gamma \in A\} = \beta \cdot (\Sigma^*)^{-1}$, then we have seen that C is not empty. Each initial segment γ in C defines an 'end segment' δ such that $\beta = \gamma \cdot \delta$. Let $R = \{i'F'_\delta \mid \delta \in \Sigma^* \text{ and } \beta = \gamma \cdot \delta \text{ for some } \gamma \in C\}$, then $R \cap T' \neq \varnothing$ otherwise β would not be recognized. Let $q' \in R \cap T'$ be such that $q' = i'F'_{\delta_0}$ and suppose that $\beta = \gamma_0 \cdot \delta_0$ with $\gamma_0 \in C$. Then $\gamma_0 \in A$ and $\delta_0 \in B$, hence $\beta \in A \cdot B$ as required. □

Theorem 5.3.3

Let $A \subseteq \Sigma^*$. If A is recognizable then so is A^*.

Proof Let $\mathfrak{M} = (\mathcal{M}, i, T)$ where $\mathcal{M} = (Q, \Sigma, F)$ and $|\mathfrak{M}| = A$. Define $\mathcal{M}^* = (\mathcal{P}(Q), \Sigma, F^*)$ where

$$PF_\sigma^* = \begin{cases} PF_\sigma & \text{if } PF_\sigma \cap T = \varnothing \\ PF_\sigma \cup \{i\} & \text{otherwise} \end{cases}$$

for $P \in \mathcal{P}(Q)$, $\sigma \in \Sigma$.

Let $T^* = \{P \in \mathcal{P}(Q) \mid P \cap T \neq \varnothing\}$ and put $\mathfrak{M}^* = (\mathcal{M}^*, \{i\}, T^*)$. If $\alpha \in A^*$ then $\alpha = a_1 \ldots a_n$ for some $n \in \mathcal{N}$ and $a_i \in A$, $1 \leq i \leq n$. Now

$$\begin{aligned}
\{i\}F^*_\alpha &= \{i\}F^*_{a_1 \ldots a_n} \\
&= (\{iF_{a_1}\} \cup \{i\})F^*_{a_2 \ldots a_n} \\
&= (\{iF_{a_1 a_2}, iF_{a_2}\} \cup \{i\})F^*_{a_3 \ldots a_n} \\
&\ \ \vdots \\
&= \{iF_\alpha, iF_{a_2 \ldots a_n}, \ldots, iF_{a_n}, i\} \in T^*
\end{aligned}$$

since $iF_{a_n} \in T$. Therefore $A^* \subseteq |\mathfrak{M}^*|$. The inequality $|\mathfrak{M}^*| \subseteq A$ will be left as an exercise. □

Theorem 5.3.4

Let $A \subseteq \Sigma^*$ be a recognizable set, then $\Sigma^* \backslash A$ is also recognizable.

Proof If $\mathfrak{M} = (\mathcal{M}, i, T)$ where $\mathcal{M} = (Q, \Sigma, F)$ is such that $|\mathfrak{M}| = A$ then $\bar{\mathfrak{M}} = (\mathcal{M}, i, Q \backslash T)$ is such that $|\bar{\mathfrak{M}}| = \Sigma^* \backslash A$. □

Theorem 5.3.5

If A and B are recognizable subsets of Σ^* then so is $A \cap B$.

Proof See exercises. □

So far we have established that there are a considerable number of recognizable sets but we have yet to meet a subset of Σ^* that is not recognizable. We will, shortly, develop techniques for testing the recognizability of certain subsets of Σ^*, but in the meantime we will briefly examine a subset which is not recognizable.

Example 5.4

Let $\Sigma = \{0, 1\}$ and put $A = \{0^n 1^n \mid n \in N\}$. Suppose that $\mathfrak{M} = (\mathcal{M}, i, T)$ is such that $A = |\mathfrak{M}|$. If $\mathcal{M} = (Q, \Sigma, F)$, as usual, then $iF_{0^n 1^n} \in T$ for each $n \in N$. Let $q_n = iF_{0^n}$ and suppose that $q_n = q_m$ where $m \in \mathcal{N}$, then $iF_{0^m 1^n} = q_n F_{1^n} = iF_{0^n 1^n} \in T$ and so $0^m 1^n \in A$. Therefore $0^m 1^n = 0^n 1^n$ which implies $m = n$. Consequently the set of states $q_1, q_2, \ldots, q_n, \ldots$ is infinite and $\mathfrak{M}$ cannot then be a recognizer. Hence we have a contradiction to A being recognizable.

Theorem 5.3.6

Let Σ, Γ be finite non-empty sets and $f: \Sigma^* \to \Gamma^*$ a function satisfying the condition $f^{-1}(\Lambda_\Gamma) = \Lambda_\Sigma$ where Λ_Γ and Λ_Σ are the empty words in Γ^* and Σ^* respectively. If $A \subseteq \Sigma^*$ is recognizable then so is $f(A)$.

Proof This is left as an exercise. □

5.4 The syntactic monoid

Suppose that $\mathcal{M} = (Q, \Sigma, F)$ is a state machine and consider the relation $\sim_{\mathcal{M}}$ defined on Σ^* by

$$\alpha \sim_{\mathcal{M}} \beta \Leftrightarrow F_\alpha = F_\beta$$

where $\alpha, \beta \in \Sigma^*$. We can immediately deduce the following proposition (cf. section 2.2).

Proposition 5.4.1

If $\mathcal{M} = (Q, \Sigma, F)$ is a state machine then $\sim_{\mathcal{M}}$ is a congruence on Σ^*.

Proof Clearly $\alpha \sim_{\mathcal{M}} \beta \Leftrightarrow x\alpha y \sim_{\mathcal{M}} x\beta y$ for all $x, y \in \Sigma^*$. □

If $\mathfrak{M} = (\mathcal{M}, i, T)$ is a recognizer with $\mathcal{M} = (Q, \Sigma, F)$ we note that if $\alpha \sim \beta$ then for $x, y \in \Sigma^*$ *either* $x\alpha y$ and $x\beta y$ both belong to $|\mathfrak{M}|$ *or* $x\alpha y$ and $x\beta y$ both do not belong to $|\mathfrak{M}|$, thus

$$\alpha \sim_{\mathcal{M}} \beta \Leftrightarrow [x\alpha y \in |\mathfrak{M}| \Leftrightarrow x\beta y \in |\mathfrak{M}|, \text{ for all } x, y \in \Sigma^*].$$

It is now possible to define a relation on Σ^* based on any given subset $A \subseteq \Sigma^*$; we put

$$\alpha \approx_A \beta \Leftrightarrow [x\alpha y \in A \Leftrightarrow x\beta y \in A \text{ for all } x, y \in \Sigma^*].$$

Then we have seen that $\alpha \sim_{\mathcal{M}} \beta \Leftrightarrow \alpha \approx_{|\mathfrak{M}|} \beta$. Can we obtain a closer connection between these two relations? In general $\mathcal{M}$ may have too many equivalent functions for the relations to be identical. We can, however, replace $\mathcal{M}$ by a more efficient machine, namely the minimal complete recognizer of $|\mathfrak{M}|$.

Theorem 5.4.2

Let $A \subseteq \Sigma^*$ be a recognizable subset of Σ^* with minimal complete recognizer $\mathfrak{M}_A = (\mathcal{M}_A, i_A, T_A)$. Then for $\alpha, \beta \in \Sigma^*$ we have

$$\alpha \sim_{\mathcal{M}_A} \beta \Leftrightarrow \alpha \approx_A \beta.$$

Proof Since $A = |\mathfrak{M}_A|$ we already have $\alpha \sim_{\mathcal{M}_A} \beta \Rightarrow \alpha \approx_A \beta$. Let $\alpha \approx_A \beta$, then $x\alpha y \in A \Leftrightarrow x\beta y \in A$ for all $x, y \in \Sigma^*$. Hence $y \in (x\alpha)^{-1} \cdot A \Leftrightarrow$

$y \in (x\beta)^{-1} \cdot A$ for all $y \in \Sigma^*$. Thus $(x\alpha)^{-1} \cdot A = (x\beta)^{-1} \cdot A$ and so

$$(x^{-1} \cdot A)F_\alpha^A = (x^{-1} \cdot A)F_\beta^A \quad \text{for all } x \in \Sigma^*$$

which means that $F_\alpha^A = F_\beta^A$ and so

$$\alpha \sim_{\mathcal{M}_A} \beta. \qquad \square$$

For a given recognizable set $A \subseteq \Sigma^*$ the congruence $\approx_A$ is called the *Myhill congruence of A*. If this congruence is factored out of the monoid Σ^* we obtain the *syntactic monoid* of A, this is given by $\Sigma^*/\approx_A = \Sigma^*/\sim_{\mathcal{M}_A} \cong \mathbf{M}(\mathcal{M}_A)$, the monoid of the minimal complete state machine $\mathcal{M}_A$. See chapter 2.

Since the congruence $\approx_A$ can be defined with respect to any subset $A \subseteq \Sigma^*$ it is of interest to see what happens when A is not recognizable. This is explored in the next result.

Theorem 5.4.3

Let $A \subseteq \Sigma^*$. The following statements are equivalent:

(i) A is recognizable.

(ii) $\Sigma^*/\approx_A$ is finite.

(iii) A is the union of congruence classes of a congruence on Σ^* of finite index.

Proof (i)$\Rightarrow$(ii). A is recognizable implies that a minimal complete recognizer $\mathfrak{M}_A$ exists and $\mathbf{M}(\mathcal{M}_A)$ is finite so that $\Sigma^*/\approx_A$ is also finite.

(ii)$\Rightarrow$(iii). If $\Sigma^*/\approx_A$ is finite then the congruence $\approx_A$ on Σ^* is of finite index. Let $[\alpha]$ denote the congruence class containing α where $\alpha \in \Sigma^*$. Now put

$$B = \cup\{[\alpha] \mid i_A F_\alpha^A \in T_A\}$$
$$= \cup\{[\alpha] \mid \alpha^{-1} \cdot A = a^{-1} \cdot A \text{ for some } a \in A\}.$$

Clearly each $\alpha \in A$ since $i_A F_\alpha^A \in T_A$ and so $B \subseteq A$. Now let $a \in A$, then $i_A F_a^A \in T_A$ and so $[a] \subseteq B$. Hence $B = A$.

(iii)$\Rightarrow$(i). Suppose that $\sim$ is a congruence of finite index on Σ^* and let $A = \cup\{[\alpha_i] \mid i = 1, \ldots, n\}$ where $\alpha_i \in \Sigma^*$ and $[\alpha_i]$ is the $\sim$-congruence class containing α_i.

Let $\mathcal{M} = (Q, \Sigma, F)$ where $Q = \Sigma^*/\sim$, $F: Q \times \Sigma \to Q$ is defined by

$$[\alpha]F_\sigma = [\alpha\sigma] \quad \text{for } [\alpha] \in \Sigma^*/\sim, \sigma \in \Sigma.$$

Put $i = [\Lambda]$ and $T = \{[\alpha] \mid \alpha \in A\}$. $\mathfrak{M} = (\mathcal{M}, i, T)$ is a recognizer since Q is finite. Let $a \in A$, then

$$iF_a = [\Lambda] \cdot F_a = [\Lambda a] = [a] \in T,$$

hence $A \subseteq |\mathfrak{M}|$. If $b \in |\mathfrak{M}|$ then $iF_b \in T$ so $[\Lambda]F_b = [b] \in T$ and $b \in A$. Hence $|\mathfrak{M}| = A$ and A is thus recognizable. □

The criterion (ii) can often be used to establish that a particular set is not recognizable since it means that the Myhill congruence is then of infinite index.

Example 5.5
Let $\Sigma = \{0, 1\}$ and put

$$A = \{0^n 10^n \mid n \in \mathcal{N}\}.$$

Consider the Myhill congruence $\approx_A$ defined by A. The infinite sequence of elements $0, 0^2, \ldots, 0^m, \ldots$ must all belong to different congruence classes for if $0^p \approx_A 0^q$ then

$$x0^p y \in A \Leftrightarrow x0^q y \in A \quad \text{for all } x, y \in \Sigma^*$$

and in particular, if we assume that $p > q$, then put $x = \Lambda$, $y = 10^q$ we get

$$0^{p-q}0^q 10^q \in A \Leftrightarrow 0^q 10^q \in A,$$

that is

$$0^{p-q}0^q 10^q \in A,$$

which is false. Thus $p = q$ and so $\approx_A$ is not of finite index and A cannot be recognizable.

Example 5.6
Consider the recognizer of example 5.3(ii) where $\Sigma = \{0, 1\}$ and $A = \{0\}^* \cdot \{1\} \cdot \{0\}^*$. The state machine is isomorphic to

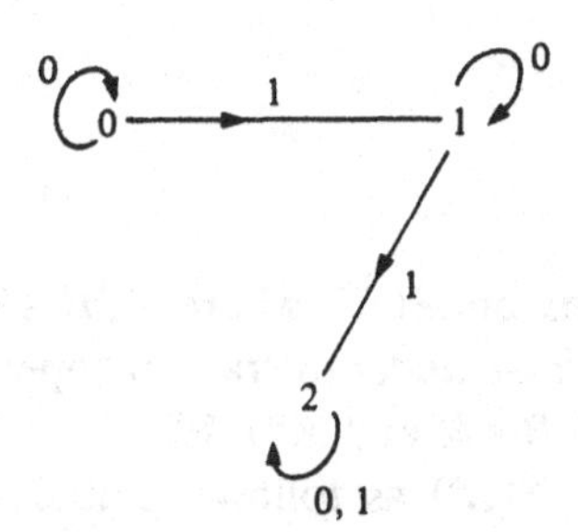

We will calculate the monoid of this machine, it is generated by $\{\Lambda, 1, 1^2\}$

with the table

	Λ	1	1^2
Λ	Λ	1	1^2
1	1	1^2	1^2
1^2	1^2	1^2	1^2

This monoid is the syntactic monoid of A. Notice that 1 is not $\approx_A$-related to 1^2 since

$$\Lambda 1 \Lambda \in A \text{ but } \Lambda 1^2 \Lambda \notin A,$$

similarly 0 is not $\approx_A$-related to 1 since $\Lambda 0 \Lambda \notin A$. There are three distinct $\approx_A$-classes and $A = [1]$.

5.5 Rational decompositions of recognizable sets

In this section we examine one of two methods of decomposing a recognizable set. This first method is the classical approach of Kleene and gives a constructive characterization of a recognizable set.

It will be recalled that any singleton word from Σ^* is recognizable, as indeed is the empty set of words. Furthermore if A and B are recognizable subsets of Σ^* then so are $A \cup B$, $A \cdot B$ and A^*. Consequently we can start with a finite collection of singleton sets of words, apply the operations of union, 'dot' product and the star operation to them a finite number of times and obtain more recognizable subsets. The question Kleene answered was whether any recognizable subsets exist that cannot be produced in this way.

Let Σ^* be the free monoid on the non-empty set Σ and consider the set $\mathscr{P}(\Sigma^*)$ consisting of all sets of words in Σ^*. We can define three operations on $\mathscr{P}(\Sigma^*)$, namely

$$A \cup B$$

$$A \cdot B$$

$$A^*.$$

They are called the *rational operations* on $\mathscr{P}(\Sigma^*)$ where $A, B \in \mathscr{P}(\Sigma^*)$. Now let $\mathscr{H} \subseteq \mathscr{P}(\Sigma^*)$, we say that $\mathscr{H}$ is *closed under* the rational operations if given $A, B \in \mathscr{H}$ then $A \cup B \in \mathscr{H}$, $A \cdot B \in \mathscr{H}$ and $A^* \in \mathscr{H}$.

We now define a subset $\text{Rat}(\Sigma) \subseteq \mathscr{P}(\Sigma^*)$ as follows. $\text{Rat}(\Sigma)$ is the smallest subset of $\mathscr{P}(\Sigma^*)$ that contains the singleton subsets and $\varnothing$, and is closed under the rational operations.

Suppose that a set $A \in \mathcal{P}(\Sigma^*)$ is either $\varnothing$ or $\{x\}$ (where $x \in \Sigma^*$) or is formed from sets of this type by a finite number of rational operations, then clearly $A \in \text{Rat}(\Sigma)$. We will call such sets *regular* sets of words. The collection of all regular words is written $\text{Reg}(\Sigma)$ and clearly

$$\text{Reg}(\Sigma) \subseteq \text{Rat}(\Sigma).$$

Notice, however, that the set $\text{Reg}(\Sigma)$ is itself closed under the rational operations, it contains the singleton subsets and the empty set, consequently it equals $\text{Rat}(\Sigma)$ which was supposed to be the smallest such set. Thus

$$\text{Reg}(\Sigma) = \text{Rat}(\Sigma)$$

and $\text{Rat}(\Sigma)$ is contained in the set of all recognizable subsets of Σ^* by 5.3.1, 5.3.2, 5.3.3 noting that $\varnothing$, $\{\Lambda\}$, $\{x\}$ $(x \in \Sigma^*)$ are all recognizable. (The first two can be found in examples 5.1 and the last one is exercise 5.1.)

Proposition 5.5.1

Let Σ and Γ be non-empty finite sets and suppose that $f : \Sigma \to \Gamma$ is a mapping. Define $f^* : \Sigma^* \to \Gamma^*$ by

$$f^*(\sigma_1 \ldots \sigma_n) = f(\sigma_1) \ldots f(\sigma_n), \quad \sigma_1 \ldots \sigma_n \in \Sigma^+$$

$$f^*(\Lambda) = \Lambda.$$

If A is a regular set of Σ^* then $f^*(A)$ is a regular set of Γ^*.

Proof If A is a singleton then $f^*(A)$ is also a singleton, similarly if A is $\varnothing$, then $f^*(A)$ is $\varnothing$. Suppose that B and C are regular sets in Σ^*, $f^*(B)$ and $f^*(C)$ are regular sets in Γ^*. Then

$$f^*(B \cup C) = f^*(B) \cup f(C) \quad \text{is regular in } \Gamma^*,$$

$$f^*(B \cdot C) = f^*(B) \cdot f^*(C) \quad \text{is regular in } \Gamma^*,$$

$$f^*(B^*) = (f^*(B))^* \quad \text{is regular in } \Gamma^*.$$

An inductive proof based on the number of regular operations in the decomposition of A will establish that $f^*(A)$ is regular in Σ^*. □

The proof of the fact that recognizable sets are regular is best examined with the help of some more abstract terminology, otherwise the details can become rather daunting.

Let S be any finite non-empty set and suppose that $\mathcal{R}$ is a relation on S. We will write

$$a\mathcal{R}a' \quad \text{to mean} \quad (a, a') \in \mathcal{R} \text{ or } a \text{ is related to } a' \text{ under } \mathcal{R}.$$

Now suppose that $\alpha = s_1 \ldots s_n \in S^+$ we call α an $\mathcal{R}$-word if

$$s_i \mathcal{R} s_{i+1} \quad \text{for all } i = 1, \ldots, n-1.$$

The empty word Λ will also be called an $\mathcal{R}$-word. Given two $\mathcal{R}$-words $\alpha = s_1 \ldots s_n$ and $\alpha' = s'_1 \ldots s'_n$ we can form further $\mathcal{R}$-words, namely

$$\alpha \cdot \alpha' = s_1 \ldots s_n \cdot s'_1 \ldots s'_n \quad \text{if } s_n \mathcal{R} s'_1.$$

Given two sets X, Y of $\mathcal{R}$-words then we define the sets

$$X \cup Y$$

$$X \cdot Y = \{x \cdot y \mid x \in X, y \in Y \text{ and } x \cdot y \text{ is an } \mathcal{R}\text{-word}\}$$

$$X^* = \{x_1 \cdot x_2 \ldots x_m \mid x_i \in X \text{ and } x_1 \cdot x_2 \ldots x_m \text{ is an } \mathcal{R}\text{-word}\}.$$

These are all sets of $\mathcal{R}$-words in S^*.

Given $s_1, s_n \in S$ we define $\mathcal{R}(s_1, s_n)$ to be the set of all $\mathcal{R}$-words in S^* of the form $s_1 \ldots s_n$.

Theorem 5.5.2

Let S be a non-empty finite set, $\mathcal{R}$ a binary relation on S and $s_1, s_n \in S$, then the set $\mathcal{R}(s_1, s_n)$ is a regular set of words of S^*.

Proof We proceed by induction on the size of the finite set S. Let $|S| = k$. Consider the case $k = 1$. Suppose that $S = \{s\}$, then we have two possibilities, either $s \mathcal{R} s$ or s is not related to s under $\mathcal{R}$. In the former case the set $\mathcal{R}(s, s) = \{\Lambda, s, s \cdot s, s \cdot s \cdot s, \ldots\} = \{s\}^*$, in the latter case $\mathcal{R}(s, s) = \{\Lambda\}$. In both cases $\mathcal{R}(s, s)$ is regular.

Not let $k > 1$ and assume that the result is true for all finite sets S of order less than m. Consider a set S of order m and put $S' = S \backslash \{s_1\}$. Let α be an $\mathcal{R}$-word belonging to $\mathcal{R}(s_1, s_n)$. Then $\alpha = s_1 \cdot \alpha' \cdot s_n$ for some $\mathcal{R}$-word $\alpha' \in S^*$. We can write α in the following form. Either

$$\alpha = s_1^{n_1} \cdot \beta_1 \cdot s_1^{n_2} \cdot \beta_2 \ldots s_1^{n_r} \cdot \beta_r \cdot s_n$$

or

$$\alpha = s_1^{n_1} \cdot \beta_1 \cdot s_1^{n_2} \cdot \beta_2 \ldots s_1^{n_r} \cdot \beta_r \cdot s_1^{n_{r+1}} \cdot s_n$$

where the $\mathcal{R}$-words $\beta_1, \ldots, \beta_r$ do not contain the symbol s_1, and are not the empty word. It is clear that $\beta_1, \ldots, \beta_r$ are $\mathcal{R}'$-words in $(S')^*$ if we consider the restriction $\mathcal{R}'$ of the relation $\mathcal{R}$ to the set S'.

Now let $\beta_1 = \gamma_{11} \ldots \gamma_{1t_1}$ where $\gamma_{11}, \ldots, \gamma_{1t_1} \in S'$ then $\beta_1 \in \mathcal{R}'(\gamma_{11}, \gamma_{1t_1})$ which is a regular set in $(S')^*$. If $\beta_1 \in S'$ then $\beta_1 \in \{\beta_1\}$ which is also regular in $(S')^*$. Similarly for $\beta_2, \ldots, \beta_r$. Hence

$$\alpha \in \{s_1\}^* \cdot A_1 \cdot \{s_1\}^* \cdot A_2 \ldots \{s_1\}^* \cdot A_r \cdot \{s_n\}$$

or

$$\alpha \in \{s_1\}^* \cdot A_1 \cdot \{s_1\}^* \cdot A_2 \ldots \{s_1\}^* \cdot A_r \cdot \{s_1\}^* \cdot \{s_n\}$$

where the $A_1, \ldots, A_r$ are all regular sets. If

$$B = \left[\bigcup_{\gamma, \gamma' \in S'} \mathcal{R}'(\gamma, \gamma') \right] \cup \left[\bigcup_{\gamma \in S'} \{\gamma\} \right]$$

then B is also a regular set in $(S')^*$, $A_i \subseteq B$ and so either

$$\alpha \in \{s_1\}^* \cdot B \cdot \{s_1\}^* \cdot B \ldots \{s_1\}^* \cdot B \cdot \{s_n\}$$

or

$$\alpha \in \{s_1\}^* \cdot B \cdot \{s_1\}^* \cdot B \ldots \{s_1\}^* \cdot B \cdot \{s_1\}^* \cdot \{s_n\}.$$

It is also clear that if

$$\alpha_1 \in \{s_1\}^* \cdot B \cdot \{s_1\}^* \cdot B \ldots \{s_1\}^* \cdot B \cdot \{s_n\}$$

or

$$\alpha_1 \in \{s_1\}^* \cdot B \cdot \{s_1\}^* \cdot B \ldots \{s_1\}^* \cdot B \cdot \{s_1\}^* \cdot \{s_n\}$$

then

$$\alpha_1 \in \mathcal{R}(s_1, s_n)$$

and hence

$$\begin{aligned} \mathcal{R}(s_1, s_n) = {} & [\{s_1\}^* \cdot B \cdot \{s_1\}^* \cdot B \ldots \{s_1\}^* \cdot B \cdot \{s_n\}] \\ & \cup [\{s_1\}^* \cdot B \cdot \{s_1\}^* \cdot B_1 \ldots \\ & \ldots \{s_1\}^* \cdot B \cdot \{s_1\}^* \cdot \{s_n\}] \end{aligned}$$

which is a regular set in S^*; and this completes the inductive proof. □

Theorem 5.5.3

If A is a recognizable set of Σ^* then A is regular.

Proof Let $\mathfrak{M} = (\mathcal{M}, i, T)$, $\mathcal{M} = (Q, \Sigma, F)$ be such that $A = |\mathfrak{M}|$. Put $S = Q \times \Sigma$ and consider the set of words S^*. Define the relation $\mathcal{R}$ on S by

$$(q, \sigma)\mathcal{R}(q', \sigma') \Leftrightarrow q' = qF_\sigma \quad \text{for } q, q' \in Q, \sigma, \sigma' \in \Sigma.$$

Now let $\alpha \in A$ then $\alpha \in \Sigma^*$ and $iF_\alpha \in T$. If $\alpha = \sigma_1 \ldots \sigma_n$ then the sequence of states i, iF_{σ_1}, $iF_{\sigma_1\sigma_2}, \ldots, iF_{\sigma_1 \ldots \sigma_n}$ defines an $\mathcal{R}$-word in S^* namely:

$$(i, \sigma_1) \cdot (iF_{\sigma_1}, \sigma_2) \ldots (iF_{\sigma_1 \ldots \sigma_{n-1}}, \sigma_n)$$

which belongs to $\mathcal{R}((i, \sigma_1), (iF_{\sigma_1 \ldots \sigma_{n-1}}, \sigma_n))$ which is a regular set of S^*. Let

$$A' = \bigcup \{\mathcal{R}((i, \sigma), (q, \sigma')) \mid \sigma, \sigma' \in \Sigma, q \in Q, qF_{\sigma'} \in T\},$$

then $\alpha \in A'$. Conversely let $\beta \in A'$, then $\beta \in \mathcal{R}((i, \sigma), (q, \sigma'))$ for some $\sigma, \sigma' \in \Sigma$, $q \in Q$, and where $qF_{\sigma'} \in T$. If $\beta = (i, \sigma)(q_1, \sigma_1) \ldots (q_n, \sigma_n)(q, \sigma')$ then $\sigma\sigma_1 \ldots \sigma_n\sigma' \in A$. Define a function $f: S^* \to \Sigma^*$ by

$$f((q_1, \sigma_1) \ldots (q_n, \sigma_n)) = \sigma_1 \ldots \sigma_n,$$
$$f(\Lambda) = \Lambda,$$

then $f(A') = A$. Furthermore A' is regular by construction and theorem 5.5.2, and by using proposition 5.5.1 we see that A is also regular. □

We will now reformulate theorem 5.5.3 along with our results from section 5.3 to obtain:

Theorem 5.5.4

(Kleene) Let Σ be a finite non-empty set. The class of recognizable sets of Σ^* equals the class $\text{Reg}(\Sigma)$ of all regular sets of Σ^*.

This result then tells us that the only recognizable sets are those sets constructed from the singletons and $\varnothing$ using the rational operations.

5.6 Prefix decompositions of recognizable sets

The other decomposition of recognizable sets is based on an analysis of the type of sets that are recognized by recognizers with single final states.

Let $\mathfrak{M} = (\mathcal{M}, i, T)$ be a recognizer such that T is a singleton; we call $\mathfrak{M}$ a *direct recognizer*.

A recognizable set $A \subseteq \Sigma^*$ is called *unitary* if the minimal complete recognizer $\mathfrak{M}_A$ is direct.

Theorem 5.6.1

Let $A \subseteq \Sigma^*$ be a recognizable set. Then A is unitary if and only if $A \neq \varnothing$ and $\alpha^{-1} \cdot A = \beta^{-1} \cdot A$ for all $\alpha, \beta \in A$.

Proof Let $\mathfrak{M}_A = (\mathcal{M}_A, i_A, T_A)$ be the minimal complete recognizer for A and suppose that $T_A = \{t_A\}$. Let $\alpha, \beta \in A$, then

$$i_A F^A_\alpha = i_A F^A_\beta = t_A$$

and so

$$\alpha^{-1} \cdot A = \beta^{-1} \cdot A.$$

Clearly $A \neq \varnothing$ as $\mathfrak{M}_A$ is accessible. Now let $\alpha^{-1} \cdot A = \beta^{-1} \cdot A$ for all

$\alpha, \beta \in A$, then

$$T_A = \{\gamma^{-1} \cdot A \in Q_A \mid \gamma \in A\} = \{\alpha^{-1} \cdot A\}$$

and so T_A is a singleton. □

Theorem 5.6.2

Let $A \subseteq \Sigma^*$ be recognizable with $A \neq \emptyset$, then $A = \bigcup_{j=1}^{r} A_j$ where the A_j are unitary and $A_j \cap A_k = \emptyset$ if $j \neq k$.

Proof Let $\mathfrak{M}_A = (\mathcal{M}_A, i_A, T_A)$ be the minimal complete recognizer and suppose that $T_A = \{t_1, \ldots, t_r\}$.

Now each $t_j \in T_A$ is of the form $\alpha_j^{-1} \cdot A$ for some $\alpha_j \in A$. Let $A_j = \{\beta \in A \mid \beta^{-1} \cdot A = \alpha_j^{-1} \cdot A\}$ for $j = 1, \ldots, r$. Then A_j is the behaviour of the recognizer $\mathfrak{M}_j = (\mathcal{M}_A, i_A, \{t_j\})$ and so A_j is recognizable; furthermore if $\beta \in A_j$ then

$$i_A F_\beta^A = t_j$$

and so

$$\beta^{-1} \cdot A_j = \alpha_j^{-1} \cdot A$$

and this holds for all $\beta \in A_j$, and thus A_j is unitary. Since $a \in A$ if and only if $i_A F_a^A = t_j$ for some $j \in \{1, \ldots, r\}$ we have $a \in \bigcup_{j=1}^{r} A_j$ and thus $A = \bigcup_{j=1}^{r} A_j$. If $\gamma \in A_j \cap A_k$ with $j \neq k$ then

$$i_A F_\gamma^A = t_j \quad \text{and} \quad i_A F_\gamma^A = t_k$$

which is clearly false. Thus $A_j \cap A_k = \emptyset$. □

We call the sets A_j $(j = 1, \ldots, r)$ the *unitary components* of A.

Example 5.7

Let $\Sigma = \{0, 1\}$ and $A = \{0\} \cdot \{10\}^* \cup \{01\}^+$, then A is recognizable and the minimal complete recognizer is given by

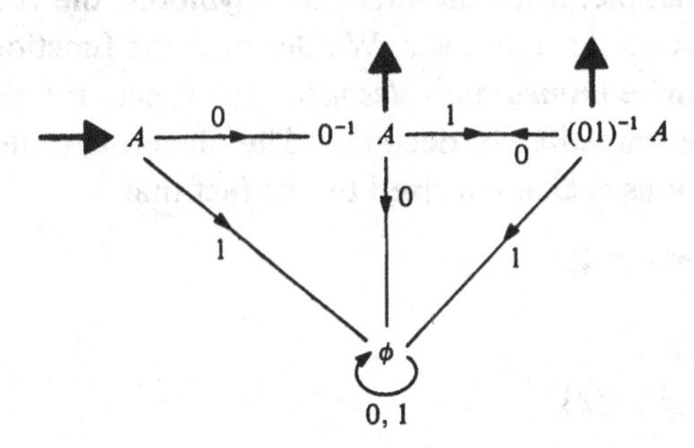

Let $A_1 = \{0\} \cdot \{10\}^*$, $A_2 = \{01\}^+$, then $A = A_1 \cup A_2$ where A_1 and A_2 are

both unitary sets. Now put $B_1=\{0\}$, $B_2=\{010\}\cdot\{10\}^*$, $B_3=\{01\}^+$; these are all unitary sets and $A=B_1\cup B_2\cup B_3$. Thus we see that the unitary decomposition may not be unique.

Let $A\subseteq\Sigma^*$ and suppose that $\alpha^{-1}\cdot A=\{\Lambda\}$ for all $\alpha\in A$. We call A a *prefix*. A prefix A then has the property that if $\alpha\in A$ the word α cannot be the start of another word from A, that is $\alpha x\notin A$ for all $x\in\Sigma^*$ except $x=\Lambda$. This concept is of considerable interest in coding theory. Here words are encoded by various methods so that transmission of messages across noisy channels can be achieved with as little distortion of the message as possible.

Example 5.8

Let $\Gamma=\{a,b,c,d,e\}$, $\Sigma=\{0,1\}$. We will encode a message, that is a word in Γ^*, into a word in Σ^* by specifying a function $f:\Gamma\to\Sigma^*$. Let $f(a)=1$, $f(b)=01$, $f(c)=001$, $f(d)=0001$, $f(e)=00001$, then the message

cbcdea

is encoded to

001010010001000011.

Now consider another coding function $f':\Gamma\to\Sigma^*$ given by $f'(a)=1$, $f'(b)=10$, $f'(c)=100$, $f'(d)=1000$, $f'(e)=10000$; then the message

cbcdea

is encoded to

100101001000100001.

The receiver will attempt to decode this as it is received and after three symbols all it can decide is that the first decoded symbol is not a or b. In our earlier example, after the first three symbols, the receiver knows that the first decoded symbol is c. We describe the function f as defining a code that can be *immediately decoded*. The function f' defines a code that cannot be immediately decoded. The algebraic difference between the two functions is characterized by the fact that

$$f(\Gamma)=\{0^n1\mid 0\le n\le 4\}$$

is a prefix whereas

$$f'(\Gamma)=\{10^n\mid 0\le n\le 4\}$$

is not a prefix.

It is immediate from theorem 5.6.1 that a prefix is unitary. Furthermore if A is recognizable and a prefix we can characterize the type of recognizer that recognizes A.

Theorem 5.6.3

Let $A \subseteq \Sigma^*$ be a recognizable set. Then A is a prefix if and only if the minimal complete recognizer $\mathfrak{M}_A$ is direct and $T_A * \Sigma = \varnothing$.

Proof If A is a prefix then $\alpha^{-1}A = \{\Lambda\}$ for all $\alpha \in A$ and so $T_A = \{\alpha^{-1} \cdot A \mid \alpha \in A\} = \{\{\Lambda\}\}$. Furthermore for $\sigma \in \Sigma$, $(\alpha^{-1} \cdot A)F_\sigma^A = (\alpha\sigma)^{-1} \cdot A$ and if $\beta \in (\alpha\sigma)^{-1} \cdot A$ then $\alpha\sigma\beta \in A$ which implies that $\sigma\beta \in \alpha^{-1} \cdot A = \{\Lambda\}$. Thus $(\alpha\sigma)^{-1} \cdot A = \varnothing$.

Conversely if A is recognizable and $\mathfrak{M}_A$ has the stated properties let $\alpha \in A$, then $i_A F_\alpha^A = \alpha^{-1} \cdot A \in T_A$ and so $\alpha^{-1} \cdot A = t_A$ where $T_A = \{t_A\}$. Suppose that $\beta \in \alpha^{-1} \cdot A$ with $\beta \neq \Lambda$, then $\alpha\beta \in A$. Let $\beta = \sigma\gamma$ where $\gamma \in \Sigma^*$, then $\alpha\sigma\gamma \in A$ and so $\gamma \in (\alpha\sigma)^{-1} \cdot A = t_A F_\sigma^A = \varnothing$ which is a contradiction. Hence $\alpha^{-1} \cdot A = \{\Lambda\}$ and A is a prefix. □

We have seen that a set A is a prefix if there are no words of the form $\alpha = \beta\gamma$ where both α and β belong to A. It is easy to construct the prefix part of any subset of Σ^* by removing all such words.

Let $A \subseteq \Sigma^*$ be recognizable, define the *prefix part of* A to be $A_p = A \backslash A \cdot \Sigma^+$. It is immediate that a recognizable subset A will be a prefix if and only if $A = A_p$. (Exercise 5.9 is concerned with the task of verifying that A_P is recognizable.)

We have seen that for a given recognizable subset A the prefix part A_P has some special properties. What can be said of the remainder of A? We first examine an example.

Example 5.9

Let $\Sigma = \{0, 1\}$ and $A = \{0\}^*\{1\} \cdot \{0\}^*$, then

$$A_P = A \backslash A \cdot \Sigma^+ = A \backslash \{0^n 1 0^m \mid m > 0\} = \{0\}^* \cdot \{1\}.$$

Notice that A is a unitary set and any element $\alpha \in A$ can be written in the form $\beta \cdot \gamma$ where $\beta \in A_P$ and $\gamma \in \{0\}^*$. Thus

$$A = A_P \cdot \{0\}^*.$$

We will now investigate the properties of $\{0\}^*$. Notice firstly that $\{0\}^*$ is a monoid, and secondly $\gamma^{-1} \cdot \{0\}^* = \{0\}^*$ for all $\gamma \in \{0\}^*$, that is $\{0\}^*$ is a unitary subset (it is clearly recognizable). We call $\{0\}^*$ a *unitary monoid*.

Our basic aim is the decomposition of a recognizable set into subsets of the form $A \cdot M$ where A is a prefix and M is a unitary monoid.

Let $B \subseteq \Sigma^*$; we call B a *unitary monoid* if

(i) B is a unitary subset of Σ^* (B is thus recognizable);

(ii) B is a submonoid of Σ^*.

Since B is a submonoid of Σ^* we see that $\Lambda \in B$ and thus $\gamma^{-1} \cdot B = B$ for all $\gamma \in B$.

Theorem 5.6.4

Given any unitary subset A of Σ^* the set

$$A_M = A^{-1} \cdot A = \{\gamma \in \Sigma^* \mid \alpha\gamma \in A \text{ for some } \alpha \in A\}$$

is a unitary monoid and $A = A_P \cdot A_M$.

Proof Since A is unitary the minimal complete recognizer $\mathfrak{M}_A = (\mathcal{M}_A, i_A, T_A)$ is direct. Let $T_A = \{t_A\}$ and consider the recognizer $\bar{\mathfrak{M}}_A = (\mathcal{M}_A, t_A, T_A)$. Now

$$\begin{aligned} \beta \in |\bar{\mathfrak{M}}_A| &\Leftrightarrow t_A F^A_\beta = t_A \\ &\Leftrightarrow i_A F^A_{\alpha\beta} = t_A \quad \text{for any } \alpha \in A \\ &\Leftrightarrow \alpha\beta \in A \quad \text{for any } \alpha \in A \\ &\Leftrightarrow \beta \in \alpha^{-1} \cdot A \quad \text{for any } \alpha \in A \\ &\Leftrightarrow \beta \in A^{-1} \cdot A. \end{aligned}$$

Thus $A^{-1} \cdot A$ is recognizable and unitary. Furthermore, if $\beta, \beta' \in A^{-1} \cdot A$ then clearly $t_A F^A_{\beta\beta'} = t_A$ and so $\beta\beta' \in A^{-1} \cdot A$ and $A^{-1} \cdot A$ is a unitary monoid. Now let $\alpha \in A$, then $i_A F^A_\alpha = t_A$. The sequence of states defined by α may contain t_A several times. If we put $\alpha = \gamma\beta$ when $\gamma \in A$ and $\beta \in A^{-1} \cdot A$ such that the path from i_A to t_A labelled by γ contains only one occurrence of t_A, namely the last one, then $\gamma \in A_P = A \backslash A\Sigma^+$. Thus $A \subseteq A_P \cdot A_M$ and the reverse inclusion is obvious. □

Theorem 5.6.5

Let A be a recognizable subset of Σ^*. Then

$$A = B_1C_1 \cup B_2C_2 \cup \ldots \cup B_rC_r$$

where B_iC_i are unitary subsets, B_i are prefixes and C_i are unitary monoids for $i = 1, 2, \ldots, r$.

Proof Using theorem 5.6.2 we have $A = A_1 \cup A_2 \cup \ldots \cup A_r$ where each A_i is a unitary subset. Now let $B_i = (A_i)_P = A_i \backslash A_i\Sigma^+$, $C_i =$

$(A_i)_M = (A_i)^{-1} \cdot A_i$ for $i = 1, \ldots, r$, then $A_i = B_iC_i$ and B_i is a prefix and C_i is a unitary monoid. □

This decomposition is called the *unitary–prefix decomposition.* We finish our discussion with some examples.

Example 5.10

Let $\Sigma = \{0, 1\}$ and $A = \{0\}^* \cdot \{\{10\} \cup \{0\}\}^* \cdot \{0\}^*$. This is recognizable by construction. Let $B = \{\{10\} \cup \{0\}\}^* \cdot \{0\}^*$. Now $0^{-1} \cdot A = A$, $1^{-1} \cdot A = 0B$, $(01)^{-1} \cdot A = 1^{-1} \cdot A = 0B$, $(10)^{-1} \cdot A = B$, $(11)^{-1} \cdot A = \varnothing$, $(100)^{-1} \cdot A = B$, $(101)^{-1} \cdot A = 0B$. The minimal complete recognizer is given by:

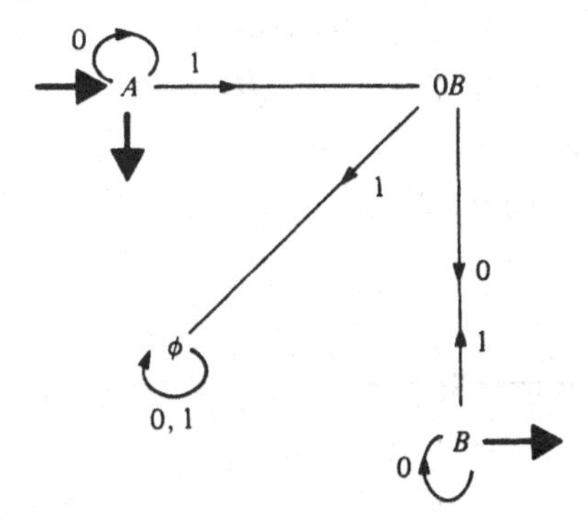

with $t_A = \{A, B\}$. The unitary decomposition is $A = \{0\}^* \cup \{0\}^*\{10\}B$. Now $\{0\}^* = \{\Lambda\} \cdot \{0\}^*$ where $\{\Lambda\}$ is a prefix and $\{0\}^*$ is a unitary monoid. Also $\{0\}^*\{10\}$ is a prefix and B is a unitary monoid.

Example 5.11

Let $\Sigma = \{0, 1\}$ and $A = (\{01\}^+ \cdot \{10\}) \cup (\{10\}^+ \cdot \{110\})$. Then $0^{-1} \cdot A = \{1\} \cdot \{01\}^* \cdot \{10\}$, $1^{-1} \cdot A = \{0\} \cdot \{10\}^* \cdot \{110\}$, $(01)^{-1} \cdot A = \{01\}^* \cdot \{10\}$, $(10)^{-1} \cdot A = \{10\}^* \cdot \{110\}$, $(011)^{-1} \cdot A = \{0\}$, $\{101\}^{-1} \cdot A = 1^{-1} \cdot A \cup \{10\}$, $(010)^{-1} \cdot A = 0^{-1} \cdot A$, $(0110)^{-1} \cdot A = \{\Lambda\} = (10110)^{-1} \cdot A$ etc.

The minimal complete recognizer is:

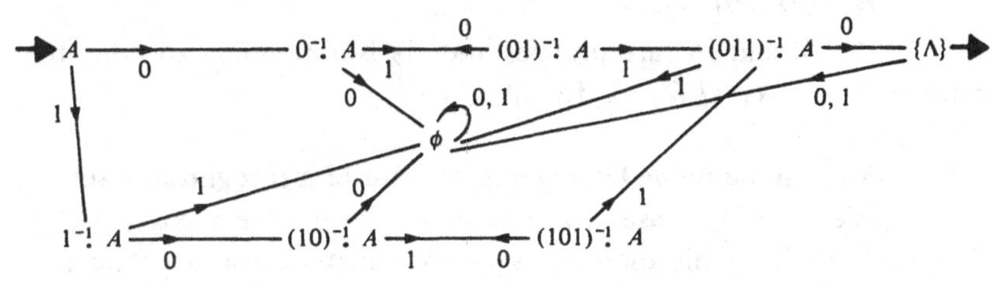

From the diagram we note that $A = A_P$ is a prefix and so $A = A \cdot \{\Lambda\}$ is the unitary decomposition.

Example 5.12

Let $\Sigma = \{0, 1\}$ and suppose that A is the set of words of Σ^* containing an equal number of 0s and 1s. Let A_1 be the set of words of A containing an odd number of 0s and A_2 the set of words of A containing an even number of 0s. Then $A = A_1 \cup A_2$. Let us try to construct the minimal complete recognizer for A.

Let

$$B_j = \{\alpha \in \Sigma^* \text{ such that } \alpha \text{ has } j \text{ more 1s than 0s}\}$$

$$C_j = \{\alpha \in \Sigma^* \text{ such that } \alpha \text{ has } j \text{ more 0s than 1s}\}$$

Then $\quad 0^{-1} \cdot A = B_1, (01)^{-1} \cdot A = A, (00)^{-1} \cdot A = B_2, \ldots$

$\quad 1^{-1} \cdot A = C_1, (10)^{-1} \cdot A = A, (11)^{-1} \cdot A = C_2, \ldots$

The 'machine' will have a graph of the following form:

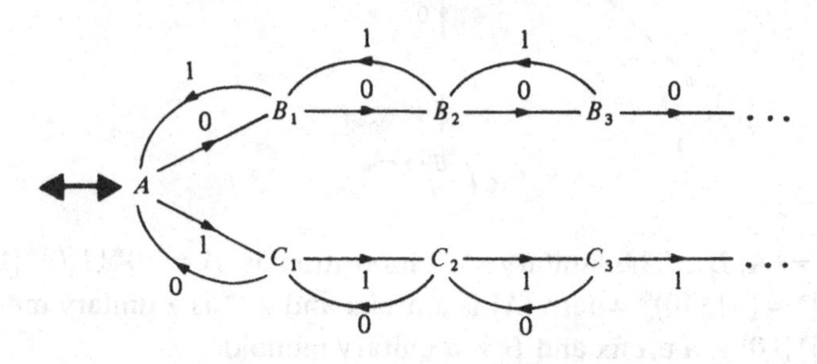

and it is clear that the set of states will have to be infinite. In fact it can easily be shown that A is not recognizable. However it is possible to devise a decomposition for A that is similar, in some respects, to the unitary–prefix decomposition. Notice that A_2 is a monoid and satisfies the condition $\alpha^{-1} \cdot A_2 = \beta^{-1} \cdot A_2 = A_2$ for any $\alpha, \beta \in A_2$. However A_2 is not a unitary monoid since it is not recognizable. Similarly $A_1 = \{01, 10\} \cdot A_2$ and $\{01, 10\}$ is a prefix. Thus

$$A = \{01, 10\} \cdot A_2 \cup \{\Lambda\} \cdot A_2$$

where $\{01, 10\}$ and $\{\Lambda\}$ are prefixes and A_2 is a monoid satisfying the condition $\alpha^{-1} \cdot A_2 = \beta^{-1} \cdot A_2$ for all $\alpha, \beta \in A_2$.

5.7 The pumping lemma and the size of a recognizable set

We examine a useful technique for testing the recognizability of subsets of Σ^*. This then leads us to a method for deciding if a recognizable subset is finite. Following this we investigate the size of an infinite recognizable set.

Lemma 5.7.1

(Pumping lemma) Let $A \subseteq \Sigma^*$ be recognizable and suppose that $n = |Q_A|$, the number of states in the minimal complete recognizer of A. If $\alpha \in A$ and the length of α is greater than or equal to n then

$$\alpha = \beta\gamma\delta$$

such that

(i) $\gamma \neq \Lambda$,
(ii) $\{\beta\} \cdot \{\gamma\}^* \cdot \{\delta\} \subseteq A$.

Proof Suppose that $\alpha \in A$, then $\alpha^{-1}A \in T_A$. The sequence of states $i_A = q_0, q_1, \ldots, q_r = \alpha^{-1} \cdot A$ defined by the word α is of length $n+1$. There must therefore be repetitions so that $q_j = q_k$ with $j \neq k$. Consider the word $\gamma \in \Sigma^*$ obtained by passing along the path defined by α between q_j and q_k. Then clearly a word $\beta \in \Sigma^*$ and a word $\delta \in \Sigma^*$ exist such that

$$i_A F_\beta^A = q_j,\ q_j F_\gamma^A = q_j,\ q_j F_\delta^A = \alpha^{-1} \cdot A = i_A F_\alpha^A.$$

Then $\alpha = \beta\gamma\delta$, $\gamma \neq \Lambda$ and any word of the form $\beta\gamma^m\delta$ is recognized. □

Example 5.13

Consider the recognizer in example 5.10. Here $n = 4$ and if $\alpha = 00010100$ we see that $\beta = 000$, $\gamma = 1010$, $\delta = 0$ gives a suitable decomposition $\alpha = \beta\gamma\delta$. Others exist, for example $\beta' = 0$, $\gamma' = 00$, $\delta' = 10100$. Notice that $\{\beta\} \cdot \{\gamma\}^* \cdot \{\delta\} \neq \{\beta'\} \cdot \{\gamma'\}^* \cdot \{\delta'\}$.

We see then that the existence in the recognizable set of a word of length at least n will guarantee that the set is infinite. If $A \subseteq \Sigma^*$ is a finite recognizable set and $n = |Q_A|$ then no words of length n can exist in A.

If $A \subseteq \Sigma^*$ let us define $A^{(n)}$ to be the set of all words of A that are of length n for $n = 0, 1, \ldots$. Then

$$A = \bigcup_{n=0}^{\infty} A^{(n)}.$$

For a finite set A we will have $A^{(n)} = A^{(n+1)} = \ldots = \varnothing$ for some value of n. For an infinite set each $A^{(n)}$ is finite, in fact

$$|A^{(n)}| \leq k^n \quad \text{where } k = |\Sigma|.$$

Our next task is to find some information about the size of the sets $A^{(n)}$ when A is a recognizable subset of Σ^*.

First let $\mathcal{M} = (Q, \Sigma, F)$ be a complete finite state machine and let $|Q| = m$. The machine $\mathcal{M}$ can be described by a set of $m \times m$ matrices that effectively define the action of F.

First we let $Q = \{q_1, \ldots, q_m\}$ and then for each $\sigma \in \Sigma$ define the matrix

$$\mathscr{f}_\sigma = (f_{ij}^\sigma) \quad \text{where } f_{ij}^\sigma = \begin{cases} 1 & \text{if } q_i F = q_j \\ 0 & \text{otherwise} \end{cases}$$

for $i, j \in \{1, \ldots, m\}$.

Each row of the matrix $\mathscr{f}_\sigma$ will consist of one 1 and $(m-1)$ 0s. Each state q_j will be represented by a $1 \times m$ row vector e_j of the form $(0 \ldots 010 \ldots 0)$ with a 1 in the j-th position. So that $q_j F_\sigma = q_k$ will be replaced by the matrix equation $e_j \cdot \mathscr{f}_\sigma = e_k$.

Given $\alpha = \sigma_1 \ldots \sigma_n \in \Sigma^*$ we define $\mathscr{f}_\alpha = \mathscr{f}_{\sigma_1} \ldots \mathscr{f}_{\sigma_n}$ and notice that

$$q_j F_\alpha = q_k \Leftrightarrow e_j \cdot \mathscr{f}_\alpha = e_k.$$

Finally we put $\mathscr{f}_\Lambda = I_m$, the $m \times m$ identity matrix. If $\mathfrak{M} = (\mathcal{M}, i, T)$ is a recognizer, let $i = q_1$ and define

$$\mathscr{E} = \{e_j \mid q_j \in T\},$$

then for each $\alpha \in |\mathfrak{M}|$ we have $e_1 \cdot \mathscr{f}_\alpha \in \mathscr{E}$ and clearly

$$|\mathfrak{M}| = \{\alpha \in \Sigma^* \mid e_1 \cdot \mathscr{f}_\alpha \in \mathscr{E}\}.$$

Let $\mathscr{F} = \sum_{\sigma \in \Sigma} \mathscr{f}_\sigma$, which is again an $m \times m$ matrix (it belongs to the set of all $m \times m$ matrices over the integers); we call $\mathscr{F}$ the *matrix* of $\mathcal{M}$. For any subset $R \subseteq Q$ we define

$$\mathscr{E}(R) = \{e_j \mid q_j \in R\}$$

and consider

$$\mathfrak{E}(R) = \sum_{e_j \in \mathscr{E}(R)} = e_j^T.$$

(e_j^T is the transpose of e_j and thus $\mathfrak{E}(R)$ is a column vector.)

Theorem 5.7.2

Let $\mathfrak{M} = (\mathcal{M}, i, T)$ be a recognizer with matrix $\mathscr{F}$. Let R be a set of states of $\mathcal{M}$ and $k \geq 0$, then the number of words of Σ^* of length k which send the initial state i to a state in R is given by

$$e_1 \cdot (\mathscr{F})^k \cdot \mathfrak{E}(R).$$

Proof Let Σ^k be the set of words of Σ^* of length k. Then $\mathscr{F}^k = \sum_{\alpha \in \Sigma^k} \mathscr{f}_\alpha$ and $e_1 \cdot (\mathscr{F})^k = \sum_{\alpha \in \Sigma^k} e_1 \cdot \mathscr{f}_\alpha = (a_{11}, \ldots, a_{1m})$ where m is the number of states in $\mathcal{M}$ and a_{ij} is the number of words of length k

that send state i to state q_j. The number of words of length k that send state i to a state in R is then given by

$$a_{11}b_1 + \ldots + a_{1m}b_m \quad \text{where } b_j = 1 \text{ if } q_j \in R$$
$$\text{and} \quad b_j = 0 \text{ if } q_j \notin R.$$

This is just $(a_{11}, \ldots, a_{1m}) \cdot \mathfrak{E}(R)$. □

Corollary 5.7.3

The number of words in $|\mathfrak{M}|$ of length k is given by

$$e_1 \cdot (\mathcal{F})^k \cdot \mathfrak{E}(T).$$

The total number of words in $|\mathfrak{M}|$ is given by

$$e_1 \cdot (\mathcal{F}^0 + \mathcal{F}^1 + \mathcal{F}^2 + \ldots + \mathcal{F}^k + \ldots) \cdot \mathfrak{E}(T) = e_1 \cdot (I - \mathcal{F})^{-1} \cdot \mathfrak{E}(T).$$

In the case of an infinite set $|\mathfrak{M}|$ the matrix $I - \mathcal{F}$ will be singular and so great care must be taken with this notation.

Example 5.14

Let $\Sigma = \{0, 1\}$ and consider the state machine $\mathcal{M}$ given by

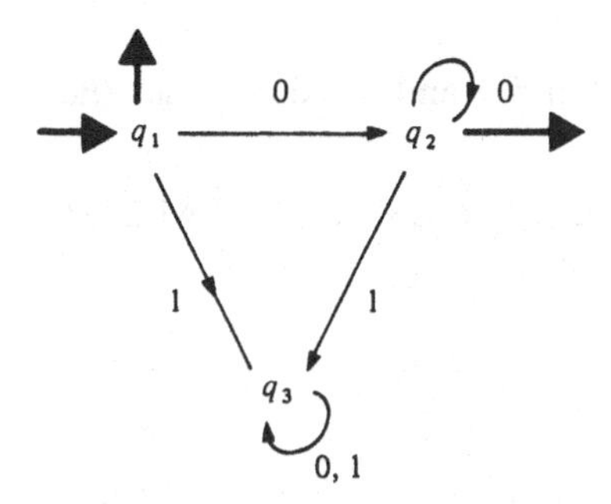

Let $i = q_1$, $T = \{q_1, q_2\}$. Then if $\mathfrak{M} = (\mathcal{M}, i, T)$, $|\mathfrak{M}| = \{0\}^*$.

$$f_0 = \begin{pmatrix} 0 & 1 & 0 \\ 0 & 1 & 0 \\ 0 & 0 & 1 \end{pmatrix}, \quad f_1 = \begin{pmatrix} 0 & 0 & 1 \\ 0 & 0 & 1 \\ 0 & 0 & 1 \end{pmatrix}, \quad \mathcal{F} = \begin{pmatrix} 0 & 1 & 1 \\ 0 & 1 & 1 \\ 0 & 0 & 2 \end{pmatrix},$$

$$e_1 = (1, 0, 0), \qquad \mathfrak{E}(T) = \begin{pmatrix} 1 \\ 1 \\ 0 \end{pmatrix}$$

and the number of words of $|\mathfrak{M}|$ of length 2 is given by

$$e_1 \cdot \mathcal{F}^2 \cdot \mathfrak{E}(T) = (1, 0, 0) \cdot \begin{pmatrix} 0 & 1 & 3 \\ 0 & 1 & 3 \\ 0 & 0 & 4 \end{pmatrix} \cdot \begin{pmatrix} 1 \\ 1 \\ 0 \end{pmatrix} = 1.$$

The use of matrix theory leads to considerable insights into the behaviour of recognizable sets. However the interested reader is referred to the literature for further details (e.g. Cohn [1975]).

5.8 Exercises

5.1 Let $x \in \Sigma^*$. Prove that $\{x\}$ is recognizable.

5.2 Let $\mathfrak{M} = (\mathcal{M}, i, T)$, $\mathcal{M} = (Q, \Sigma, F)$ and $q^{-1} \circ T = q_1^{-1} \circ T \Rightarrow q = q_1$ for all $q, q_1 \in Q$. Prove that if $A = |\mathfrak{M}|$ then $\mathfrak{M} \cong \mathfrak{M}_A$.

5.3 Let $\mathfrak{M} = (\mathcal{M}, i, T)$, $\mathcal{M} = (Q, \Sigma, F)$. Define a relation E on Q by

$$qEq_1 \Leftrightarrow q^{-1} \circ T = q_1^{-1} \circ T \quad \text{for } q, q_1 \in Q.$$

Consider $\mathcal{M}/E = (Q/E, \Sigma, \bar{F})$ where $\bar{F}$ is defined by

$$[q]\bar{F}_\sigma = [qF_\sigma] \quad \text{for } [q] \in Q/E, \sigma \in \Sigma.$$

Show that $\mathcal{M}/E$ is well-defined and if

$$\mathfrak{M}/E = (\mathcal{M}/E, [i], \{[t]/t \in T\})$$

then

$$|\mathfrak{M}/E| = |\mathfrak{M}|.$$

5.4 With $\mathfrak{M}$ and $\mathcal{M}$ as defined in 5.3 and $n \geq 0$ consider the relation E_n on Q given by

$$qE_nq_1 \Leftrightarrow \{q\alpha \in T \Leftrightarrow q_1\alpha \in T \text{ for all } \alpha \in \Sigma^* \text{ with } |\alpha| \leq n\}.$$

Prove that $E = \bigcap_{n \geq 0} E_n$.

5.5 Complete the proof of theorem 5.3.3.

5.6 Prove theorem 5.3.5.

5.7 Prove theorem 5.3.6.

5.8 If $A, B, C \subseteq \Sigma^*$ prove that

$$(A \cdot B)^{-1} \cdot C = B^{-1}(A^{-1} \cdot C)$$
$$(A^{-1} \cdot B) \cdot C \subseteq A^{-1} \cdot (B \cdot C).$$

5.9 If $A \subseteq \Sigma^*$ is recognizable then so is $A_P = A \backslash A\Sigma^+$.

5.10 *If* π is an admissible partition on $\mathcal{M}$ what does

$$\mathfrak{M}/\pi = (\mathcal{M}/\pi, [i], [T])$$

recognize?

6

Sequential machines and functions

Mealy machines were briefly introduced in chapter 2 to provide a motivational basis for the discussion of products of state machines and transformation semigroups. In this chapter, Mealy machines and their associated functions will be examined in their own right and some of the results from earlier chapters will be applied to them.

6.1 Mealy machines again

Recall that a Mealy machine, as defined in section 2.5, is a quintuple $\hat{\mathcal{M}} = (Q, \Sigma, \Theta, F, G)$ where Q, Σ and Θ are finite sets, and

$$F: Q \times \Sigma \to Q,$$

$$G: Q \times \Sigma \to \Theta$$

are functions.

Thus $\mathcal{M} = (Q, \Sigma, F)$ is a complete state machine. It is now reasonable to extend our concept slightly by including the possibilities that either $F: Q \times \Sigma \to Q$ or $G: Q \times \Sigma \to \Theta$ are partial functions rather than functions. Thus $\mathcal{M} = (Q, \Sigma, F)$ may not be complete.

A *Mealy machine* is now understood to be a quintuple $\hat{\mathcal{M}} = (Q, \Sigma, \Theta, F, G)$ where Q, Σ, Θ are finite sets (Σ and Θ being non-empty) and $F: Q \times \Sigma \to Q$, $G: Q \times \Sigma \to \Theta$ partial functions. If F and G are both functions we say that $\hat{\mathcal{M}}$ is a *complete* Mealy machine.

It is now inappropriate to describe Mealy machines by directed graphs, we will have to use tables.

Example 6.1

Let

$Q = \{q_1, q_2, q_3\}$ $\quad\quad$ $\Sigma = \{0, 1\}$ $\quad\quad$ $\Theta = \{a, b\}$.

We define $F: Q \times \Sigma \to Q$ and $G: Q \times \Sigma \to \Theta$ by the table

$\hat{\mathcal{M}}$		q_1	q_2	q_3
F	0	q_1	$\varnothing$	q_3
	1	q_2	q_3	$\varnothing$
G	0	a	a	a
	1	$\varnothing$	b	$\varnothing$

Thus

$$q_1F_0 = q_1, \quad q_2F_0 \text{ is undefined,}$$

$$q_1G_0 = a, \quad q_1G_1 \text{ is undefined etc.}$$

(Some authors use a dash – instead of the symbol $\varnothing$ in such tables.)

If $\hat{\mathcal{M}} = (Q, \Sigma, \Theta, F, G)$ satisfies the property that

$$qF_\sigma = \varnothing \Leftrightarrow qG_\sigma = \varnothing$$

we can use the directed graph method of describing $\hat{\mathcal{M}}$. Such Mealy machines are called *normal.*

$\hat{\mathcal{M}}$ is called *state complete* if F is a function and *output complete* if G is a function.

Example 6.2

$Q = \{q_1, q_2, q_3\}$, $\Sigma = \{0, 1\}$, $\Theta = \{a, b\}$.

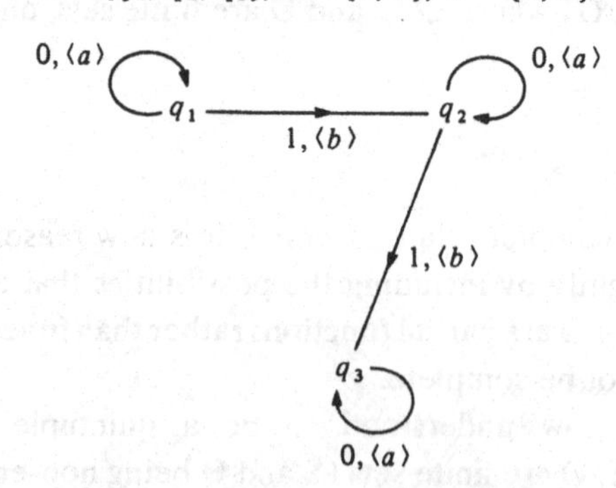

represents the normal Mealy machine:

$\hat{\mathcal{M}}$		q_1	q_2	q_3
F	0	q_1	q_2	q_3
	1	q_2	q_3	$\varnothing$
G	0	a	a	a
	1	b	b	$\varnothing$

The output of a Mealy machine clearly depends on the set of states that are traversed in the process of the operation of the machine. We have seen that for complete machines, if $\sigma_1\sigma_2\ldots\sigma_k \in \Sigma$ and $q \in Q$ then the output word of Θ^* obtained when $\sigma_1\sigma_2\ldots\sigma_k$ is the input word and q is the initial state is given by

$$\theta_1\theta_2\ldots\theta_k$$

where

$$\theta_1 = qG_{\sigma_1}$$
$$\theta_2 = qF_{\sigma_1}G_{\sigma_2}$$
$$\vdots$$
$$\theta_k = qF_{\sigma_1\ldots\sigma_{k-1}}G_{\sigma_k}.$$

The state q defines a function $f_q : \Sigma^* \to \Theta^*$ described by

$$f_q(\sigma_1\sigma_2\ldots\sigma_k) = \theta_1\theta_2\ldots\theta_k, \quad \text{for } \sigma_1\sigma_2\ldots\sigma_k \in \Sigma^+.$$

It is clear that f_q satisfies the following properties

$$f_q(\sigma) = qG_\sigma$$
$$f_q(x\sigma) = (f_q(x))(qF_xG_\sigma) = f_q(x)f_{qF_x}(\sigma), \quad \text{for } \sigma \in \Sigma, x \in \Sigma^+,$$

and these are enough to define f_q.

We will also ask that f_q satisfies the property $f_q(\Lambda) = \Lambda$.

Proposition 6.1.1

If $\hat{\mathcal{M}} = (Q, \Sigma, \Theta, F, G)$ is a complete Mealy machine and $q \in Q$ then for $x, y \in \Sigma^*$

$$f_q(xy) = f_q(x)f_{qF_x}(y).$$

Proof We proceed by induction on the length of y. If $y = \sigma \in \Sigma$ then

$$f_q(x\sigma) = f_q(x)qF_xG_\sigma$$
$$= f_q(x)f_{qF_x}(\sigma).$$

We assume now that $f_q(xy) = f_q(x)f_{qF_x}(y)$ holds for all words $y \in \Sigma^*$ of length less than t, and states $q \in Q$.

Let $y = \sigma_1\ldots\sigma_{t-1}\sigma_t$ and put $z = \sigma_1\ldots\sigma_{t-1}$, then

$$f_q(xz) = f_q(x)f_{qF_x}(z).$$

Now

$$\begin{aligned} f_q(xy) &= f_q(xz\sigma_t) \\ &= f_q(xz)qF_{xz}G_{\sigma_t} \\ &= f_q(x)f_{qF_x}(z)qF_{xz}G_{\sigma_t} \\ &= f_q(x)f_{qF_x}(z)qF_xF_zG_{\sigma_t} \\ &= f_q(x)f_{qF_x}(z\sigma_t) \\ &= f_q(x)f_{qF_x}(y). \end{aligned}$$

For $x = \Lambda$ or $y = \Lambda$ the result is immediate. □

Next we turn to the case where $\hat{\mathcal{M}}$ is not complete. For $q \in Q$ we can define a partial function $f_q : \Sigma^* \to \Theta^*$ by

$$f_q(\sigma_1\sigma_2 \ldots \sigma_k) = \theta_1\theta_2 \ldots \theta_k \quad \text{for } \sigma_1\sigma_2 \ldots \sigma_k \in \Sigma^*$$

if

$$qF_{\sigma_1}, qG_{\sigma_1}, qF_{\sigma_1\sigma_2}, qF_{\sigma_1}G_{\sigma_2}, \ldots, qF_{\sigma_1\ldots\sigma_k}, qF_{\sigma_1\ldots\sigma_{k-1}}G_{\sigma_k}$$

are all defined and $f_q(\Lambda) = \Lambda$.

However if $qF_{\sigma_1}, qF_{\sigma_1\sigma_2}, \ldots, qF_{\sigma_1\sigma_2\ldots\sigma_{i-1}}$ are defined but $qF_{\sigma_1\sigma_2\ldots\sigma_i} = \varnothing$ we note that the machine stops completely and no more output symbols can be printed.

The case where $qF_{\sigma_1}, \ldots, qF_{\sigma_1\ldots\sigma_{i-1}}$ are all defined but $qG_{\sigma_1}, \ldots, qF_{\sigma_1\ldots\sigma_{i-1}}G_{\sigma_i}$ are not all defined can be dealt with as follows. If $qF_{\sigma_1\ldots\sigma_{i-1}}G_{\sigma_i} = \varnothing$ we regard the output as a blank space on the output tape. This will be denoted by □. Consequently the output word could take the form

$$\theta_1 \ldots \theta_i \,\square\, \theta_{i+1} \ldots \theta_k.$$

(A slightly different interpretation will be used in sections 6.4 and 6.5.) We should not confuse □ with Λ. A simple example will illustrate some of these points.

Example 6.3
$\hat{\mathcal{M}}$ is defined by the table:

$\hat{\mathcal{M}}$		q_1	q_2
F	0	$\varnothing$	q_1
	1	q_2	q_2
G	0	a	$\varnothing$
	1	b	b

so that

$$q_1F_0 = \varnothing, q_1F_1G_0 = \varnothing \text{ etc.}$$

Now $f_{q_1}(0) = a$, $f_{q_1}(10) = b\Box$, $f_{q_1}(101) = b\Box b$, $f_{q_1}(100) = b\Box a$, and $f_{q_1}(01) = \varnothing$ since $q_1F_0 = \varnothing$. But what is $f_{q_1}(1001)$? Either $f_{q_1}(1001) = \varnothing$ since the machine stops before all the input word has been fed in, that is after 100 in this case, or we put $f_{q_1}(1001) = b\Box a$, that is $f_{q_1}(1001) = f_{q_1}(100)$ where 100 is the smallest initial segment of 1001 for which the machine produces a complete output.

To avoid these problems we will only consider $f_q(x)$ to be defined if

$$qF_{\sigma_1}, qF_{\sigma_1\sigma_2}, \ldots, qF_{\sigma_1\ldots\sigma_{i-1}}$$

are all defined where $x = \sigma_1\sigma_2 \ldots \sigma_i \in \Sigma^*$. We describe this by saying that x is *applicable* to q. This will guarantee that the image $f_q(x)$ is of the same length as the length of x whenever x is applicable to q. Notice that the length of $\Box$ is 1, whereas the length of Λ is 0 and clearly in our example $f_{q_2}(\Lambda) = \Lambda$ (no tape goes in and no tape comes out!) whereas $f_{q_2}(0) = \Box$ (a blank tape comes out of length 1).

If xy is applicable to q then the conclusion of 6.1.1, namely $f_q(xy) = f_q(x)f_{qF_x}(y)$ is valid. Clearly if $x \in \Sigma^*$ is not applicable to q then neither is xy for any $y \in \Sigma^*$.

We can, to a certain extent, overcome some of the difficulties concerned with the applicability of inputs by moving to the completion of the Mealy machine.

Let $\hat{\mathcal{M}} = (Q, \Sigma, \Theta, F, G)$ be a Mealy machine such that the state machine $\mathcal{M} = (Q, \Sigma, F)$ is incomplete. Let $\mathcal{M}^c = (Q \cup \{z\}, \Sigma, F')$ be the completion of $\mathcal{M}$ and define $G' : (Q \cup \{z\}) \times \Sigma \to \Theta$ by

$$\begin{aligned} G'(q, \sigma) &= G(q, \sigma) && \text{for } q \in Q, \sigma \in \Sigma, \\ G'(z, \sigma) &= \varnothing && \text{for } \sigma \in \Sigma. \end{aligned}$$

Then $\hat{\mathcal{M}}^c = (Q \cup \{z\}, \Sigma, \Theta, F', G')$ is called the *state completion* of $\hat{\mathcal{M}}$.

We now notice that every $x \in \Sigma^+$ is applicable to any state $q' \in Q \cup \{z\}$ and so $f'_{q'}$ is defined as a function although from Σ^* to $(\Theta \cup \{\Box\})^*$.

If $\sigma_1\sigma_2 \ldots \sigma_k$ is not applicable to q in the original machine $\hat{\mathcal{M}}$ but $\sigma_1\sigma_2 \ldots \sigma_{k-1}$ is applicable to q, the output obtained by applying $\sigma_1\sigma_2 \ldots \sigma_k$ to q in the state completion is given by $f_q(\sigma_1 \ldots \sigma_{k-1})\Box$.

Note that $f'_q(x) = f_q(x)$ if x is applicable to q.

In general if $\hat{\mathcal{M}}$ is an arbitrary Mealy machine and q is a state of $\hat{\mathcal{M}}$ then q defines a partial function $f_q : \Sigma^* \to (\Theta \cup \{\Box\})^*$ where $f_q(x)$ is defined only when x is applicable to q with $x \in \Sigma^*$.

Put $\Theta_1 = \Theta \cup \{\square\}$. Let $x, y \in \Theta_1^*$, say

$$x = \alpha_1\alpha_2 \ldots \alpha_k, \; y = \alpha'_1\alpha'_2 \ldots \alpha'_k \quad \text{where } \alpha_i, \alpha'_i \in \Theta_1.$$

We say that x *covers* y if, for each $1 \leq i \leq k$ we have either $\alpha_i = \alpha'_i$ or $\alpha'_i = \square$. This is written as $x \# y$. We say that x and y are *compatible*, written $x \| y$ if, for each $1 \leq i \leq k$, we have either $\alpha_i = \alpha'_i$ or $\alpha_i = \square$ or $\alpha'_i = \square$.

Clearly $x \# y$ implies $x \| y$.

Thus for $\Theta = \{0, 1\}$ we have $0\square110 \# \square\square1\square0$ and $0\square110 \| \square\square1\square0$, also $\square\square1\square0 \| 0110\square$. Notice however that $0\square110 \| 0110\square$ is false and so compatibility is not a transitive relation. The relation 'covers' is not even symmetric.

Turning from compatibility amongst words to a related concept for states we proceed to the following definition. Let $q, q_1 \in Q$, we say that q and q_1 are *compatible* (or output compatible) if

$$f_q(x) \| f_{q_1}(x) \quad \text{for all } x \in \Sigma^* \text{ and } x \text{ applicable to } q \text{ and } q_1,$$

and write $q \| q_1$. If two states are compatible and the machine is started in either of these states then the output words will not be 'noticeably different', they may not be identical but where they do differ one word will have a blank space at that position.

One basic aim is to construct a Mealy machine with a state set of minimal size that will behave in the same way as a given Mealy machine. This involves looking at the partial functions f_q for each state q in the original machine.

6.2 Minimizing Mealy machines

We first consider a complete Mealy machine $\hat{\mathcal{M}} = (Q, \Sigma, \Theta, F, G)$. Define a relation $\sim$ on Q by

$$q \sim q_1 \Leftrightarrow f_q = f_{q_1}, \quad \text{where } q, q_1 \in Q.$$

A machine $\hat{\mathcal{M}}/\!\sim\, = (Q/\!\sim, \Sigma, \Theta, F', G')$ can now be constructed by defining

$$\left.\begin{aligned}[q]F'_\sigma &= [qF_\sigma] \\ [q]G'_\sigma &= qG_\sigma\end{aligned}\right\} \text{for } q \in Q, \sigma \in \Sigma,$$

where $[q]$ denotes the $\sim$-class containing q. This definition is meaningful since $\sim$ is an equivalence relation and if $q \sim q_1$ and $\sigma \in \Sigma$, $x \in \Sigma^*$ then

$$f_q(\sigma x) = f_q(\sigma) f_{qF_\sigma}(x) \quad \text{by 6.1.1}$$

and

$$f_{q_1}(\sigma x) = f_{q_1}(\sigma) \cdot f_{q_1F_\sigma}(x) = f_q(\sigma) \cdot f_{q_1F_\sigma}(x)$$

so that

$$f_{qF_\sigma} = f_{q_1F_\sigma} \quad \text{and thus } qF_\sigma \sim q_1F_\sigma.$$

Furthermore

$$qG_\sigma = f_q(\sigma) = f_{q_1}(\sigma) = q_1G_\sigma \quad \text{for } \sigma \in \Sigma.$$

We will see that $f'_{[q]} : \Sigma^* \to \Theta^*$, the function defined by $\hat{\mathcal{M}}/\sim$ in state $[q]$ equals $f_q : \Sigma^* \to \Theta^*$.

We call $\hat{\mathcal{M}}/\sim$ the *minimal Mealy machine* of $\hat{\mathcal{M}}$. The reason for this name is to be found in 6.2.2.

Theorem 6.2.1

Let $\hat{\mathcal{M}} = (Q, \Sigma, \Theta, F, G)$ be a complete Mealy machine. If $\hat{\mathcal{M}}/\sim$ is the minimal Mealy machine of $\hat{\mathcal{M}}$ then

(i) the surjective function $\psi : Q \to Q/\sim$ defined by

$\psi(q) = [q]$ satisfies the conditions

$\psi(q)F'_\sigma = \psi(qF_\sigma)$

$\psi(q)G'_\sigma = qG_\sigma \quad$ for $q \in Q, \sigma \in \Sigma$

(ii) for each $q \in Q$, $f'_{[q]} = f_q$.

Proof (i) For $q \in Q$, $\sigma \in \Sigma$ the definition of $\hat{\mathcal{M}}/\sim$ yields

$$\psi(q)F'_\sigma = [q]F'_\sigma = [qF_\sigma] = \psi(qF_\sigma)$$

and

$$\psi(q)G'_\sigma = [q]G'_\sigma = qG_\sigma.$$

(ii) Let $\sigma \in \Sigma$, then

$$f'_{[q]}(\sigma) = [q]G'_\sigma = qG_\sigma = f_q(\sigma).$$

Assume that for words $x \in \Sigma^*$ of length less than n we have $f'_{[q]}(x) = f_q(x)$ and let $y \in \Sigma^*$ be of length n, so that $y = x\sigma$ for some $x \in \Sigma^*$ and $\sigma \in \Sigma$, $q \in Q$, then $f'_{[q]}(y) = f'_{[q]}(x)[q]F'_xG'_\sigma = f_q(x)\psi(qF_x)G'_\sigma = f_q(x)qF_xG_\sigma = f_q(x\sigma) = f_q(y)$. Hence the result follows by induction.

(We note that $\psi(q)F'_\sigma = \psi(qF_\sigma)$ can easily be extended by induction to $\psi(q)F'_x = \psi(qF_x)$ where $x \in \Sigma^*$.) □

Corollary 6.2.2

Let $\hat{\mathcal{M}} = (Q, \Sigma, \Theta, F, G)$ be a complete Mealy machine and $\hat{\mathcal{M}}/\sim = (Q/\sim, \Sigma, \Theta, F', G')$ the minimal Mealy machine of $\hat{\mathcal{M}}$. Suppose that $\hat{\mathcal{M}}_1 = (Q_1, \Sigma, \Theta, F_1, G_1)$ is a complete Mealy machine and $\phi : Q \to Q_1$ is a surjective function satisfying

(i) $\phi(q)(F_1)_\sigma = \phi(qF_\sigma) \quad$ for all $q \in Q, \sigma \in \Sigma$

(ii) $(f_1)_{\phi(q)} = f_q \quad$ for all $q \in Q$

then a surjective function $\xi: Q_1 \to Q/\sim$ exists such that

(i)′ $\xi(q_1)F'_\sigma = \xi(q_1(F_1)_\sigma)$ for all $q_1 \in Q_1, \sigma \in \Sigma$

(ii)′ $f'_{\xi(q_1)} = (f_1)_{q_1}$ for all $q_1 \in Q_1$.

Proof Exercise 6.2. □

The method of actually calculating the minimal Mealy machine depends on finding the relation $\sim$. This can be done by a series of approximations to the relation. For each positive integer i define a relation $\sim_i$ on Q by

$$q \sim_i q' \Leftrightarrow f_q(x) = f_{q'}(x) \quad \text{for all } x \in \Sigma^+ \text{ of length less than or equal to } i.$$

Clearly $q \sim q' \Leftrightarrow q \sim_i q'$ for all $i > 0$.

Proposition 6.2.3

For $i > 1$, $q \sim_i q'$ if and only if

$$q \sim_1 q' \text{ and } qF_\sigma \sim_{i-1} q'F_\sigma \quad \text{for all } \sigma \in \Sigma.$$

Proof Suppose $q \sim_1 q'$ and $qF_\sigma \sim_{i-1} q'F_\sigma$ and let $x \in \Sigma^+$ be of length i. Then $x = \sigma y$ for some $y \in \Sigma^+$ of length $i-1$ and $\sigma \in \Sigma$. Now

$$\begin{aligned} f_q(x) &= f_q(\sigma y) \\ &= f_q(\sigma) \cdot f_{qF_\sigma}(y) \\ &= f_{q'}(\sigma) \cdot f_{q'F_\sigma}(y) \\ &= f_{q'}(\sigma y) \\ &= f_{q'}(x). \end{aligned}$$

The converse is now obvious. □

Each equivalence relation $\sim_i$ defines a partition π_i of the set Q and it is clear that

$$\pi_1 \supseteq \pi_2 \supseteq \ldots$$

Suppose that H_1 is a π_1-block and $q, q' \in H_1$; if $\sigma \in \Sigma$ is such that qF_σ and $q'F_\sigma$ belong to different π_1-blocks then proposition 6.2.3 tells us that $q \sim_2 q'$ cannot hold. More generally if q and q' belong to the same π_i-block but qF_σ and $q'F_\sigma$ belong to different π_i-blocks then q and q' cannot belong to the same π_{i+1}-block. In the language of state machines this means that if π_i is an admissible partition then $\pi_{i+1} = \pi_i$.

Proposition 6.2.4

For $i \geq 1$, $\pi_{i+1} = \pi_i$ if and only if π_i is an admissible partition on $\mathcal{M} = (Q, \Sigma, F)$.

Proof Suppose that $\pi_{i+1} = \pi_i$ and $q, q' \in Q$ are such that $q \sim_i q'$. Then for $\sigma \in \Sigma$, $qF_\sigma \sim_i q'F_\sigma$ since $q \sim_{i+1} q'$.

Conversely suppose that π_i is admissible and let $q, q' \in Q$ with $q \sim_i q'$ but $q \not\sim_{i+1} q'$. By 6.2.3 either $q \not\sim_1 q'$ or some $\sigma \in \Sigma$ exists such that $qF_\sigma \not\sim_i q'F_\sigma$, but this is impossible. □

We are now in a position to calculate the minimal Mealy machine since we can easily establish that $\pi_i = \pi_{i+1} \Rightarrow \pi_i = \pi_{i+k}$ for $k \geq 0$ and hence the relations $\sim_i$ and $\sim$ coincide.

Example 6.4

Let $\Sigma = \Theta = \{0, 1\}$ and $\hat{\mathcal{M}} = (Q, \Sigma, \Theta, F, G)$ be given by

	$\hat{\mathcal{M}}$	q_1	q_2	q_3	q_4	q_5
F	0	q_2	q_4	q_2	q_1	q_5
	1	q_5	q_5	q_5	q_3	q_4
G	0	0	1	0	0	1
	1	1	1	1	1	1

Then

$$\pi_1 = \{\{q_1, q_3, q_4\}, \{q_2, q_5\}\},$$
$$\pi_2 = \{\{q_1, q_3\}, \{q_4\}, \{q_2\}, \{q_5\}\}$$

which is admissible and hence $\pi_2 = \pi_3$ etc. The minimal machine is thus:

	$\hat{\mathcal{M}}/\sim$	$[q_1]$	$[q_2]$	$[q_4]$	$[q_5]$
F'	0	$[q_2]$	$[q_4]$	$[q_1]$	$[q_5]$
	1	$[q_2]$	$[q_2]$	$[q_4]$	$[q_4]$
G'	0	0	1	0	1
	1	1	1	1	1

Turning to the incomplete case we will first examine the problem of minimizing a state complete Mealy machine. If $\hat{\mathcal{M}} = (Q, \Sigma, \Theta, F, G)$ is state complete, the relation $\|$ on Q may not be an equivalence relation,

since transitivity may fail. We can, however, still define a sequence of relations on Q as follows:
for each positive integer i and $q, q' \in Q$ define

$$q \parallel_i q' \Leftrightarrow f_q(x) \parallel f_{q'}(x)$$

for all $x \in \Sigma^*$ of length less than or equal to i. Then

$$q \parallel q' \Leftrightarrow q \parallel_i q' \quad \text{for all } i > 0.$$

Proposition 6.2.5
For $i > 1$, $q \parallel_i q'$ if and only if $q \parallel_1 q'$ and $qF_\sigma \parallel_{i-1} q'F_\sigma$ for all $\sigma \in \Sigma$.

Proof Suppose that $q \parallel_1 q'$ and $qF_\sigma \parallel_{i-1} q'F_\sigma$ and let $x \in \Sigma^*$ be of length i. Then $x = \sigma y$ for some $y \in \Sigma^*$ of length $i-1$ and $\sigma \in \Sigma$. Now

$$f_q(x) = f_q(\sigma y) = f_q(\sigma) f_{qF_\sigma}(y)$$

$$f_{q'}(x) = f_{q'}(\sigma y) = f_{q'}(\sigma) f_{q'F_\sigma}(y).$$

We have

$$f_q(\sigma) \parallel f_{q'}(\sigma) \text{ and } f_{qF_\sigma}(y) \parallel f_{q'F_\sigma}(y)$$

and clearly this means $f_q(x) \parallel f_{q'}(x)$, that is $q \parallel_i q'$. The converse is easily checked. □

For each relation $\parallel_{i+1}$ on Q, we examine the relation $\parallel_i$ on Q and see what the state maps F_σ ($\sigma \in \Sigma$) do to the pairs of states (q, q') satisfying $q \parallel_i q'$. If $qF_\sigma \parallel_i q'F_\sigma$ is false for some $\sigma \in \Sigma$ then $q \parallel_{i+1} q'$ is false. As before we eventually must reach a position where the relations $\parallel_n$ and $\parallel_{n+1}$ are identical. Then $\parallel_n$ equals the relation $\parallel$.

For each $q \in Q$, define

$$A(q) = \{q' \mid q \parallel q'\}.$$

Clearly $q \in A(q)$. The collection $\mathscr{X}$ of *distinct* $A(q)$ ($q \in Q$) forms a set of subsets of Q but not generally a partition, i.e. we could have $A(q) \cap A(q') \neq \varnothing$ and $A(q) \neq A(q')$, we could also have $A(q) \subsetneq A(q')$, $q, q' \in Q$. It is clear that if $q \parallel q'$ then $qF_\sigma \parallel q'F_\sigma$ for $\sigma \in \Sigma$, and so for all $q' \in A(q)$ we have $q'F_\sigma \in A(qF_\sigma)$ and thus $A(q)F_\sigma \subseteq A(qF_\sigma)$ for $q \in Q$, $\sigma \in \Sigma$.

The subsets $A(q) \in \mathscr{X}$ may have the following unfortunate property; namely that if $q', q'' \in A(q)$ then $q' \parallel q''$ is false. We now search for an *admissible* subset system $\pi = \{H_i\}_{i \in I}$ of Q satisfying the following conditions:

(i) Given $i \in I$, there exists $q \in Q$ such that

$$H_i \subseteq A(q).$$

(ii) If $q', q'' \in H_i$ then

$$q' \| q''.$$

It is always possible to find such an admissible subset system for any machine $\hat{\mathcal{M}}$, since 1_Q clearly satisfies the conditions. We call such an admissible subset system a *compatible subset system*. In general it may not be a partition of Q.

If $\pi = \{H_i\}_{i \in I}$ is a compatible subset system then a Mealy machine $\hat{\mathcal{M}}/\pi$ can be defined as follows:

$$\hat{\mathcal{M}}/\pi = (\{H_i\}_{i \in I}, \Sigma, \Theta, F^\pi, G^\pi)$$

where

$$H_i F_\sigma^\pi = H_j \quad \text{where } j \in I \text{ is chosen so that } H_i F_\sigma \subseteq H_j,$$

$$H_i G_\sigma^\pi = \begin{cases} qG_\sigma & \text{if a } q \in H_i \text{ exists such that } qG_\sigma \neq \varnothing \\ \varnothing & \text{otherwise.} \end{cases}$$

Since π is an admissible subset system, rather than a partition, in general there may be many possibilities for the definition of F^π and we will assume that a particular choice has been made (see chapter 4 for a similar definition) and then F^π is well-defined. Since π is compatible it is clear that G^π is also well-defined.

Now let $q \in Q$, there then exists an $i \in I$ such that $q \in H_i$; we now establish a connection between the sequential function f_q defined with respect to $\hat{\mathcal{M}}$ and the sequential function $f_{H_i}^\pi$ defined with respect to $\hat{\mathcal{M}}/\pi$.

Theorem 6.2.6

Let $\hat{\mathcal{M}} = (Q, \Sigma, \Theta, F, G)$ be a state complete Mealy machine and $\pi = \{H_i\}_{i \in I}$ a compatible subset system on Q. Let $q \in Q$ then $q \in H_i$ for some $i \in I$ and if $f_{H_i} : \Sigma^* \to (\Theta \cup \square)^*$ is the sequential function of $\hat{\mathcal{M}}/\pi$ in state H_i then for each $x \in \Sigma^*$,

$$f_{H_i}^\pi(x) \# f_q(x).$$

Proof Let $\sigma \in \Sigma$, then $f_q(\sigma) = qG_\sigma$. If $q'G_\sigma = \varnothing$ for all $q' \in H_i$ then $qG_\sigma = H_i G_\sigma^\pi = \varnothing$. If $q'G_\sigma \neq \varnothing$ for some $q' \in H_i$ then $H_i G_\sigma^\pi = q'G_\sigma$. Since $q \| q'$ either $qG_\sigma = q'G_\sigma$ or $qG_\sigma = \varnothing$. In all cases $H_i G_\sigma^\pi \# qG_\sigma$ so $f_{H_i}^\pi(\sigma) \# f_q(\sigma)$. Now suppose that $f_{H_i}^\pi(x) \# f_q(x)$ for all $x \in \Sigma^*$ of length less than n and let $y \in \Sigma^*$ be of length n. Writing $y = x\sigma$ for $x \in \Sigma^*$, $\sigma \in \Sigma$ we see that

$$f_q(y) = f_q(x) \cdot qF_x G_\sigma$$

and

$$f_{H_i}^\pi(y) = f_{H_i}^\pi(x) \cdot H_i F_x^\pi G_\sigma^\pi.$$

By the inductive assumption $f^{\pi}_{H_i}(x) \# f_q(x)$. Let $H_iF^{\pi}_x = H_j$, where $H_iF_x \subseteq H_j$, then $qF_x \in H_i$. Now $H_jG^{\pi}_{\sigma} = q'G_{\sigma}$ where $q' \in H_j$ and since $q' \| qF_x$ we see that $H_jG^{\pi}_{\sigma} \# qF_xG_{\sigma}$ and so $f^{\pi}_{H_i}(y) \# f_q(y)$. The result follows by induction. □

In many ways the Mealy machine $\hat{\mathcal{M}}/\pi$ performs similar tasks to the original machine $\hat{\mathcal{M}}$, but it may not be the smallest such machine. The size of $\hat{\mathcal{M}}/\pi$ equals the number of subsets in the compatible subset system $\pi = \{H_i\}_{i \in I}$ and we would naturally ask for this to be as small as possible. A compatible subset system π is called *maximal* if no non-trivial compatible subset system τ exists such that $\pi < \tau$. We regard $\{Q\}$ as a trivial compatible subset system.

The Mealy machine $\hat{\mathcal{M}}/\pi$, where π is a maximal compatible subset system, will be called a *minimal cover for* $\hat{\mathcal{M}}$. There is no unique minimal cover in general for a Mealy machine $\hat{\mathcal{M}}$, and in fact different minimal covers for a particular machine $\hat{\mathcal{M}}$ can have rather different properties. The task of constructing the minimal covers will not be discussed in any detail here; it amounts to the calculation of the maximal compatible subset systems and this in general is done using *ad hoc* methods.

Example 6.5

Consider the machine $\hat{\mathcal{M}} = (Q, \Sigma, \Theta, F, G)$, where $\Sigma = \Theta = \{0, 1\}$, given by:

$\hat{\mathcal{M}}$		q_1	q_2	q_3	q_4	q_5
F	0	q_2	q_1	q_2	q_1	q_5
	1	q_5	q_5	q_5	q_3	q_4
G	0	0	$\varnothing$	$\varnothing$	0	1
	1	1	1	$\varnothing$	1	$\varnothing$

To calculate the relation $\|$ on Q we proceed as follows. First we describe the relation $\|_1$ by writing $(i, j)_1$ to denote $q_i \|_1 q_j$ and recall that the relation is symmetric. Thus

$$(1, 2)_1, (1, 3)_1, (1, 4)_1, (2, 3)_1, (2, 4)_1, (2, 5)_1, (3, 4)_1, (3, 5)_1.$$

To determine the relation $\|_2$ we examine $(q_iF_{\sigma}, q_jF_{\sigma})$ for each pair $(i, j) \in \|_1$ and if $q_iF_{\sigma} = q_k, q_jF_{\sigma} = q_l$ and $(k, l) \notin \|_1$ then by 6.2.4 we know that $(i, j) \notin \|_2$. This leads to

$$(1, 2)_2, (1, 3)_2, (1, 4)_2, (2, 3)_2, (2, 4)_2, (3, 4)_2$$

and then

$$(1, 2)_3, (1, 3)_3, (2, 3)_3$$

and

$$(1, 2)_4, (1, 3)_4, (2, 3)_4.$$

Therefore $\|$ is the same as $\|_3$ and the set $\mathscr{X} = \{\{q_1, q_2, q_3\}, \{q_4\}, \{q_5\}\}$ which is, in this case, a compatible subset system, and a partition. A minimal cover is thus given by:

$\hat{\mathscr{M}}/\pi$		H_1	H_2	H_3
F^π	0	H_1	H_1	H_3
	1	H_3	H_1	H_2
G^π	0	0	0	1
	1	1	1	$\varnothing$

where $\pi = \{H_1, H_2, H_3\}$ and $H_1 = \{q_1, q_2, q_3\}$, $H_2 = \{q_4\}$, $H_3 = \{q_5\}$.

Example 6.6
Let $\hat{\mathscr{M}} = (Q, \Sigma, \Theta, F, G)$ be defined by

$\hat{\mathscr{M}}$		q_1	q_2	q_3	q_4	q_5
F	0	q_1	q_1	q_5	q_3	q_1
	1	q_3	q_1	q_1	q_5	q_3
G	0	0	$\varnothing$	0	0	0
	1	0	1	$\varnothing$	1	$\varnothing$

where $\Sigma = \Theta = \{0, 1\}$. Then $\|_1$ is given by

$$(1, 3)_1, (1, 5)_1, (2, 3)_1, (2, 4)_1, (2, 5)_1, (3, 4)_1, (3, 5)_1, (4, 5)_1$$

and we also obtain

$$(1, 3)_2, (1, 5)_2, (2, 3)_2, (2, 4)_2, (2, 5)_2, (3, 4)_2, (3, 5)_2, (4, 5)_2.$$

Thus

$$\mathscr{X} = \{\{q_1, q_3, q_5\}, \{q_2, q_3, q_4, q_5\}, Q\}.$$

If $H_1 = \{q_1, q_3, q_5\}$ and $H_2 = \{q_2, q_4\}$ then $\pi = \{H_1, H_2\}$ is an admissible subset system which is also compatible. Then $\hat{\mathcal{M}}/\pi$ is given by

$\hat{\mathcal{M}}/\pi$		H_1	H_2
F^π	0	H_1	H_1
	1	H_1	H_1
G^π	0	0	0
	1	0	1

and this is a minimal cover for $\hat{\mathcal{M}}$. Here again π was a partition of Q even though $\mathscr{X}$ was not.

Example 6.7
Let $\hat{\mathcal{M}} = (Q, \Sigma, \Theta, F, G)$ be given by

$\hat{\mathcal{M}}$		q_1	q_2	q_3	q_4	q_5
F	a	q_1	q_3	q_2	q_5	q_4
	b	q_1	q_1	q_3	q_4	q_4
	c	q_1	q_3	q_2	q_5	q_4
G	a	0	1	$\varnothing$	0	$\varnothing$
	b	$\varnothing$	0	1	1	0
	c	1	$\varnothing$	0	$\varnothing$	1

where $\Sigma = \{a, b, c\}$ and $\Theta = \{0, 1\}$. Then $\|_1$ is given by $(1, 4)_1$, $(1, 5)_1$, $(2, 5)_1$, $(3, 4)_1$ which is also $\|$.

$$\mathscr{X} = \{\{q_1, q_4, q_5\}, \{q_2, q_5\}, \{q_3, q_4\}, \{q_1, q_3, q_4\}, \{q_1, q_2, q_5\}\}.$$

Let $H_1 = \{q_1, q_4\}$, $H_2 = \{q_2, q_5\}$, $H_3 = \{q_3, q_4\}$, $H_4 = \{q_1, q_5\}$, then $\pi = \{H_1, H_2, H_3, H_4\}$ is an admissible and compatible subset system which is not a partition. A machine, $\hat{\mathcal{M}}/\pi$, which is a minimal cover for $\hat{\mathcal{M}}$ is given by

$\hat{\mathcal{M}}/\pi$		H_1	H_2	H_3	H_4
F^π	a	H_2	H_3	H_2	H_1
	b	H_1	H_1	H_3	H_1
	c	H_2	H_3	H_2	H_1
G^π	a	0	1	$\varnothing$	0
	b	$\varnothing$	0	1	0
	c	1	$\varnothing$	0	1

More general covers can be introduced as follows. First let $\hat{\mathcal{M}} = (Q, \Sigma, \Theta, F, G)$ be a Mealy machine. If $\hat{\mathcal{M}}' = (Q', \Sigma, \Theta, F', G')$ is another state complete Mealy machine and $\phi: Q \to Q'$ is a function then we say that ϕ is a *covering* of $\hat{\mathcal{M}}$ by $\hat{\mathcal{M}}'$ if, for each $q \in Q$,

$$f'_{\phi(q)}(x) \# f_q(x) \quad \text{for all } x \in \Sigma^*,$$

where f_q and $f'_{\phi(q)}$ are the partial functions associated with $\hat{\mathcal{M}}$ in state q and $\hat{\mathcal{M}}'$ in state $\phi(q)$ respectively. We write

$$\hat{\mathcal{M}} \leq \hat{\mathcal{M}}'.$$

This means that machine $\hat{\mathcal{M}}'$ will do all that $\hat{\mathcal{M}}$ can do, and possibly more. In the case where $\hat{\mathcal{M}}$ is state complete and $\hat{\mathcal{M}}' = \hat{\mathcal{M}}/\pi$ for some compatible subset system π then

$$\hat{\mathcal{M}} \leq \hat{\mathcal{M}}/\pi.$$

It is now necessary to extend our concepts of compatibility to Mealy machines that may not be state complete.

Let $\hat{\mathcal{M}} = (Q, \Sigma, \Theta, F, G)$ be a general Mealy machine and let $q, q_1 \in Q$. We say that q and q_1 are *compatible* if, whenever $x \in \Sigma^*$ is applicable to both q and q_1, then

$$f_q(x) \| f_{q_1}(x).$$

As before we may define the relations $\|_i$ on Q for each positive integer i. The subsets

$$A(q) = \{q' \mid q \| q'\}$$

may be formed for each $q \in Q$ and also the collection $\mathscr{X}$ of the distinct $A(q)$. Using our new compatibility definition we can now look for admissible subset systems $\pi = \{H_i\}_{i \in I}$ of Q satisfying

(i) for each $i \in I$ there exists a $q \in Q$ such that $H_i \subseteq A(q)$,
(ii) if $q', q'' \in H_i$ then $q' \| q''$.

We call π a *compatible subset system* as before. Define a Mealy machine $\hat{\mathcal{M}}/\pi = (\{H_i\}_{i \in I}, \Sigma, \Theta, F^\pi, G^\pi)$ as follows:

$$H_i F_\sigma^\pi = \begin{cases} H_j & \text{if } \exists\, j \in I \text{ such that } \varnothing \neq H_i F_\sigma \subseteq H_j \\ \varnothing & \text{otherwise} \end{cases}$$

$$H_i G_\sigma^\pi = \begin{cases} qG_\sigma & \text{if a } q \in H_i \text{ exists satisfying } qG_\sigma \neq \varnothing \\ \varnothing & \text{otherwise.} \end{cases}$$

As before we make a choice for the definition of F^π. The compatibility of π ensures that G^π is well-defined.

Theorem 6.2.7

Let $\hat{\mathcal{M}} = (Q, \Sigma, \Theta, F, G)$ be a Mealy machine which is not state complete and suppose that $\hat{\mathcal{M}}^c = (Q \cup \{z\}, \Sigma, \Theta, F', G')$ is the state completion of $\hat{\mathcal{M}}$. If $\pi = \{H_i\}_{i \in I}$ is a compatible subset system for $\hat{\mathcal{M}}$ then

$$\pi^c = \{H_i \cup \{z\}\}_{i \in I}$$

is a compatible subset system for $\hat{\mathcal{M}}^c$. Conversely, let $\tau = \{K_j\}_{j \in J}$ be a compatible subset system for $\hat{\mathcal{M}}^c$ then

$$\tau^* = \{K_j \backslash \{z\}\}_{j \in J}$$

is a compatible subset system for $\hat{\mathcal{M}}$.

Proof Let $\pi = \{H_i\}_{i \in I}$ be a compatible subset system for $\hat{\mathcal{M}}$. Clearly $\pi^c = \{H_i \cup \{z\}\}_{i \in I}$ is an admissible subset system for $\hat{\mathcal{M}}^c$ for if $\sigma \in \Sigma$ then

$$H_i F_\sigma \subseteq H_j \quad \text{for some } j \in I$$

and

$$(H_i \cup \{z\})F'_\sigma \subseteq H_j \cup \{zF_\sigma\} = H_j \cup \{z\} \in \pi^c.$$

Now let $H_i \in A(q)$. Since $z \| q$ in $\hat{\mathcal{M}}^c$ we see that $H_i \cup \{z\} \subseteq A^c(q)$ where $A^c(q)$ denotes the set of states of $\hat{\mathcal{M}}^c$ compatible with q.

Finally for each $q', q'' \in H_i$ we have $q' \| q''$ in $\hat{\mathcal{M}}$. Clearly $q' \| z$ and $q'' \| z$ in $\hat{\mathcal{M}}^c$ and so π^c is a compatible subset system.

Now suppose that $\tau = \{K_j\}_{j \in J}$ is a compatible subset system for $\hat{\mathcal{M}}^c$. First note that the non-empty subsets of the form $K_j \backslash \{z\}$ $(j \in J)$ form an admissible subset system for $\hat{\mathcal{M}}$, since

$$(K_j \backslash \{z\})F_\sigma = K_j F_\sigma \backslash \{z\} \subseteq K_l \backslash \{z\}$$

for some $l \in J$, where $\sigma \in \Sigma$.

Now let $K_j \subseteq A^c(q)$ for some $q \in Q \cup \{z\}$. We may assume that $q \neq z$ since $A^c(z) = Q \cup \{z\}$. Let $q' \in K_j \backslash \{z\}$, then $q' \| q$ in $\hat{\mathcal{M}}^c$, where $q \in Q$. Suppose that $x \in \Sigma^*$ is such that x is applicable to both q' and q in $\hat{\mathcal{M}}$, then $f_{q'}(x)$ and $f_q(x)$ exist (in $\hat{\mathcal{M}}$) and since $f_{q'}(x) \| f_q(x)$ in $\hat{\mathcal{M}}^c$ we have $f_{q'}(x) \| f_q(x)$ in $\hat{\mathcal{M}}$. Hence $q' \| q$ in $\hat{\mathcal{M}}$.

Finally let $q', q'' \in K_j \backslash \{z\}$. By a similar argument we see that $q' \| q''$ in $\hat{\mathcal{M}}^c$ and thus $q' \| q''$ in $\hat{\mathcal{M}}$. Therefore τ^* is a compatible subset system for $\hat{\mathcal{M}}$. □

As before a compatible subset system π is called *maximal* if no non-trivial compatible subset system τ exists such that $\pi < \tau$.

Theorem 6.2.8

Let $\hat{\mathcal{M}} = (Q, \Sigma, \Theta, F, G)$ be a Mealy machine which is not state complete and $\hat{\mathcal{M}}^c = (Q \cup \{z\}, \Sigma, \Theta, F', G')$ its state completion. If $\pi = \{H_i\}_{i \in I}$ is a maximal compatible subset system for $\hat{\mathcal{M}}$ then $\pi^c = \{H_i \cup \{z\}\}_{i \in I}$ is a maximal compatible subset system for $\hat{\mathcal{M}}^c$.

Conversely let $\tau = \{K_j\}_{j \in J}$ be a maximal compatible subset system for $\hat{\mathcal{M}}^c$, then

$$\tau^* = \{K_j \backslash \{z\}\}_{j \in J}$$

is a maximal compatible subset system for $\hat{\mathcal{M}}$.

Proof Assume first that $\pi = \{H_i\}_{i \in I}$ is maximal and let $\pi^c < \tau$ where τ is a compatible subset system of $\hat{\mathcal{M}}^c$. Consider the subset system

$$\tau^* = \{K_j \backslash \{z\}\}_{j \in J}$$

where the system $\tau = \{K_j\}_{j \in J}$. From the previous result we see that τ^* is a compatible subset system for $\hat{\mathcal{M}}$ and clearly $\pi \leq \tau^*$, if $\pi = \tau^*$ then we must have $\pi^c = (\tau^*)^c$, but $\pi^c < \tau$ implies that $z \in K_j$ for all $j \in J$ and so $(\tau^*)^c = \tau$. Thus we obtain a contradiction and so $\pi < \tau^*$. This means that $\tau^* = \{Q\}$ since π is maximal. Then $\tau = (\tau^*)^c = \{Q \cup \{z\}\}$ and so π^c is maximal in $\hat{\mathcal{M}}^c$.

Now let $\tau = \{K_j\}_{j \in J}$ be maximal in $\hat{\mathcal{M}}^c$ and suppose that $\tau^* < \rho$ where ρ is a compatible subset system for $\hat{\mathcal{M}}$. Then $(\tau^*)^c \leq \rho^c$ and clearly

$$\tau^* = ((\tau^*)^c)^* \leq (\rho^c)^* = \rho$$

which implies that $(\tau^*)^c < \rho^c$. If $\rho = \{L_t\}_{t \in T}$ then for each $j \in J$, $K_j \backslash \{z\} \subseteq L_t$ for some $t \in T$ and so $K_j \subseteq L_t \cup \{z\} \in \rho^c$ for each $j \in J$, even when $K_j = \{z\}$. Therefore $\tau < \rho^c$ and the maximality of τ forces $\rho^c = \{Q \cup \{z\}\}$ and thus $\rho = \{Q\}$. □

These two results enable us to obtain minimal covering machines for incomplete Mealy machines directly from the covering machines of their state completions.

Let $\hat{\mathcal{M}}$ be a Mealy machine. If $\hat{\mathcal{M}}$ is not state complete, consider the state completion $\hat{\mathcal{M}}^c$ and construct a maximal compatible subset system π for $\hat{\mathcal{M}}^c$. Then the compatible subset system π^* for $\hat{\mathcal{M}}$ is maximal and any Mealy machine of the form $\hat{\mathcal{M}}/\pi^*$ will be a minimal cover for $\hat{\mathcal{M}}$.

The justification for this terminology is obtained if we generalize our notion of Mealy machine covering to include incomplete Mealy machines.

For any arbitrary Mealy machines $\hat{\mathcal{M}} = (Q, \Sigma, \Theta, F, G)$ and $\hat{\mathcal{M}}' = (Q', \Sigma, \Theta, F', G')$ we say that $\hat{\mathcal{M}}'$ *covers* $\hat{\mathcal{M}}$, written $\hat{\mathcal{M}}' \geq \hat{\mathcal{M}}$, if there exists a function $\phi : Q \to Q'$ such that for each $q \in Q$

$$f'_{\phi(q)}(x) \# f_q(x)$$

for all $x \in \Sigma^*$ applicable to q.

Theorem 6.2.9

Let $\hat{\mathcal{M}}$ be a Mealy machine and π a compatible subset system for $\hat{\mathcal{M}}$. Then

$$\hat{\mathcal{M}} \leq \hat{\mathcal{M}}/\pi.$$

Proof Let $x \in \Sigma^*$ be applicable to the state $q \in Q$, then, if $x = \sigma_1 \ldots \sigma_k$, all of $qF_{\sigma_1}, \ldots, qF_{\sigma_1 \ldots \sigma_{k-1}}$ are defined. If $q \in H_i$ for $i \in I$ we have

$$qF_{\sigma_1} \in H_i F_{\sigma_1}, \ldots, qF_{\sigma_1 \ldots \sigma_{k-1}} \in H_i F_{\sigma_1 \ldots \sigma_{k-1}}$$

and so x is applicable in $\hat{\mathcal{M}}/\pi$ to H_i.

Putting $\phi : Q \to \{H_i\}_{i \in I}$ to be any function satisfying $q \in \phi(q)$, $q \in Q$ we see that a similar proof to 6.2.6 will yield

$$f^{\pi}_{\phi(q)}(x) \# f_q(x). \qquad \square$$

Example 6.8

Let $\hat{\mathcal{M}} = (Q, \Sigma, \Theta, F, G)$ be given by

$\hat{\mathcal{M}}$		q_1	q_2	q_3	q_4	q_5
F	a	q_1	$\varnothing$	q_2	q_5	q_4
	b	q_1	$\varnothing$	q_3	q_4	$\varnothing$
	c	q_1	q_3	q_2	q_5	q_4
G	a	0	1	$\varnothing$	0	$\varnothing$
	b	$\varnothing$	0	1	1	0
	c	1	$\varnothing$	0	$\varnothing$	1

where $\Sigma = \{a, b, c\}$, $\Theta = \{0, 1\}$. (This is an incomplete version of example 6.7.)

Then $\hat{\mathcal{M}}^c$ is given by

$\hat{\mathcal{M}}^c$		q_1	q_2	q_3	q_4	q_5	z
F	a	q_1	z	q_2	q_5	q_4	z
	b	q_1	z	q_3	q_4	z	z
	c	q_1	q_3	q_2	q_5	q_4	z
G	a	0	1	$\varnothing$	0	$\varnothing$	$\varnothing$
	b	$\varnothing$	0	1	1	0	$\varnothing$
	c	1	$\varnothing$	0	$\varnothing$	1	$\varnothing$

We calculate the relation $\|$ for $\hat{\mathcal{M}}^c$.

Now $\|_1$ is given by

$$(1,4)_1, (1,5)_1, (2,5)_1, (3,4)_1, (1,z)_1, (2,z)_1, (3,z)_1, (4,z)_1, (5,z)_1.$$

The relation $\|$ is given by

$$(1,4)_2, (1,5)_2, (1,z)_2, (2,5)_2, (2,z)_2, (3,4)_2, (3,z)_2, (4,z)_2, (5,z)_2.$$

So

$$\mathcal{X} = \{\{q_1, q_4, z\}, \{q_2, q_5, z\}, \{q_3, q_4, z\}, \{q_1, q_3, q_4, z\},$$
$$\{q_1, q_2, q_5, z\}, \{q_1, q_2, q_3, q_4, q_5, z\}\}.$$

Let $H_1 = \{q_1, q_4, z\}, H_2 = \{q_2, q_5, z\}, H_3 = \{q_3, q_4, z\}, H_4 = \{q_1, q_5, z\}$, then $\pi = \{H_1, H_2, H_3, H_4\}$ is a compatible subset system for $\hat{\mathcal{M}}^c$. It is a maximal compatible subset system for $\hat{\mathcal{M}}^c$ and so $\pi^* = \{H_1\backslash\{z\}, H_2\backslash\{z\}, H_3\backslash\{z\}, H_4\backslash\{z\}\}$ is a maximal compatible subset system for $\hat{\mathcal{M}}$.

Now $\hat{\mathcal{M}}/\pi^*$ could take the form, for example

$\hat{\mathcal{M}}/\pi^*$		H_1^*	H_2^*	H_3^*	H_4^*
F^*	a	H_2^*	H_3^*	H_2^*	H_1^*
	b	H_1^*	$\varnothing$	H_3^*	H_1^*
	c	H_2^*	H_3^*	H_2^*	H_1^*
G^*	a	0	1	$\varnothing$	0
	b	$\varnothing$	0	1	0
	c	1	$\varnothing$	0	1

where $H_1^* = \{q_1, q_4\}$, $H_2^* = \{q_2, q_5\}$, $H_3^* = \{q_3, q_4\}$, $H_4^* = \{q_1, q_5\}$.

This is 'almost isomorphic' to the Mealy machine $\hat{\mathcal{M}}/\pi$ obtained in example 6.7, but this should come as no surprise since the machine $\hat{\mathcal{M}}$ in 6.7 clearly covers the machine considered here and so we would expect some close connection between their minimal covers.

We close with the remark that our approach to the minimization of a Mealy machine actually makes use of the fact that the machine may not be completely defined. The entries $\varnothing$ in the tables specifying the machine's output are sometimes called 'don't care' entries since their value is of no consequence. We can take advantage of this freedom to generate much smaller covering machines than if we were to complete the output function in a similar way to the completion of the state function. For this reason we have chosen a rather general form of the concept of machine covering.

6.3 Two sorts of covering

The purpose of this section is to examine the relationship between the covering of one Mealy machine by another and the connections between their state machines. To examine this problem in general it is necessary to extend the definition of Mealy machine covering to include the case where the input and output alphabets do not coincide.

Let $\hat{\mathcal{M}} = (Q, \Sigma, \Theta, F, G)$ and $\hat{\mathcal{M}}' = (Q, '\Sigma', \Theta', F', G')$ be Mealy machines, not necessarily state complete.

Let $\xi : \Sigma \to \Sigma'$, $\rho : \Theta \to \Theta'$ be functions and suppose that a function $\psi : Q \to Q'$ exists such that for each $q \in Q$ we have

$$f'_{\psi(q)}(\xi(x)) \# \rho(f_q(x))$$

for all $x \in \Sigma^*$ applicable to q and such that $\xi(x)$ is applicable to $\psi(q)$. (The functions ξ and ρ are of course assumed to have been extended to the free monoids Σ^* and Θ^* respectively.)

As usual we will write $\hat{\mathcal{M}} \leq \hat{\mathcal{M}}'$. If $\mathcal{M} = (Q, \Sigma, F)$ we will call $\mathcal{M}$ the state machine of $\hat{\mathcal{M}}$.

Theorem 6.3.1

Let $\hat{\mathcal{M}} = (Q, \Sigma, \Theta, F, G)$ be a Mealy machine and suppose that $\mathcal{M}' = (Q', \Sigma', F')$ is a state machine satisfying $\mathcal{M} \leq \mathcal{M}'$, then there exists a Mealy machine $\hat{\mathcal{M}}' = (Q', \Sigma', \Theta', F', G')$ such that for each $q \in Q$ and $x \in \Sigma^*$ applicable to q

$$f_q(x) = f'_{\psi(q)}(\xi(x)), \quad \text{for some function } \psi : Q \to Q'.$$

Proof We are given a function $\xi : \Sigma \to \Sigma'$ and a surjective partial function $\eta : Q' \to Q$ such that $\eta(q')F_x \subseteq \eta(q'F'_{\xi(x)})$ for each $q' \in Q'$ and $x \in \Sigma^*$.

Put $\Theta' = \Theta$ and define $G' : Q' \times \Sigma' \to \Theta$ by

$$G'(q', \sigma') = \begin{cases} G(\eta(q'), \sigma) & \text{if } \sigma' = \xi(\sigma) \text{ for some } \sigma \in \Sigma \text{ and } \eta(q') \neq \varnothing, \\ \varnothing & \text{otherwise.} \end{cases}$$

Now let $\psi: Q \to Q'$ be a function satisfying the condition $\eta \circ \psi = 1_Q$; such a function must exist since η is surjective. We show that ψ defines a covering of Mealy machines $\hat{\mathcal{M}} \leq \hat{\mathcal{M}}'$ where $\hat{\mathcal{M}}' = (Q', \Sigma', \Theta, F', G')$. Choose any $q \in Q$. Let $x \in \Sigma^*$ be applicable to q in $\hat{\mathcal{M}}$ and suppose that $x = \sigma_1 \ldots \sigma_k$. Then

$$qF_{\sigma_1}, \ldots, qF_{\sigma_1 \ldots \sigma_{k-1}}$$

are all defined. Since $q = \eta(\psi(q))$ we see that

$$qF_{\sigma_1} = \eta(\psi(q))F_{\sigma_1} \subseteq \eta(\psi(q)F'_{\xi(\sigma_1)})$$
$$\vdots$$
$$qF_{\sigma_1 \ldots \sigma_{k-1}} = \eta(\psi(q))F_{\sigma_1 \ldots \sigma_{k-1}} \subseteq \eta(\psi(q)F'_{\xi(\sigma_1 \ldots \sigma_{k-1})})$$

since η is a state machine covering. Thus $\xi(x)$ is applicable to $\psi(q)$ in $\hat{\mathcal{M}}'$. Now for $\sigma \in \Sigma$,

$$f_q(\sigma) = qG_\sigma = \eta(\psi(q))G_\sigma = \psi(q)G'_{\xi(\sigma)} = f'_{\psi(q)}(\xi(\sigma)).$$

Assume that $f_q(x) = f'_{\psi(q)}(\xi(x))$ for all words $x \in \Sigma^*$ of length less than n which are applicable to q.

Now let $y = x\sigma$ where y is of length n and y is applicable to q. Then

$$f_q(y) = f_q(x) \cdot qF_xG_\sigma$$
$$= f'_{\psi(q)}\xi(x) \cdot qF_xG_\sigma.$$

Now

$$qF_xG_\sigma = \eta(\psi(q))F_xG_\sigma$$
$$\subseteq \eta(\psi(q)F'_{\xi(x)})G_\sigma$$
$$= \psi(q)F'_{\xi(x)}G'_{\xi(\sigma)} \quad \text{by the definition of } G'.$$

Since

$$qF_x \neq \varnothing$$

we have

$$qF_xF_\sigma = \psi(q)F'_{\xi(x)}G'_{\xi(\sigma)}$$

and so

$$f_q(y) = f'_{\psi(q)}(\xi(y)).$$ □

Corollary 6.3.2
In the situation of 6.3.1 we have
$\hat{\mathcal{M}} \leq \hat{\mathcal{M}}'$.

One conclusion that we may draw from this result is that whereas the concept of covering of Mealy machines developed in section 6.2 and above is suitable for the problem of minimizing incomplete Mealy machines, when we come to examine the relationship of Mealy machine covering with state machine covering it is too general. Our aim in this chapter is to apply the results of chapters 3 and 4 on state machines to the theory of Mealy machines, and to achieve this we will introduce a special form of Mealy machine covering more suitable for this task. When the machines are complete there is no difference in the two concepts.

Let $\hat{\mathcal{M}} = (Q, \Sigma, \Theta, F, G)$ and $\hat{\mathcal{M}}' = (Q', \Sigma', \Theta', F', G')$ be Mealy machines. Suppose that $\xi : \Sigma \to \Sigma'$, $\rho : \Theta \to \Theta'$ are functions, and a function $\psi : Q \to Q'$ exists such that for each $q \in Q$ and $x \in \Sigma^*$, x is applicable to q if and only if $\xi(x)$ is applicable to $\psi(q)$ and

$$\rho(f_q(x)) = f'_{\psi(q)}(\xi(x)).$$

We say that $\hat{\mathcal{M}}'$ *strongly covers* $\hat{\mathcal{M}}$, or that ψ is a strongly covering function, and write

$$\hat{\mathcal{M}} \ll \hat{\mathcal{M}}'.$$

Clearly $\hat{\mathcal{M}} \ll \hat{\mathcal{M}}'$ implies $\hat{\mathcal{M}} \leq \hat{\mathcal{M}}'$.

To make progress in the other direction we need the following concept. A Mealy machine $\hat{\mathcal{M}} = (Q, \Sigma, \Theta, F, G)$ is called *reduced* if given distinct states q, q_1 then there exists $x \in \Sigma^*$ such that $f_q(x) \neq f_{q_1}(x)$, with x applicable to both q and q_1.

Theorem 6.3.3

(Ginzburg [1968]) Let $\hat{\mathcal{M}} = (Q, \Sigma, \Theta, F, G)$ be a reduced Mealy machine and suppose that $\hat{\mathcal{M}} \ll \hat{\mathcal{M}}'$ where $\hat{\mathcal{M}}' = (Q', \Sigma, \Theta, F', G')$. Then $\mathcal{M} \leq \mathcal{M}'$ as state machines.

Proof Let $\psi : Q \to Q'$ be given such that, for $q \in Q$,

$$f_q(x) = f'_{\psi(q)}(x)$$

for all $x \in \Sigma^*$ applicable to q.

We first note that ψ is a one-one function, for if q, $q_1 \in Q$ and $\psi(q) = \psi(q_1)$ then $f'_{\psi(q)}(x) = f_q(x) = f_{q'}(x)$ for all $x \in \Sigma^*$ applicable to q and q_1 and so $q = q_1$ since $\hat{\mathcal{M}}$ is reduced.

We wish to construct a surjective partial function $\eta : Q' \to Q$ such that

$$\eta(q')F_x \subseteq \eta(q'F'_x) \quad \text{for all } q' \in Q',\ x \in \Sigma^*.$$

First note that there exists a unique function $\chi : \psi(Q) \to Q$ defined by

$$\chi(\psi(q)) = q \quad \text{for all } \psi(q) \in \psi(Q).$$

This is well-defined since ψ is one-one. Thus χ defines a surjective partial function from Q' to Q, but it does not necessarily satisfy the requirement for it to be a state machine covering since

$$\psi(Q)F'_x \subseteq \psi(Q)$$

may not hold for all $x \in \Sigma^*$.

For $x \in \Sigma^*$ define a partial function

$$\psi_x : Q \to Q'$$

by

$$\psi_x(q) = (\psi(q))F'_x \quad \text{for } q \in Q.$$

Now choose a partial function $\alpha_x : Q' \to Q$ such that

$$\mathfrak{D}(\alpha_x) = \psi(Q)F'_x$$

and

$$\psi_x\alpha_x(q') = q' \quad \text{for all } q' \in \psi(Q)F'_x$$

thus

$$(\psi(\alpha_x(q'))F'_x = q' \quad \text{for all } q' \in \psi(Q)F'_x.$$

Define a partial function

$$\eta_x : Q' \to Q$$

by

$$\eta_x(q') = (\alpha_x(q'))F_x \quad \text{for all } q' \in \psi(Q)F'_x.$$

Notice that $\psi_\Lambda = \psi$ and $\eta_\Lambda(\psi(q)) = \alpha_\Lambda(\psi(q))$ for all $\psi(q) \in \psi(Q)$, and since ψ is one-one and $\psi\alpha_\Lambda(\psi(q)) = \psi(q)$ we have

$$\alpha_\Lambda(\psi(q)) = q.$$

Consider now the relation

$$\eta = \bigcup_{x \in \Sigma^*} \eta_x : Q' \to Q.$$

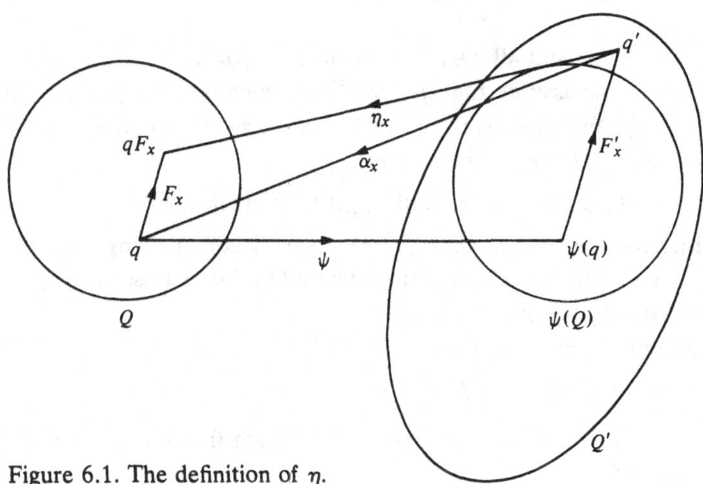

Figure 6.1. The definition of η.

Then $\mathfrak{D}(\eta)=\bigcup_{x\in\Sigma^*}\psi(Q)F'_x$. We establish that η is a partial surjective function. Since $\eta_\Lambda\subseteq\eta$ it is clear that η is surjective. See figure 6.1. We must now show that η is a partial function.

First let $q\in Q$, and $\alpha,\beta\in\Sigma^*$ with $\alpha\beta$ applicable to q. Then $f_q(\alpha\beta)=f'_{\psi(q)}(\alpha\beta)$, which by 6.1.1 gives

$$f_q(\alpha)\cdot f_{qF_\alpha}(\beta)=f'_{\psi(q)}(\alpha)\cdot f'_{\psi(q)F'_\alpha}(\beta)$$

and so

$$f_{qF_\alpha}(\beta)=f'_{\psi(q)F'_\alpha}(\beta)$$

since $f_q(\alpha)=f'_{\psi(q)}(\alpha)$.

Now

$$f_{qF_\alpha}(\beta)=f'_{\psi(qF_\alpha)}(\beta)$$

since β is applicable to qF_α and thus

$$f'_{\psi(qF_\alpha)}(\beta)=f'_{\psi(q)F'_\alpha}(\beta)=f_{qF_\alpha}(\beta).$$

Now let $z\in\Sigma^+$ be applicable to $\eta_x(q')$ where $q'\in\mathfrak{D}(\eta_x)$.

Then

$$\begin{aligned}f_{\eta_x(q')}(z)&=f_{\alpha_x(q')F_x}(z)\\&=f'_{\psi(\alpha_x(q')F_x)}(z)\\&=f'_{\psi(\alpha_x(q'))F'_x}(z)\\&=f'_{\psi_x(\alpha_x(q'))}(z)\\&=f'_{q'}(z).\end{aligned}$$

Thus

$$f_{\eta(q')}(z)=\bigcup_{x\in\Sigma^*}\{f_{\eta_x(q')}(z)\}=f'_{q'}(z)$$

for $q'\in\mathfrak{D}(\eta)$ and all $z\in\Sigma^+$ applicable to $\eta(q')$.

If we now assume that $q_1,q_2\in Q$ are such that $q_1,q_2\in\eta(q')$ for some $q'\in\mathfrak{D}(\eta)$ and $q_1\neq q_2$, then there exists a $z\in\Sigma^*$ applicable to q_1 and q_2 such that $f_{q_1}(z)\neq f_{q_2}(z)$. Then

$$f_{q'}(z)\in\{f'_{q'}(z)\}\quad\text{and}\quad f_{q_2}(z)\in\{f'_{q'}(z)\}$$

which implies that $f_{q_1}(z)=f_{q_2}(z)$ since $\{f'_{q'}(z)\}$ is a singleton element of Θ^*. This contradicts the assumption that $q_1\neq q_2$. Consequently $\eta:Q'\to Q$ is a partial surjective function.

We now show that for $q'\in Q'$, $t\in\Sigma^*$

$$\eta(q')F_t\subseteq\eta(q'F'_t).$$

If $\eta(q')F_t\neq\varnothing$ then there exists $z\in\Sigma^*$ such that $f_{\eta(q')F_t}(z)\neq\varnothing$ since $\hat{\mathcal{M}}$ is reduced.

Now

$$f_{\eta(q')}(tz)=f'_{q'}(tz)$$

and so

$$f_{\eta(q')}(t)f_{\eta(q')F_t}(z)=f'_{q'}(t)f'_{q'F'_t}(z)$$

which implies

$$f_{\eta(q')F_t}(z)=f'_{q'F'_t}(z)\neq\varnothing.$$

Hence $q'F'_t\neq\varnothing$.

For $x\in\Sigma^*$ applicable to $q\in Q$ we have

$$f_q(x)=f'_{\psi(q)}(x)$$

and for $x\in\Sigma^*$ applicable to qF_x we have

$$f_{qF_x}(z)=f'_{\psi(qF_x)}(z)$$

and

$$f_{\eta_x(\psi(qF_x))}(z)=f'_{\psi(qF_x)}(z).$$

But

$$f_qF_x(z)=f'_{\psi(qF_x)}(z)=f'_{\psi(q)F'_x}(z)$$

and thus

$$f_{\eta_x(\psi(q)F'_x)}(z)=f'_{\psi(q)F'_x}(z)=f_{qF_x}(z).$$

Hence $\eta_x(\psi(q)F'_x)=qF_x$ for all $x\in\Sigma^*$ applicable to q.

Finally let $\eta(q')$ be defined, then

$$q'=\psi(q)F'_x\in\psi(Q)F'_x\quad\text{for some } x\in\Sigma^*,\ q\in Q$$

and

$$\begin{aligned}\eta(q')F_t\subseteq\eta_x(\psi(q)F'_x)F_t&=qF_xF_t=qF_{xt}\\&=\eta_{xt}(\psi(q)F'_{xt})=\eta(q'F'_t)\end{aligned}$$

as required. □

We can now piece together some of our earlier results. Let $\hat{\mathcal{M}}$ be a Mealy machine and suppose that $\mathcal{M}$ is the state machine of $\hat{\mathcal{M}}$. From chapters 3 and 4 we can obtain a decomposition

$$\mathcal{M}\leq\mathcal{A}_1\circ\mathcal{A}_2\circ\ldots\circ\mathcal{A}_n$$

and then by 6.3.1 the state machine $\mathcal{A}_1\circ\mathcal{A}_2\circ\ldots\circ\mathcal{A}_n$ can be provided with outputs to turn it into a Mealy machine that covers the original machine. It follows that in general a Mealy machine can be replaced by a minimal covering machine which, in turn, can then be replaced by a series of machines connected up in series and parallel which have, as underlying state machines, group machines and reset machines. This is a very significant result.

6.4 Sequential functions

For this and the next section we will use a slightly different interpretation of the behaviour of an incomplete Mealy machine. We will only consider normal Mealy machines, and the difference between their operation here and in the previous sections is concerned with the appearance of blanks on the output tape.

Let $\hat{\mathcal{M}} = (Q, \Sigma, \Theta, F, G)$ be a normal Mealy machine and let $q \in Q$, $x \in \Sigma^*$. If $x = \sigma_1\sigma_2 \ldots \sigma_k$ and $qF_{\sigma_1}, qF_{\sigma_1\sigma_2}, \ldots, qF_{\sigma_1\sigma_2\ldots\sigma_k}$ are all defined then the output word $f_q(x)$ is completely defined and is an element of Θ^*. We define a partial function $\bar{f}_q : \Sigma^* \to \Theta^*$ by

$$\bar{f}_q(x) = \begin{cases} f_q(x) & \text{if } qF_{\sigma_1}, qF_{\sigma_1\sigma_2}, \ldots, qF_{\sigma_1\sigma_2\ldots\sigma_k} \text{ are all defined} \\ & \text{where } x = \sigma_1\sigma_2 \ldots \sigma_k. \\ \varnothing & \text{otherwise} \end{cases}$$

This adaptation of the function f_q satisfies several properties. Clearly blanks cannot occur in $\bar{f}_q(x)$ for any $x \in \Sigma^*$. Another point of interest is that for $\bar{f}_q(x) \neq \varnothing$ the machine must stop in a defined state, i.e. qF_x must exist. Thus $\bar{f}_q(x) \neq \varnothing$ if and only if x is applicable to q and $qF_x \neq \varnothing$. In general $\bar{f}_q : \Sigma^* \to \Theta^*$ is a partial function according to this interpretation, and will be a function if $\hat{\mathcal{M}}$ is complete, in this case $\bar{f}_q = f_q$.

We can now state some simple consequences of this interpretation which are really analogues of some earlier results, namely 6.1.1, 6.2.1, 6.2.2 and 6.2.3.

Proposition 6.4.1

Let $\hat{\mathcal{M}} = (Q, \Sigma, \Theta, F, G)$ be a normal Mealy machine and $q \in Q$, then for $x, y \in \Sigma^*$

$$\bar{f}_q(xy) = \bar{f}_q(x) \cdot \bar{f}_{qF_x}(y).$$

Proof We need only note that if $\bar{f}_q(xy) = \varnothing$ then either $\bar{f}_q(x) = \varnothing$ or $\bar{f}_{qF_x}(y) = \varnothing$ which means that $\bar{f}_q(x) \cdot \bar{f}_{qF_x}(y) = \varnothing$. □

Theorem 6.4.2

Let $\hat{\mathcal{M}} = (Q, \Sigma, \Theta, F, G)$ be a normal Mealy machine. The relation defined on Q by

$$q \sim q_1 \Leftrightarrow \bar{f}_q = \bar{f}_{q_1}$$

is an equivalence relation. If $\hat{\mathcal{M}}/\sim = (Q/\sim, \Sigma, \Theta, F', G')$ is defined by

$$[q]F'_\sigma = [qF_\sigma]$$
$$[q]G'_\sigma = [qG_\sigma]$$

for $q \in Q$, $\sigma \in \Sigma$ then

(i) the function $\phi : Q \to Q/\sim$ defined by $\phi(q) = [q]$, $q \in Q$, satisfies

$$\left.\begin{aligned} \phi(q)F'_\sigma &= [qF_\sigma] \\ \phi(q)G'_\sigma &= qG_\sigma \end{aligned}\right\} \text{ for } q \in Q, \sigma \in \Sigma;$$

(ii) $\bar{f}'_{[q]} = \bar{f}_q$ for each $q \in Q$.

We call $\hat{\mathcal{M}}/\sim$ the *minimal machine of* $\hat{\mathcal{M}}$. (An analogue of 6.2.4 also holds here.) The machine $\hat{\mathcal{M}}/\sim$ has the property that if $[q]$, $[q_1] \in Q/\sim$ there exists a word $x \in \Sigma^*$ applicable to both $[q]$ and $[q_1]$ such that

$$\bar{f}'_{[q]}(x) \neq \bar{f}'_{[q_1]}(x).$$

This property will be described by saying that $\hat{\mathcal{M}}/\sim$ is *sequentially reduced*, or *s-reduced* for short.

Let $f : \Sigma^* \to \Theta^*$ be a partial function. We call f a *sequential partial function* if there exists a normal Mealy machine $\hat{\mathcal{M}} = (Q, \Sigma, \Theta, F, G)$ and a state $q \in Q$ such that $f(x) = \bar{f}_q(x)$ for all $x \in \Sigma^*$.

Naturally the machine $\hat{\mathcal{M}}$ may not be unique and one of our aims is to find a minimal Mealy machine satisfying the required conditions. This can be set into the more general problem of minimizing an arbitrary Mealy machine, and the minimization procedure will yield a machine with as few states as possible.

Example 6.9

Let $\Sigma = \Theta = \{0, 1\}$. The function $f : \Sigma^* \to \Theta^*$ is defined by

$$f(x) = x \quad \text{for all } x \in \Sigma^*.$$

If we construct the Mealy machine:

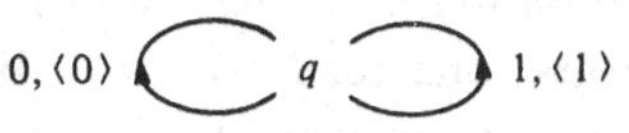

0, ⟨0⟩ q 1, ⟨1⟩

then

$$f_q = f.$$

The Mealy machine:

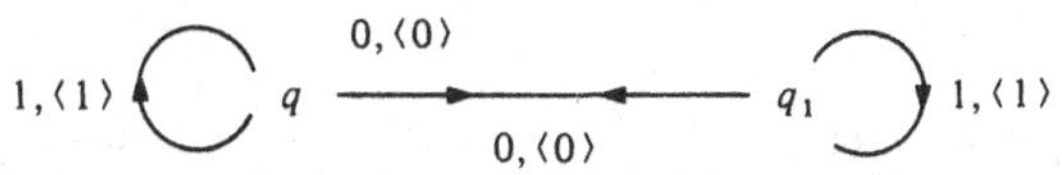

1, ⟨1⟩ q 0, ⟨0⟩ 0, ⟨0⟩ q_1 1, ⟨1⟩

also satisfies the property

$$f_q = f.$$

Furthermore

$$f_{q_1} = f.$$

This second machine is in some sense less efficient than the first, it has more states but can do nothing more than the first machine. Both machines are complete.

Example 6.10

Let $\Sigma = \{0\}$, $\Theta = \{0, 1\}$. Define $f : \Sigma^* \to \Theta^*$ by

$$f(\Lambda) = \Lambda, f(0) = 0, f(00) = 00, f(0^{n+2}) = 001^n \ (n > 0).$$

To see that f is sequential we construct the Mealy machine:

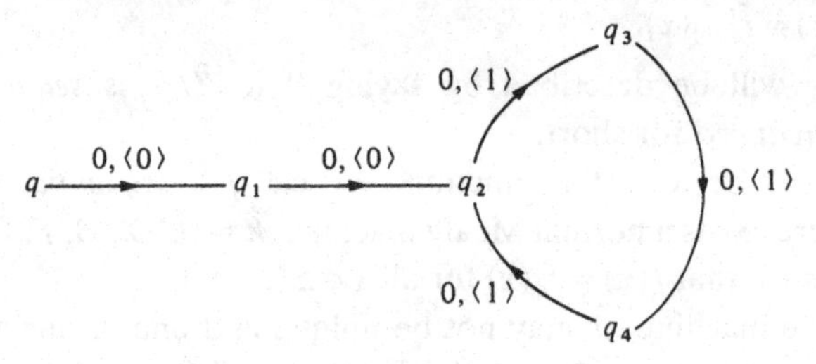

and note that $f_q = f$. This machine is also complete.

Example 6.11

Let $\Sigma = \{0, 1\}$ and suppose that $x, y \in \Sigma^*$ then x and y represent binary numbers, and we will define a machine that adds them together and gives the result as a binary number. Recall that if $x = \sigma_1\sigma_2 \ldots \sigma_k$ then x can represent a positive integer by using the expansion

$$x = \sigma_1 + \sigma_2 \cdot 2 + \sigma_3 \cdot 2^2 + \ldots \sigma_k \cdot 2^{k-1}.$$

We have written this expansion out in the reverse order to what is normal; this is caused by our convention that the tapes enter machines so that the left-most symbol is the first one read. When adding two numbers normally we look first at the right-most symbols, so these are the symbols that we must input first. Thus $2 \equiv 01$, $3 \equiv 11$, $4 \equiv 001$, $5 \equiv 101$, $6 \equiv 011$, $7 \equiv 111$, $8 \equiv 0001$ etc. Define the machine $\hat{\mathcal{M}} = (Q, \Sigma, \Theta, F, G)$, where $\Sigma = \{0, 1\} \times \{0, 1\}$, $\Theta = \{0, 1\}$,

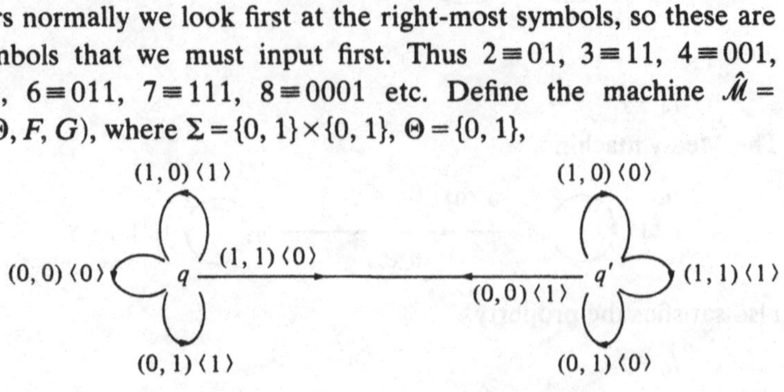

(1, 0) ⟨1⟩ (1, 0) ⟨0⟩

(0, 0) ⟨0⟩ q (1, 1) ⟨0⟩ (0, 0) ⟨1⟩ q′ (1, 1) ⟨1⟩

(0, 1) ⟨1⟩ (0, 1) ⟨0⟩

Now let x, y be binary numbers, we first ensure that they are of the same length by adding a succession of 0s to the *right* hand side of the shortest word until they are of equal length. Now we have two words $\sigma_1 \dots \sigma_k$ and $\sigma'_1 \dots \sigma'_k$ representing the binary numbers x and y. If we add them together the binary representation of $x+y$ is either of length k or $k+1$. Since our sequential machine cannot convert two words of length k into a word of length $k+1$ we must make sure that our original inputs are of length $k+1$ by adding a further 0 to the right of each word $\sigma_1 \dots \sigma_k$ and $\sigma'_1 \dots \sigma'_k$. Now we input the word $(\sigma_1, \sigma'_1)(\sigma_2, \sigma'_2) \dots (\sigma_k, \sigma'_k)(0, 0) \in \Sigma^*$ into the machine in state q. The resulting output

$$f_q((\sigma_1, \sigma'_1) \dots (\sigma_k, \sigma'_k)(0, 0))$$

will represent the sum $x+y$ (in our reverse binary representation).

For example $2 \equiv 01$ and so the input $(0, 0)(1, 1)(0, 0)$ will result in the sum $2+2$ which can be read off from the machine diagram as $001 \equiv 4$. Similarly $5+8$ is obtained with the input $(1, 0)(0, 0)(1, 0)(0, 1)(0, 0)$ which gives $10110 \equiv 13$, and so on. The sequential function f_q is thus a *binary adder.* The final input $(0, 0)$, which must be incorporated in any input word, is called a *carry* and ensures that the final state is q and that no part of the binary sum has been 'left at' q'.

We now consider a sequential partial function $f: \Sigma^* \to \Theta^*$. Thus a normal Mealy machine $\hat{\mathcal{M}} = (Q, \Sigma, \Theta, F, G)$ exists such that $f = \bar{f}_q$ for some $q \in Q$. Because of 6.4.2 we can replace $\hat{\mathcal{M}}$ by the minimal machine $\hat{\mathcal{M}}/\sim$ and then $f = \bar{f}'_{[q]}$. This means that given a sequential partial function we can find an s-reduced machine to represent the partial function. To ensure that the machine is the most efficient possible we remove all states that cannot be reached from the initial state.

Let $f: \Sigma^* \to \Theta^*$ be a sequential partial function and $\hat{\mathcal{M}} = (Q, \Sigma, \Theta, F, G)$ the minimal, s-reduced Mealy machine such that $f = \bar{f}_q$ for some $q \in Q$.

Form the set $Q_f = \{qF_x \mid x \in \Sigma^*\}$ and define the Mealy machine

$$\hat{\mathcal{M}}_f = (Q_f, \Sigma, \Theta, F_1, G_1)$$

where

$$q'(F_1)_\sigma = q'F_\sigma$$

and

$$q'(G_1)_\sigma = q'G_\sigma$$

for $\sigma \in \Sigma$, $q' \in Q_f$.

The pair $(\hat{\mathcal{M}}_f, q)$ is called the *minimal sequential machine* for f. Our terminology implies that $(\hat{\mathcal{M}}_f, q)$ is unique, but to be more precise it is only unique up to isomorphism, where this is defined next.

Let $\hat{\mathcal{M}} = (Q, \Sigma, \Theta, F, G)$ and $\hat{\mathcal{M}}' = (Q', \Sigma, \Theta, F', G')$ be complete Mealy machines. A function $\psi : Q \to Q'$ is called a *Mealy machine homomorphism* if

$$\psi(qF_\sigma) = \psi(q)F'_\sigma$$
$$qG_\sigma = \psi(q)G'_\sigma$$

for $q \in Q$, $\sigma \in \Sigma$.

If ψ is a bijective function then ψ is called an *isomorphism*.

Theorem 6.4.3

Let $f : \Sigma^* \to \Theta^*$ be a sequential partial function. Suppose that s-reduced Mealy machines

$$\hat{\mathcal{M}} = (Q, \Sigma, \Theta, F, G) \quad \text{and} \quad \hat{\mathcal{M}}' = (Q', \Sigma, \Theta, F', G')$$

exist such that $f = \bar{f}_q$ for some $q \in Q$ and $f = \bar{f}'_{q'}$ for some $q' \in Q'$. If $(\hat{\mathcal{M}}_f, q)$ and $(\hat{\mathcal{M}}_{f'}, q')$ are minimal sequential machines of f then there exists an isomorphism

$$\psi : \hat{\mathcal{M}}_f \to \hat{\mathcal{M}}'_{f'} \quad \text{such that } \psi(q) = q'.$$

Proof Let $x, z \in \Sigma^*$. Since $\bar{f}_q = \bar{f}'_{q'}$ we have $\bar{f}_q(xz) = \bar{f}'_{q'}(xz)$ and so

$$\bar{f}_q(x) \cdot \bar{f}_{qF_x}(z) = \bar{f}'_{q'}(x) \cdot \bar{f}'_{q'F'_x}(z)$$

which implies that $\bar{f}_{qF_x}(z) = \bar{f}'_{q'F'_x}(z)$ and so

$$\bar{f}_{qF_x} = \bar{f}_{q'F'_x}.$$

Define $\psi : \hat{\mathcal{M}}_f \to \hat{\mathcal{M}}'_f$ by

$$\psi(qF_x) = q'F'_x \quad (x \in \Sigma^*).$$

This is well-defined for if $qF_x = qF_y$ $(x, y \in \Sigma^*)$ then

$$\bar{f}_{qF_x} = \bar{f}_{qF_y} \quad \text{and also} \quad \bar{f}_{qF_x} = \bar{f}'_{q'F'_x},$$

$\bar{f}_{qF_y} = \bar{f}'_{q'F'_y}$ which implies that $\bar{f}'_{q'F'_x} = \bar{f}'_{q'F'_y}$. By the s-reduced nature of $\hat{\mathcal{M}}'$ we must then have $q'F'_x = q'F'_y$. A similar argument yields the fact that ψ is one-one. It is clearly onto and for $x \in \Sigma^*$

$$\psi(qF_x) = \psi(q)F'_x \quad \text{since } q' = \psi(q).$$

Given $\sigma \in \Sigma$ we have

$$\psi(qF_xF_\sigma) = \psi(qF_x)F'_\sigma$$

and

$$qF_xG_\sigma = \bar{f}_{qF_x}(\sigma) = \bar{f}'_{q'F'_x}(\sigma) = (qF_x)G'_\sigma.$$

Hence ψ is an isomorphism. □

Theorem 6.4.4

Let $\psi : \hat{\mathcal{M}} \to \hat{\mathcal{M}}'$ be a Mealy machine homomorphism and suppose that $q \in Q$, the state set of $\hat{\mathcal{M}}$. Then

$$\bar{f}_q = \bar{f}'_{\psi(q)}.$$

Proof This is left as an exercise. □

Given a partial sequential function $f : \Sigma^* \to \Theta^*$ we can now consider associating with it a minimal machine $(\hat{\mathcal{M}}_f, q)$ in an essentially unique way, any other minimal machine will be isomorphic by 6.4.3. This justifies calling $(\hat{\mathcal{M}}_f, q)$ *the minimal machine of f*. By the construction of $\hat{\mathcal{M}}_f$ it is accessible in the sense that any state of $\hat{\mathcal{M}}_f$ occurs as an image of q under a suitable input. This ensures that there are no 'redundant' states in $\hat{\mathcal{M}}_f$.

Let $f : \Sigma^* \to \Theta^*$ and $g : \Theta^* \to \Gamma^*$ be partial sequential functions and suppose that $(\hat{\mathcal{M}}_f, q)$, $(\hat{\mathcal{M}}'_g, q')$ are the minimal machines of f and g. Writing

$$\hat{\mathcal{M}}_f = (Q, \Sigma, \Theta, F, G) \quad \text{and} \quad \hat{\mathcal{M}}'_g = (Q', \Theta, \Gamma, F', G')$$

we can now form the Mealy machine

$$\hat{\mathcal{M}}'_g \omega \hat{\mathcal{M}}_f = (Q' \times Q, \Sigma, \Gamma, F^\omega, G^\omega)$$

where

$$(q'_1, q_1)F^\omega_\sigma = (q'_1 F'_{\omega(q,\sigma)}, q_1 F_\sigma)$$

and

$$(q'_1, q_1)G^\omega_\sigma = q'_1 G'_{\omega(q_1,\sigma)} \quad \text{for } q_1 \in Q_1,\ q'_1 \in Q'_1,\ \sigma \in \Sigma,$$

and where

$$\omega : Q \times \Sigma \to \Theta$$

is defined by

$$\omega(q, \sigma) = G(q, \sigma) \quad \text{for } q \in Q,\ \sigma \in \Sigma.$$

Consider the partial sequential function $h : \Sigma^* \to \Gamma^*$ defined by this machine in state (q', q). Let $\sigma \in \Sigma$, then

$$\begin{aligned} h(\sigma) &= q' G'_{\omega(q,\sigma)} \\ &= q' G'_{qG_\sigma} \\ &= q' G'_{f(\sigma)} \\ &= g(f(\sigma)). \end{aligned}$$

For $x \in \Sigma^*, \sigma \in \Sigma$

$$\begin{aligned} h(x\sigma) &= h(x) \cdot (q', q) F_x^{\omega} G_{\sigma}^{\omega} \\ &= h(x) \cdot (q' F'_{\omega^+(q,x)}, qF_x) G_{\sigma}^{\omega} \quad \text{using the notation of 2.6} \\ &= h(x) \cdot (q' F'_{\omega^+(q,x)}) G'_{\omega(qF_x,\sigma)} \\ &= h(x) \cdot q' F'_{\omega^+(q,x)} G'_{qF_x G_{\sigma}} \\ &= h(x) \cdot q' F'_{f(x)} G'_{qF_x G_{\sigma}} \\ &= g(f(x) q G_x G_{\sigma}) \\ &= g(f(x\sigma)) \end{aligned}$$

providing that we can establish the identity

$$\omega^+(q_1, x) = \bar{f}_{q_1}(x) \quad \text{for } q_1 \in Q_1, x \in \Sigma^*.$$

Now for $\sigma \in \Sigma$, $\omega^+(q, \sigma) = \omega(q, \sigma) = qG_{\sigma} = f(\sigma)$. Let us consider $\alpha \in \Sigma^*$ and

$$\begin{aligned} \omega^+(q_1, \sigma\alpha) &= \omega(q_1, \sigma)\omega^+(q_1 F_{\sigma}, \alpha) \\ &= \bar{f}_{q_1}(\sigma) \cdot \bar{f}_{q_1 F_{\sigma}}(\alpha) \\ &= \bar{f}_{q_1}(\sigma\alpha) \end{aligned}$$

by the usual inductive process. Hence $\omega^+(q_1, x) = \bar{f}_{q_1}(x)$ as required. Consequently the machine $\hat{\mathcal{M}}'_g \omega \hat{\mathcal{M}}_f$ in state (q', q) defines the partial sequential function $g \circ f : \Sigma^* \to \Gamma^*$.

Thus the composition of two partial sequential functions is again a partial sequential function.

The other 'products' defined between Mealy machines give rise to natural operations on the corresponding partial sequential functions. For example, given Mealy machines $\hat{\mathcal{M}} = (Q, \Sigma, \Theta, F, G)$ and $\hat{\mathcal{M}}' = (Q', \Sigma', \Theta', F', G')$ we can define the product

$$\hat{\mathcal{M}} \times \hat{\mathcal{M}}' = (Q \times Q', \Sigma \times \Sigma', \Theta \times \Theta', F \times F', G \times G')$$

which will then define a partial sequential function

$$\overline{(f \times f')}_{(q,q')} : (\Sigma \times \Sigma')^* \to (\Theta \times \Theta')^*$$

by

$$\overline{(f \times f')}_{(q,q')}(x, x') = (\bar{f}_q(x), \bar{f}'_{q'}(x'))$$

for $x \in \Sigma^*, x' \in (\Sigma')^*, q \in Q, q' \in Q'$.

6.5 Decompositions of sequential functions

In this final section we will apply some of our earlier results to problems associated with sequential functions.

Let $f:\Sigma^* \to \Theta^*$ be a partial sequential function and suppose then $(\hat{\mathcal{M}}_f, q)$ is the minimal machine of f. If $\hat{\mathcal{M}}_f = (Q_f, \Sigma, \Theta, F, G)$, then (Q_f, Σ, F) is a state machine, and so we may construct the transformation semigroup $\mathbf{TS}(Q_f, \Sigma, F)$ and we call this the *syntactic transformation semigroup* of f. It will be convenient to write this as

$$\mathcal{A}_f = (Q_f, S_f).$$

Suppose that $\mathcal{A}_f$ has a decomposition of the form

$$\mathcal{A}_f \leq \mathcal{B}_1 \circ \ldots \circ \mathcal{B}_n,$$

what can we say about the sequential function f?

Theorem 6.5.1

(Eilenberg [1976]) Let $f:\Sigma^* \to \Theta^*$ be a partial sequential function and let $\mathcal{A}_f \leq \mathcal{B}_1 \circ \mathcal{B}_2$. Then there exist partial sequential functions $g_1:\Sigma^* \to \Gamma^*$, $g_2:\Gamma^* \to \Theta^*$ such that

$$f \subseteq g_1 \circ g_2$$

and

$$\mathcal{A}_{g_1} \leq \mathcal{B}_1, \quad \mathcal{A}_{g_2} \leq \mathcal{B}_2.$$

Proof Consider the syntactic transformation semigroup $\mathcal{A}_f = (Q_f, S_f)$ of the minimal machine $(\hat{\mathcal{M}}_f, q)$. For each $x \in \Sigma^+$ we have $F_x \in S_f$. If $\mathcal{B}_1 = (P_1, T_1)$ and $\mathcal{B}_2 = (P_2, T_2)$ then $\mathcal{A} \leq \mathcal{B}_1 \circ \mathcal{B}_2$ implies that a partial function $\psi: P_1 \times P_2 \to Q_f$ exists such that for each $s \in S_f$ there exists $t_2^s \in T_2$, $h^s: P_2 \to T_1$ such that

$$\psi(p_1, p_2)s \subseteq \psi(p_1 h^s(p_2), p_2 t_2^s)$$

for $(p_1, p_2) \in P_1 \times P_2$. Let $(i_1, i_2) \in P_1 \times P_2$ such that $\psi(i_1, i_2) = q$.

Put $\Gamma = P_2 \times \Sigma \times T_1$ and define

$$\hat{\mathcal{M}}_2 = (P_2, \Sigma, \Gamma, F^2, G^2)$$

by

$$p_2 F^2_\sigma = p_2 t_2^s \quad \text{where } F_\sigma = s \in S_f$$
$$p_2 G^2_\sigma = (p_2, \sigma, h^s(p_2)).$$

Then consider the partial sequential function

$$g_2: \Sigma^* \to \Gamma^*$$

defined by $\hat{\mathcal{M}}_2$ in state i_2.

Now put $\hat{\mathcal{M}}_1 = (P_1, \Gamma, \Theta, F^1, G^1)$ where

$$p_1 F^1_\gamma = p_1 t_1 \quad \text{if } \gamma = (p_2, \sigma, t_1)$$

and

$$p_1G^1_\gamma = \begin{cases} \psi(p_1, p_2)G_\sigma & \text{if } \psi(p_1, p_2) \neq \varnothing \neq p_1t_1 \\ \varnothing & \text{if } p_1t_1 = \varnothing \\ \text{arbitrary} & \text{otherwise.} \end{cases}$$

Now $g_1 : \Gamma^* \to \Theta^*$ is defined by $\hat{\mathcal{M}}_1$ in state i_1. The partial function $g_1 \circ g_2 : \Sigma^* \to \Theta^*$ is then defined by $\hat{\mathcal{M}}_1 \omega \hat{\mathcal{M}}_2$ in state (i_1, i_2).

For $p_1 \in P_1$, $p_2 \in P_2$, $\sigma \in \Sigma$ there exists $s = F_\sigma \in S_f$ and then

$$\begin{aligned}(p_1, p_2)\bar{G}_\sigma &= p_1G^1_{G^2(p_2,\sigma)} \\ &= p_1G^1_\gamma \quad \text{where } \gamma = (p_2, \sigma, h^s(p_2)) \\ &= \psi(p_1, p_2)G_\sigma\end{aligned}$$

whenever $\psi(p_1, p_2) \neq \varnothing \neq p_1h^s(p_2)$. (Here $\bar{G}$ is the output function from $\hat{\mathcal{M}}_1 \omega \hat{\mathcal{M}}_2$.) If $p_1h^s(p_2) = \varnothing$ then $(p_1, p_2)(h^s, t^s) = \varnothing$ and so $\psi(p_1, p_2)s = \varnothing$. Consequently $f \subseteq g_1 \circ g_2$ as required. □

Theorem 6.5.2

Let $f : \Sigma^* \to \Theta^*$ be a partial sequential function and let

$$\mathcal{A}_f \leq \mathcal{B}_1 \times \mathcal{B}_2,$$

then there exist partial sequential functions

$$g_1 : \Sigma^* \to \Gamma_1^*$$
$$g_2 : \Sigma^* \to \Gamma_2^*$$

and a function $\beta : \Gamma_1 \times \Gamma_2 \to \Theta$ such that

$$f(x) = (g_1 \wedge g_2)(\beta(x)),$$

where

$$(g_1 \wedge g_2)(x) = (g_1(x), g_2(x))$$

for $x \in \Sigma^*$ and

$$\mathcal{A}_{g_1} \leq \mathcal{B}_1, \quad \mathcal{A}_{g_2} \leq \mathcal{B}_2.$$

Proof This construction follows a similar argument to the previous proof. However we use the Mealy machine construction $\hat{\mathcal{M}}_1 \wedge \hat{\mathcal{M}}_2$, that is, the restricted direct product. Here the alphabets Γ_1 and Γ_2 are defined to be $P_1 \times \Sigma$ and $P_2 \times \Sigma$ respectively.

$$\beta : \Gamma_1 \times \Gamma_2 \to \Theta$$

is defined by

$$\beta(p_1, \sigma, p_2, \sigma') = \begin{cases} (\psi(p_1, p_2), \sigma)G_\sigma & \text{if } \sigma = \sigma' \text{ and } \psi(p_1, p_2) \neq \varnothing \\ \text{arbitrary} & \text{otherwise,} \end{cases}$$

where $\psi: P_1 \times P_2 \to Q$ is the covering partial function. The details are left to the reader; they will be found in Eilenberg [1976].

Example 6.12

Let $\Sigma = \{\sigma\}$, $\Theta = \{0, 1\}$ and $f: \Sigma^* \to \Theta^*$ be defined by

$$f(\sigma) = 0,\ f(\sigma^2) = 01,\ f(\sigma^n) = 010^{n-2}$$

for $n \geq 3$. Then f is a sequential function defined by the complete Mealy machine started in state a:

$$a \xrightarrow{\sigma,\langle 0\rangle} b \xrightarrow{\sigma,\langle 1\rangle} c \circlearrowright \sigma,\langle 0\rangle$$

This has state machine $\mathscr{C}_{(1,2)}$ which, by the holonomy decomposition theorem, has a covering

$$\mathscr{C}_{(1,2)} \geq \bar{\mathbf{2}} \circ \mathscr{C}.$$

The covering $\psi: P_1 \times P_2 \to Q$ (using the notation of theorem 6.5.1) is given by

$$a = \psi((b, a)) = \psi((c, a))$$
$$b = \psi((b, \{b, c\}))$$
$$c = \psi((c, \{b, c\}))$$

where $P_1 = \{b, c\}$, $P_2 = \{\{a\}, \{b, c\}\}$. If $T_1 = \{t_1, t_1'\}$, $P_1 t_1 = \{c\}$, $P_1 t_1' = \{b\}$, $T_2 = \{t_2\}$ then $h^\sigma: P_2 \to T_1$ is defined by $h^\sigma(\{a\}) = t_1'$, $h^\sigma(\{b, c\}) = t_1$ and $g_2: \Sigma^* \to \Gamma^*$ is defined by

$$g_2(\sigma) = (a, \sigma, t_1'), \quad g_2(\sigma^2) = (a, \sigma, t_1') \cdot (\{b, c\}, \sigma, t_1)$$

and generally

$$g_2(\sigma^n) = (a, \sigma, t_1')(\{b, c\}, \sigma, t_1) \ldots (\{b, c\}, \sigma, t_1)$$

for $n > 1$.

Now

$$\begin{aligned} g_1(g_2(\sigma)) &= g_1(a, \sigma, t_1') \\ &= \psi((b, a))G_\sigma = aG_\alpha = 0 \\ g_1(g_2(\sigma^2)) &= 0bF'_{(a,\sigma,t_1)}G'_\gamma \quad \text{where } \gamma = (\{b, c\}, \sigma, t_1) \\ &= 0bt_1G'_\gamma \\ &= 0\psi(bt_1, \{b, c\})G_\sigma \\ &= 0bG_\sigma \\ &= 01 \end{aligned}$$

$$\begin{aligned} g_1(g_2(\sigma^3)) &= 01(bF'_\gamma G'_\gamma) \\ &= 01(cG'_\gamma) \\ &= 01(cG_\sigma) \\ &= 010. \end{aligned}$$

Continuing we see that

$$g_1(g_2(\sigma^n)) = 010^{n-2} \quad \text{for } n \geq 3.$$

6.6 Conclusion

We have seen how the concept of a Mealy machine can be used to model a variety of discrete situations. Using the results of this chapter we can analyse the underlying state machine and transformation semigroup by means of our results from chapters 3 and 4. To recover information about the original situation we can apply the results of this chapter to give facts about Mealy machine coverings, or if the model is concerned with sequential functions we can decompose them. By choosing a suitable decomposition theory we can then highlight various properties of our original model and this may well throw light on the situation that we are modelling.

The subject discussed here is undergoing much rapid development and it is likely that over the next few years many new and useful results will appear. For those interested in reading further I would strongly recommend that the two masterful volumes by S. Eilenberg be studied.

6.7 Exercises

6.1 Let $\hat{\mathcal{M}} = (Q, \Sigma, \Theta, F, G)$ be a Mealy machine and $i \in Q$ a given initial state. Consider the partial sequential function $\bar{f}_i : \Sigma^* \to \Theta^*$. Let $\mathfrak{M} = (\mathcal{M}, i, Q)$ be a recognizer defined by $\mathcal{M} = (Q, \Gamma, H)$ where $\Gamma = \Sigma \times \Theta$, $H : Q \times \Sigma \to Q$ is given by

$$(q_1(\sigma, \theta))H = qF_\sigma \quad \text{if and only if } qG_\sigma = \theta,\ q \in Q,\ \sigma \in \Sigma,\ \theta \in \Theta.$$

Prove that $|\mathfrak{M}| = \{(\alpha, \beta) \mid \alpha \in \Sigma^*, \beta \in \Theta^*, \bar{f}_i(\alpha) = \beta\}$.
This shows that $\bar{f}_i$ is a *rational* function, i.e. one whose graph is a rational subset of $\Sigma^* \times \Theta^*$.

6.2 Prove 6.2.2.

6.3 In the notation of 6.2.8 prove the following:

$$(\pi^c)^* = \pi$$

$$(\tau^*)^c \leq \tau.$$

6.4 Examine 6.3.3 in the case where $\psi : Q \to Q'$ satisfies

$$f_q(x) \subseteq f'_{\psi(q)}(x) \quad \text{for all } x \in \Sigma^*.$$

6.5 Prove that if $\psi : \hat{\mathcal{M}} \to \hat{\mathcal{M}}'$ is an isomorphism and $\psi(q) = q'$, then $f_q = f'_{q'}$.

6.6 Prove 6.5.2.

6.7 Minimize (if possible) the following machines (where $\Sigma = \Theta = \{0, 1\}$):

(i)

	$\hat{\mathcal{M}}_1$	a	b	c	d	e	f
F	0	a	b	c	e	b	a
	1	c	d	c	b	d	f
G	0	0	1	0	1	0	1
	1	1	1	1	1	0	1

(ii)

	$\hat{\mathcal{M}}_2$	a	b	c	d	e	f	g
F	0	a	d	b	f	c	a	c
	1	c	d	e	d	a	b	d
G	0	$\varnothing$	0	$\varnothing$	0	1	$\varnothing$	1
	1	1	$\varnothing$	1	0	$\varnothing$	0	1

(iii)

	$\hat{\mathcal{M}}_3$	a	b	c	d	e	f	g
F	0	a	d	b	c	c	a	c
	1	$\varnothing$	$\varnothing$	e	a	a	b	d
G	0	$\varnothing$	0	0	$\varnothing$	0	$\varnothing$	1
	1	1	0	0	1	$\varnothing$	1	1

6.8 Describe the sequential partial functions f_a, $\bar{f}_a$ for the machine

	$\hat{\mathcal{M}}$	a	b	c
F	0	b	c	a
	1	a	$\varnothing$	$\varnothing$
G	0	1	1	1
	1	0	$\varnothing$	$\varnothing$

6.9 Describe the sequential function $f_{\langle \text{off},\text{off} \rangle}$ of example 3.2.7.

6.10 Describe the sequential function f_{off} of example 3.2.9. Find a minimal machine for this function.

6.11 Complete the details of theorem 6.5.2.

Appendix

The following program evaluates the semigroup of a state machine with up to five states and nine inputs. The semigroup elements are listed and a semigroup multiplication table constructed. The states are described by numbers 1, 2, 3, 4, 5 and the inputs by letters $A, B, C, D, E, \ldots$. The next state function is described as an n-tuple ($n \leq 5$). Implementation is on an Apple or ITT 2020 microcomputer running Apple Pascal with a printer. The program was written by Dr A. W. Wickstead, Department of Pure Mathematics, Queen's University, Belfast.

```
PROGRAM SEMIGROUPS;
CONST BLANK='               ';(*15 BLANKS*)
SEPARATOR='------------------------------------------------------------------';
   (* '-' 66 TIMES*)
      MAXWORD=15;
      STACKSIZE=50;
TYPE WORD=RECORD VAL:PACKED ARRAY[1..5] OF 1..5;
                 STR:STRING[MAXWORD]
                 END;
     VALUES=PACKED ARRAY[1..5] OF 1..5;
VAR PRINT:TEXT;
    SGFILE:FILE OF WORD;
    DOMPOINT,I,J:INTEGER;
    OPTION,NUM:CHAR;
    FILENAME:STRING;

PROCEDURE CONTINUE;
BEGIN
WRITELN;
WRITE('PRESS RETURN TO CONTINUE ');
READLN;
END;

PROCEDURE GENERATE;
VAR PRINTCNT,WORDSIZE,WHICH,WORDNUM,FUNCTNUM,DOMSIZE,FUNCTSIZE:INTEGER;
    NEWONE:WORD;
    STACKEND:ARRAY[0..1] OF INTEGER;
    STACK:ARRAY[0..1,1..STACKSIZE] OF WORD;
    STARTER:ARRAY[1..9] OF WORD;
    USED:PACKED ARRAY [1..5,1..5,1..5,1..5,1..5] OF BOOLEAN;
    TEMP:STRING[1];

PROCEDURE CHECKPAGE;
VAR I:INTEGER;
BEGIN
IF PRINTCNT>59 THEN
```

```
  BEGIN
  FOR I:=1 TO 66-PRINTCNT DO WRITELN(PRINT);
  PRINTCNT:=0;
  END
END;

PROCEDURE WORDMESS(I:INTEGER);
BEGIN
WRITELN(PRINT);
WRITELN(PRINT,'WORD SIZE: ',WORDSIZE,' NUMBER OF NEW WORD(S): ',STACKEND[I]);
WRITELN(PRINT);
PRINTCNT:=PRINTCNT+3;
CHECKPAGE;
END;

PROCEDURE OUT;
BEGIN
WRITE(PRINT,'(');
FOR DOMPOINT:=1 TO DOMSIZE DO
  BEGIN
  WRITE(PRINT,NEWONE.VAL[DOMPOINT]);
  IF DOMPOINT<DOMSIZE THEN WRITE(PRINT,',') ELSE WRITE(PRINT,')  ');
  END;
WRITELN(PRINT,NEWONE.STR);
PRINTCNT:=PRINTCNT+1;
CHECKPAGE;
(*$I-*)
SGFILE^:=NEWONE;
PUT(SGFILE);
(*$I+*)
END;

PROCEDURE SETUP;
BEGIN
WRITELN('FILENAME FOR OUTPUT? (<RETURN> FOR');
WRITE('NONE ');
READLN(FILENAME);
IF (FILENAME<>'') THEN IF (FILENAME[LENGTH(FILENAME)]<>':')  THEN
     REWRITE(SGFILE,FILENAME) ELSE WRITELN('NO OUTPUT FILE OPENED');
FOR DOMPOINT:=1 TO 5 DO NEWONE.VAL[DOMPOINT]:=DOMPOINT;
NUM:=CHR(0);
PRINTCNT:=0;
TEMP:='?';
WORDSIZE:=1;
WRITE('SIZE OF DOMAIN (1..5)? ');
REPEAT UNITREAD(2,NUM,1) UNTIL NUM IN ['1'..'5'];
    (*THIS USE OF UNITREAD READS SINGLE CHARACTER NUM FROM KEYBOARD WITHOUT
                PRINTING IT ON THE VDU*)
WRITELN(NUM);
DOMSIZE:=ORD(NUM)-48;
WRITE('NUMBER OF FUNCTIONS (1..9)? ');
REPEAT UNITREAD(2,NUM,1) UNTIL NUM IN ['1'..'9'];
WRITELN(NUM);
FUNCTSIZE:=ORD(NUM)-48;
FOR FUNCTNUM:=1 TO FUNCTSIZE DO
  BEGIN
  FOR DOMPOINT:=1 TO DOMSIZE DO
    BEGIN
    WRITE('VALUE OF FUNCTION ',FUNCTNUM,' AT ',DOMPOINT,' ? ');
    REPEAT UNITREAD(2,NUM,1) UNTIL NUM IN ['1'..CHR(48+DOMSIZE)];
    WRITELN(NUM);
    NEWONE.VAL[DOMPOINT]:=ORD(NUM)-48;
    TEMP[1]:=CHR(64+FUNCTNUM);
    NEWONE.STR:=TEMP;
    END;
  STACK[0,FUNCTNUM]:=NEWONE;
  STARTER[FUNCTNUM]:=NEWONE;
  OUT;
  END;
STACKEND[0]:=FUNCTSIZE;
WRITELN(PRINT);
WRITELN(PRINT,'ORIGINAL ',FUNCTSIZE,' FUNCTION(S)');
WRITELN(PRINT);
PRINTCNT:=PRINTCNT+3;
CHECKPAGE;
WHICH:=1;
FILLCHAR(USED,SIZEOF(USED),CHR(0));
```

```
FOR FUNCTNUM:=1 TO FUNCTSIZE DO USED[STARTER[FUNCTNUM].VAL[1],
    STARTER[FUNCTNUM].VAL[2],STARTER[FUNCTNUM].VAL[3],STARTER[FUNCTNUM].VAL[4],
    STARTER[FUNCTNUM].VAL[5]]:=TRUE;
END;

PROCEDURE TOOHARD;
BEGIN
WRITELN('SEMIGROUP IS TOO BIG FOR THIS PROGRAM');
(*$I-*)
CLOSE(SGFILE,PURGE);
(*$I+*)
CONTINUE;
EXIT(GENERATE)
END;

PROCEDURE NEWLEVEL;
BEGIN
STACKEND[WHICH]:=0;
FOR WORDNUM:=1 TO STACKEND[1-WHICH] DO (*FOR EACH NEW WORD AT LAST LEVEL*)
  BEGIN
  FOR FUNCTNUM:=1 TO FUNCTSIZE DO (*FOR EACH ORIGINAL FUNCTION*)
    BEGIN
    FOR DOMPOINT:=1 TO DOMSIZE DO (*FOR EACH POINT OF DOMAIN*)
      BEGIN
      NEWONE.VAL[DOMPOINT]:=STARTER[FUNCTNUM].VAL[STACK[1-WHICH,WORDNUM].
          VAL[DOMPOINT]]; (*VALUE OF ((ORIGINAL FUNCTION))*WORD) AT DOMPOINT*)
      END;
    NEWONE.STR:=CONCAT(STACK[1-WHICH,WORDNUM].STR,STARTER[FUNCTNUM].STR);
    IF NOT USED[NEWONE.VAL[1],NEWONE.VAL[2],NEWONE.VAL[3],NEWONE.VAL[4],
                                                    NEWONE.VAL[5]] THEN
      BEGIN
      IF WORDSIZE>MAXWORD THEN TOOHARD;
      USED[NEWONE.VAL[1],NEWONE.VAL[2],NEWONE.VAL[3],NEWONE.VAL[4],
                                                    NEWONE.VAL[5]]:=TRUE;
      STACKEND[WHICH]:=STACKEND[WHICH]+1;
      IF STACKEND[WHICH]>STACKSIZE THEN TOOHARD;
      STACK[WHICH,STACKEND[WHICH]]:=NEWONE;
      OUT;
      END;
    END
  END;
WORDSIZE:=WORDSIZE+1;
IF STACKEND[WHICH]<>0 THEN WORDMESS(WHICH);
WHICH:=1-WHICH;
END;

BEGIN(*GENERATE*)
PAGE(OUTPUT);
GOTOXY(10,6);
WRITELN('SEMIGROUP GENERATION');
GOTOXY(0,10);
WRITELN('YOU MAY SPECIFY UP TO 9 FUNCTIONS ON');
WRITELN('A SET OF UP TO 5 ELEMENTS. THE PROGRAM');
WRITELN('WILL LIST THE ELEMENTS IN THE ');
WRITELN('SEMIGROUP THAT THEY GENERATE, AND A');
WRITELN('DESCRIPTION OF EACH IN TERMS OF THE');
WRITELN('ORIGINAL FUNCTIONS.');
WRITELN;
SETUP;
REPEAT NEWLEVEL UNTIL STACKEND[1-WHICH]=0;
(*$I-*)
CLOSE(SGFILE,LOCK);
(*$I+*)
FOR I:=1 TO 66-PRINTCNT DO WRITELN(PRINT);
END;

PROCEDURE MULTIPLY;
VAR PRODUCT:VALUES;
    STACK:ARRAY[0..255] OF VALUES;
    CODE:PACKED ARRAY [1..5,1..5,1..5,1..5,1..5] OF 0..255;
    PAGESWD,PAGESHT,SIZE,I,J:INTEGER;
    LIST:ARRAY[0..255] OF STRING[MAXWORD];


PROCEDURE SETUP;
BEGIN
WRITE('FILENAME FOR INPUT? ');
REPEAT READLN(FILENAME) UNTIL FILENAME<>'';
```

```
RESET(SGFILE,FILENAME);
FILLCHAR(CODE,SIZEOF(CODE),CHR(0));
SIZE:=0;
REPEAT
  BEGIN
  STACK[SIZE]:=SGFILE^.VAL;
  LIST[SIZE]:=CONCAT(SGFILE^.STR,COPY(BLANK,1,15-LENGTH(SGFILE^.STR)));
  CODE[SGFILE^.VAL[1],SGFILE^.VAL[2],SGFILE^.VAL[3],SGFILE^.VAL[4],
              SGFILE^.VAL[5]]:=SIZE;
  (*$I-*)
  GET(SGFILE);
  (*$I+*)
  SIZE:=SIZE+1;
  END
UNTIL EOF(SGFILE) OR (SIZE>255);
IF NOT EOF(SGFILE) THEN
  BEGIN
  WRITELN('SEMIGROUP IS TOO BIG TO COMPUTE TABLE');
  WRITELN('ROUTINE ABORTING');
  CONTINUE;
  CLOSE(SGFILE,LOCK);
  EXIT(MULTIPLY);
  END;
CLOSE(SGFILE,LOCK);
PAGESHT:=(SIZE-1) DIV 60;
PAGESWD:=(SIZE-1) DIV 7;
END;

PROCEDURE PRINTPAGE(I,J:INTEGER);
VAR K,L,X,Y:INTEGER;
BEGIN
WRITE(PRINT,BLANK,CHR(124));
FOR K:=0 TO 6 DO
  BEGIN
  IF 7*J+K<SIZE THEN WRITE(PRINT,' ',LIST[7*J+K]);
  END;
WRITELN(PRINT);
WRITELN(PRINT,SEPARATOR,SEPARATOR);
FOR L:=0 TO 59 DO
  BEGIN
  X:=60*I+L;
  IF X<SIZE THEN
    BEGIN
    WRITE(PRINT,LIST[X],CHR(124));
    FOR K:=0 TO 6 DO
      BEGIN
      Y:=7*J+K;
      IF Y<SIZE THEN
        BEGIN
        FOR DOMPOINT:=1 TO 5 DO
          BEGIN
          PRODUCT[DOMPOINT]:=STACK[Y,STACK[X,DOMPOINT]];
          END;
        WRITE(PRINT,' ',LIST[CODE[PRODUCT[1],PRODUCT[2],PRODUCT[3],
                                 PRODUCT[4],PRODUCT[5]]]);

        END;
      END;
    END;
  WRITELN(PRINT);
  END;
FOR L:=0 TO 3 DO WRITELN(PRINT);
END;

BEGIN(*MULTIPLY*)
PAGE(OUTPUT);
GOTOXY(7,6);
WRITELN('PRINT MULTIPLICATION TABLE');
GOTOXY(0,10);
WRITELN('THIS ROUTINE WILL PRINT THE TABLE');
WRITELN('OF A FUNCTION SEMIGROUP THAT HAS BEEN');
WRITELN('PRODUCED BY THE GENERATION OPTION');
WRITELN;
SETUP;
FOR I:=0 TO PAGESHT DO
  BEGIN
  FOR J:=0 TO PAGESWD DO PRINTPAGE(I,J)
  END;
END;
```

```
BEGIN(*PROGRAM*)
OPTION:=CHR(0);
REWRITE(PRINT,'REMOUT:');
REPEAT
  BEGIN
  PAGE(OUTPUT);
  GOTOXY(10,6);
  WRITELN('FUNCTION SEMIGROUPS');
  GOTOXY(9,8);
  WRITELN('(C) 1981 A.W. WICKSTEAD');
  GOTOXY(0,12);
  WRITELN('OPTIONS: G]ENERATE SEMIGROUP');
  WRITELN;
  WRITELN('          M]ULTIPLICATION TABLE');
  WRITELN;
  WRITELN('          Q]UIT');
  REPEAT UNITREAD(2,OPTION,1) UNTIL OPTION IN ['G','M','Q'];
  CASE OPTION OF
    'G':GENERATE;
    'M':MULTIPLY;
    'Q':PAGE(OUTPUT)
    END(*CASES*)
  END;
UNTIL OPTION='Q';
CLOSE(PRINT);
END.
```

References

Arbib, M. A. [1964] *Brains, machines and mathematics.* McGraw-Hill, New York

Arbib, M. A. [1969] 'Memory limitations of S-R models'. *Psychol. Review* **76** 507–10

Arbib, M. A. [1969A] *Theories of abstract automata.* Prentice-Hall, Englewood Cliffs, N.J.

Chittenden, B. [1978] 'Specification of software as finite state automata' in *Proceedings of the International Symposium on Mini and Micro Computers.* I.E.E.E. New York pp. 105–11

Cohn, P. M. [1975] 'Algebra and Language theory'. *Bull. L.M.S.* **7** 1–29

Dilger, E. [1976] 'On permutation-reset automata'. *Information and Control* **30** 86–95

Eilenberg, S. [1974] *Automata, languages and machines, Vol. A.* Academic Press, New York

Eilenberg, S. [1976] *Automata, languages and machines, Vol. B.* Academic Press, New York.

Fiksel, J. R. and Bower, G. H. [1976] 'Question answering by a semantic network of parallel automata'. *J. Math. Psychol.* **13** 1–45

Ginzburg, A. [1968] *Algebraic theory of automata.* Academic Press, New York

Keville, T. J. [1978] 'The decomposition of transformation semigroups'. M.Sc. thesis, Queen's University, Belfast

Kieras, D. E. [1976] 'Finite automata and S-R models'. *J. Math. Psychol.* **13** 127–47

Krohn, K., Langer, R. and Rhodes, J. L. [1967] 'Algebraic principles for the analysis of a biochemical system'. *J. Comp. and System Sci.* **1** 119–36

Krohn, K. and Rhodes, J. L. [1965] 'Algebraic theory of machines, I. Prime decomposition theorem for finite semigroups and machines'. *Trans. Amer. Math. Soc.* **116** 450–64

Minsky, M. [1967] *Computation, finite and infinite machines.* Prentice-Hall, Englewood Cliffs, N.J.

Nelson, R. J. [1975] 'Behaviorism, finite automata and stimulus response theory'. *Theory and Decision* **6** 249–67

Roedding, W. [1975] *Netzwerke endlicher Automaten als Modelle wirtshaftlicher und sozialer systeme.* Reports of the Austrian Society for Cybernetic Studies

Rosen, R. [1972] *Foundations of mathematical biology Vol. 2.* Academic Press, New York

Shibata, Y. [1972] 'On the structure of an automaton and automorphisms'. *Systems-Computers-Controls* **3** 10–15

Suppes, P. [1969] 'Stimulus response theory of finite automata'. *J. Math. Psychol.* **6** 327–55

Index of notation

Index

Printed in the United States
by Baker & Taylor Publisher Services